W0230104

Automatic Control for Food Processing Systems

ASPEN FOOD ENGINEERING SERIES

Series Editor

Gustavo V. Barbosa-Cánovas, Washington State University

Advisory Board

Jose Miguel Aguilera, Pontifica Universidad Catolica de Chile
Pedro Fito, Universidad Politecnica
Richard W. Hartel, University of Wisconsin
Jozef Kokini, Rutgers University
Michael McCarthy, University of California at Davis
Martin Okos, Purdue University
Micha Peleg, University of Massachusetts
Leo Pyle, University of Reading
Shafiur Rahman, Hort Research
M. Anandha Rao, Cornell University
Yrjo Roos, University of Helsinki
Walter L. Spiess, Bundesforschungsanstalt
Jorge Welti-Chanes, Universidad de las Américas-Puebla

Aspen Food Engineering Series

Jose M. Aguilera and David W. Stanley, *Microstructural Principles of Food Processing and Engineering*, second edition (1999)

Stella M. Alzamora, María S. Tapia, and Aurelio López-Malo, *Minimally Processed Fruits and Vegetables: Fundamental Aspects and Applications* (2000)

Gustavo Barbosa-Cánovas and Humberto Vega-Mercado, *Dehydration of Foods* (1996)

Pedro Fito, Enrique Ortega-Rodríguez, and Gustavo Barbosa-Cánovas, *Food Engineering 2000* (1997)

P. J. Fryer, D. L. Pyle, and C. D. Rielly, *Chemical Engineering for the Food Industry* (1997)

S. D. Holdsworth, *Thermal Processing of Packaged Foods* (1997)

Michael Lewis and Neil Heppell, *Continuous Thermal Processing of Foods: Pasteurization and UHT Sterilization* (2000)

Rosana G. Moreira, M. Elena Castell-Perez, and Maria A. Barrufet, *Deep-Fat Frying: Fundamentals and Applications* (1999)

M. Anandha Rao, *Rheology of Fluid and Semisolid Foods: Principles and Applications* (1999)

Automatic Control for Food Processing Systems

Rosana G. Moreira, PhD, PE
Associate Professor
Department of Agricultural Engineering
Texas A&M University
College Station, Texas

AN ASPEN PUBLICATION®
Aspen Publishers, Inc.
Gaithersburg, Maryland
2001

The author has made every effort to ensure the accuracy of the information herein. However, appropriate information sources should be consulted, especially for new or unfamiliar procedures. It is the responsibility of every practitioner to evaluate the appropriateness of a particular opinion in the context of actual clinical situations and with due considerations to new developments. The author, editors, and the publisher cannot be held responsible for any typographical or other errors found in this book.

Library of Congress Cataloging-in-Publication Data

Moreira, Rosana G.
Automatic control for food processing systems/Rosana G. Moreira
p. cm.
Includes bibliographical references and index.
ISBN 0-8342-1781-3
1. Food processing machinery—Automatic control. I. Title.
TP373.M56 2001
664'.0028'4—dc21
00-067400

Copyright © 2001 by Aspen Publishers, Inc.
A Wolters Kluwer Company
www.aspenpublishers.com
All rights reserved.

Aspen Publishers, Inc., grants permission for photocopying for limited personal or internal use. This consent does not extend to other kinds of copying, such as copying for general distribution, for advertising or promotional purposes, for creating new collective works, or for resale. For information, address Aspen Publishers, Inc., Permissions Department, 200 Orchard Ridge Drive, Suite 200, Gaithersburg, Maryland 20878.

Orders: (800) 638-8437
Customer Service: (800) 234-1660

About Aspen Publishers • For more than 40 years, Aspen has been a leading professional publisher in a variety of disciplines. Aspen's vast information resources are available in both print and electronic formats. We are committed to providing the highest quality information available in the most appropriate format for our customers. Visit Aspen's Internet site for more information resources, directories, articles, and a searchable version of Aspen's full catalog, including the most recent publications: **www.aspenpublishers.com**

Aspen Publishers, Inc. • The hallmark of quality in publishing
Member of the worldwide Wolters Kluwer group.

Editorial Services: Timothy Sniffin
Library of Congress Catalog Card Number: 00-067400
ISBN: 0-8342-1781-3

Printed in the United States of America

1 2 3 4 5

Contents

Foreword

The application of automatic control has grown rapidly in the food industry during the past decade. The sophistication of control applications has increased markedly with the development of personal computers, microchips, and biosensors. As a consequence, food engineers and scientists now require a basic understanding of the principles of automatic control, or at least a cursory knowledge of the potential and the limitations of the various controller types. Most of the available texts are written for the chemical and mechanical engineering industries, and are overly theoretical for the average food engineer. *Automatic Control for Food Processing Systems* addresses the specific control problems in the food industry and provides the food engineer and scientist with the necessary background to evaluate the potential of automatic control in food processing systems.

The control of a processing system, such as a grain dryer, is complicated by the changeably of the control variable (e.g., the moisture content) and/or of the raw product (e.g. the grain). The experience or skill of the operator is still the basis for the control of many food processing systems. This personal control may have been acceptable in the past but cannot satisfy the present demands of the food industry to maximize system capacity, product quality, energy efficiency, and personnel safety, as well as minimize operating costs. Only automatic control can establish the optimum operating conditions to fulfill these requirements.

The reader of this text will be able to design some of the classical control systems required in a food plant, and to understand the function of some of the more recently developed controllers. The aim of the author, herself well known as the developer of automatic controls for food extruders and grain dryers, is to present fundamental control theory followed by its application in specific food processing examples. The major topics of PID control, adaptive control, model-predictive control, and fuzzy control are covered, each time explaining the advantages and the limitations of a controller type for a specific food-processing operation or system of operations.

Some of the mathematics in this book will challenge the casual reader. However, even without understanding every equation in a chapter, readers will gain a better understanding of the power of automatic control in establishing the optimum operating conditions in a food processing plant for running at maximum capacity, regardless of the raw product variability and the inherently lengthy dead times.

This book will be useful to senior- and graduate-level food engineers and those engineers in the food industry who did not cover control in their university studies. Managers of food plants will benefit by obtaining a general idea of the important subject of automatic control.

—*Fred W. Bakker-Arkema, PhD*
Professor
Department of Agricultural Engineering
Michigan State Unversity
East Lansing, MI

Preface

I don't know what it is I have inside me...
something that wants to come out ...
cher maître, I swear
I will always adore two goddesses, Liberty and the Muse.

— *Arthur Rimbaud, Letters*

The idea for this book originated during my time as a graduate student working in the area of food process control. I quickly realized that design and implementation of an automatic controller for a food processing system (in my case a grain dryer) is a difficult task because of the extreme variability of raw materials and the lack of sufficient knowledge of the main physicochemical interactions that take place during processing of agricultural materials. I still remember the long hours spent in front of a computer screen hoping to see the desired setpoint moisture content. The system's highly unpredictable performance and numerous disturbances definitely presented me with a huge challenge. Since then, I have spent much of my professional life working in the applications of process control strategies for industrial food processing systems. Although excellent reviews on this subject are readily available, I felt there was still a need for a text that would describe the basics of automatic data collection and process control systems, which are critical for improving product quality and efficiency in today's food and agricultural industries.

Automatic Control for Food Processing Systems intends to give an introduction to the theory and applications of process control strategies in selected food processing systems. Its audience is either the reader who may have little experience with this subject matter or the more experienced engineer who may find this text a useful reference to refresh his or her memory in specific aspects of process control. This book may also be a helpful reference for teachers and students involved in courses addressing process control issues in the food and other related industries, since it includes a significant amount of background material on process control theory.

Throughout the book, the reader will learn about the importance of practicing automatic control for a variety of food processes (since proper control is necessary to improve process efficiency); how to decrease product quality variations; and how to reduce waste. A nice feature of this book is the presentation of specific examples of application to food processes (Chapter 2); background material about the basic components and specific requirements of model-based controllers for food processing operations, leading up to the most recent developments in modern control theory (Chapter 3); description of empirical and theoretical modeling of the process dynamics and how they improve system understanding (Chapters 4 through 6); control theories, such as adaptive control with supervision level (Chapter 7), fuzzy logic, and neural network (Chapter 8); and finally, the latest information on available equipment and instrumentation to carefully control processing parameters and evaluate product quality model-based controllers (Chapter 9).

I hope to make the reader aware of the relevant issues in selecting and applying appropriate control strategies. I also have attempted to provide an extensive reference list at the end of each chapter for those readers interested in reading the original research.

In looking back on those challenging times as a graduate student, now I realize I gained much from them. Like Rimbaud, "I have stretched ropes from steeple to steeple, garlands from window to window, golden chains from star to star."

CHAPTER 1

Introduction

In the United States, food is the largest of all manufacturing industries, exceeding $350 billion annually. It represents over 25% of all nondurable goods manufactured domestically. To remain competitive in the international market, US food companies are shifting their manufacturing technology to flexible systems that are closely coupled with corporate strategies. Although the older technology was largely based on traditional experience, the new technology is based on scientific principles and must be engineered: that is, predictable, controllable, portable, and subject to modeling and scale-up.

The Institute of Food Technologists (IFT) has identified the need to improve process design and operation efficiencies through closed-loop control strategies (IFT Research Committee, 1993). Process control that maximizes throughput while optimizing product quality and safety and extending the processing time between start-up and shutdown/ cleanup would obviously increase efficiency.

This book concerns the automatic control of process systems, particularly agricultural and food processing: drying, frying, extrusion, baking, distillation, heating and cooling, and so forth.

THE STATE OF THE FOOD INDUSTRY

The US food industries, while profitable, have not shared the level of profitability enjoyed by durable producers in the last years. Most of the economy boom has been sparked by durable goods such as housing, automobiles, appliances, planes, and computers rather than by nondurables such as food, apparel, and medical services (Morris, 1998).

Over the past several years, US food manufacturers have intensified their efforts to boost income and market share through acquisitions, overseas expansion, and cutting costs. Figure 1–1 shows that food industry mergers and acquisitions have increased sharply in the past 3 years, reaching a high value of 218 in 1997. Growing overseas investment by US companies (Figure 1–2) more than doubled since 1990, according to the US Department of Commerce. Although food industries have experienced cost reduction in the past few years, most of their moderate capital spending in 1997 has been to upgrade process control and maintenance, with the objective of improving productivity (Morris, 1998).

Compared to other manufacturing industries, the food industry has been slow in adopting new technologies that have emerged in recent years (McGrath et al, 1998).

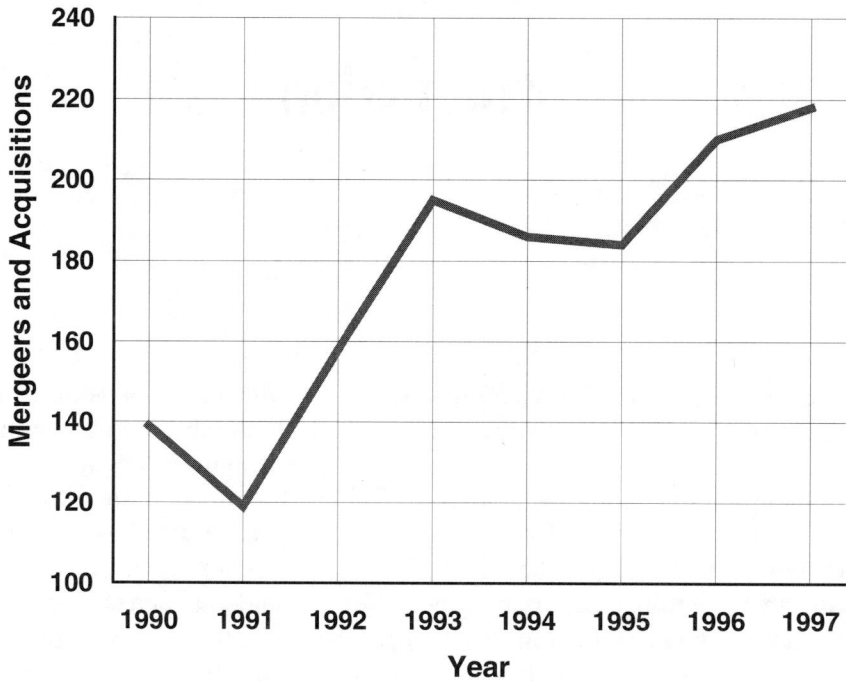

Figure 1–1 Mergers and acquisitions in the US food industry from 1990 to 1996.

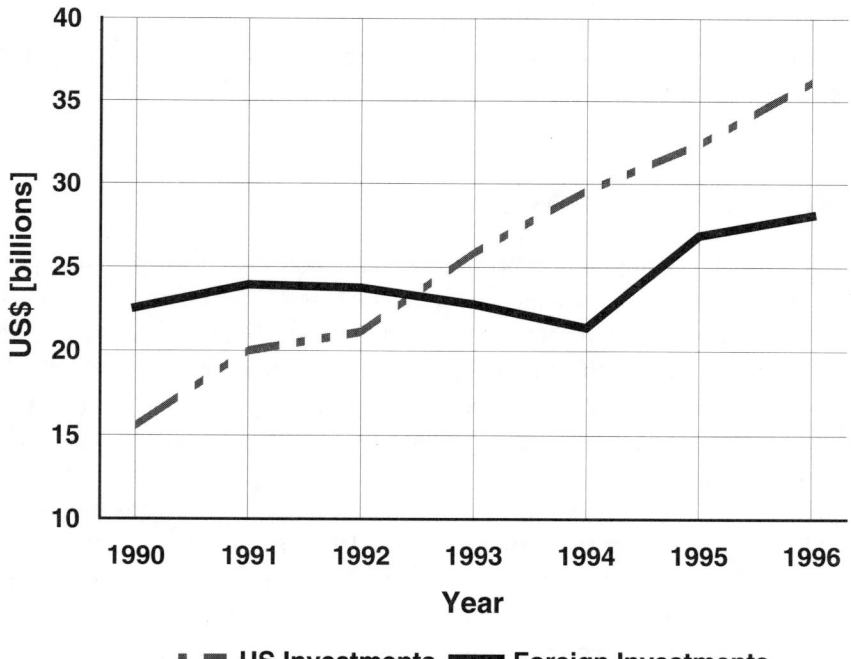

Figure 1–2 US direct investment in foreign food manufacturing versus foreign direct investment in US food manufacturing from 1990 to 1996.

According to Skjöldebrand (1991), the food industry is still characterized by a traditional art of food processing and the employment of few trained engineers. Even today, there is not enough application of scientific principles to the process, and many production process operations are still heavily dependent on the experience and skill of operators.

The level of automation in the US food industries is shown in Figure 1–3. About 52% of the companies reported scattered "islands of control," and 23% have distributed process control, but only 2% have achieved integrated top-to-bottom manufacturing control, according to the *Food Engineering* 1998 food manufacturing survey (Morris, 1998).

According to Willians (1998), the food industry has been poorly served by automation product vendors traditionally targeted to one or two main groups: (1) discrete manufacturing, as in the case of the automation industry, which requires fast sequencing with counters, timers, and many conditional logic operations and which uses programmable logic controllers (PLCs); and (2) continuous manufacturing, as in the chemical industry, which needs sophisticated continuous control for safety critical applications and uses distributed control systems (DCSs). Many food industries, however, require both continuous and sequential controls, so that the manufacturers must buy one of each control and then find a way of making them communicate with each other.

McGrath et al (1998) clearly discussed the requirements of the food industry today with respect to process control implementation. They commented that the food industry

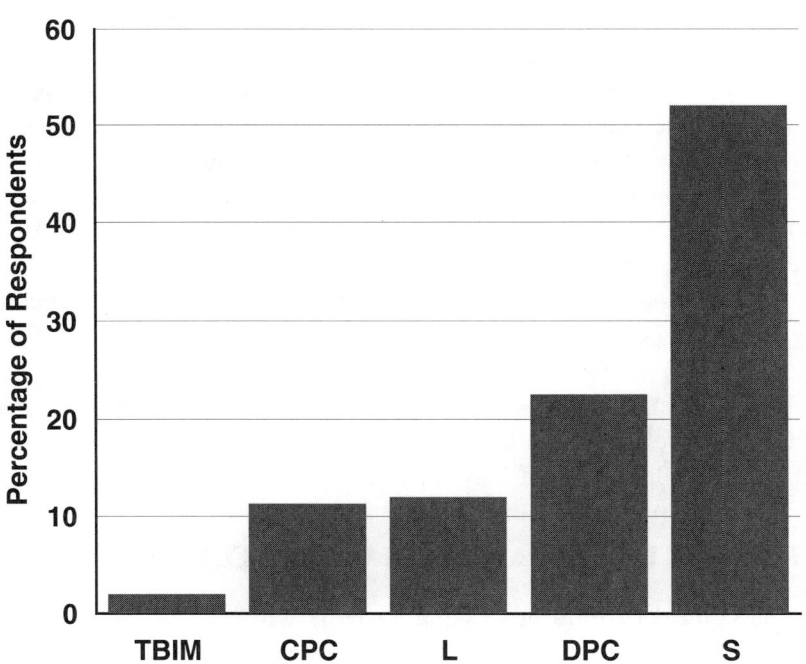

Figure 1–3 Process control integration in the food industry. *Note:* TBIM, top-to-bottom integrated manufacturing; CPC, central process control; L, little or no process control; DPC, distributed process control; S, scattered " islands of control."

needs low-cost solutions, with powerful but flexible control platforms to facilitate rapid development of tools to meet new requirements.

CONTROL SCHEMES IN THE FOOD PROCESSING INDUSTRY

In the food-processing industry, automatic control is one of the key elements that affect productivity, quality, safety, and competitiveness. The control schemes available commercially include proportional-integral-derivative (PID) control, robust control, adaptive control, and model predictive control (Seborg et al, 1989).

The *PID* controller has been around for more than 50 years and today is still the most widely used industrial controller (Ogata, 1970). It is simple, is easy to understand, and does not require a precise process model to start up or maintain. It works well if the process system is simple, but if the process dynamics change, it needs to be retuned, and it has trouble controlling complex systems—that is, processes that are nonlinear, time variant and coupled and that have parameter or structure uncertainties.

Robust control is a controller design method that focuses on the reliability (robustness) of the control algorithm. Robust control methods are suited to processes that have large uncertainty ranges and small stability margins (Takahashi et al, 1972). The main drawback of robust control systems is that they require a high level of expertise to design. They will work well without any operator attention if they are designed properly. However, if major modification or upgrades are required, they must be redesigned.

Adaptive control is defined as an intelligent feedback control system that adjusts its characteristics to changes in the environment so as to operate in an optimal manner (Isermann, 1989). In process control applications, the PID self-tuning scheme is widely implemented in commercial products. Some of the problems with the traditional adaptive control include (1) the amount of off-line training required, (2) the trade-off between the persistent excitation signals for correct identification and the steady system response for control performance, (3) the assumption of the process structure, and (4) the model convergence and system stability issues in real applications.

Model predictive control (MPC) is one of the few advanced control techniques used successfully in industrial control applications (Prett & Garcia, 1988). MPC is based on three key elements: (1) predictive model, (2) optimization in range of a temporal window, and (3) feedback correction. These steps are carried on continuously by computer programs on line. MPC works well in controlling complex process systems but requires a high level of expertise, since the design of MPC systems is very complicated.

AUTOMATIC CONTROL STRATEGIES FOR THE FOOD INDUSTRY

Automatic control of food-processing systems will help to improve final product quality, increase process efficiency, and reduce waste of raw materials. Food processes are generally multiple-input, multiple-output systems involving complex interactions between process inputs and outputs. These are generally characterized by strong relationships among mass, energy, and momentum transfer, including complex physicochemical transformations such as gelatinization of starch, denaturization of proteins, and browning reactions. Such changes are influenced by the chemical composition and physical state of the materials used and by the process conditions (Schonauer, 1995).

With limited understanding of the physical and chemical interactions occurring during food processing, it is difficult to design proper control systems for food-processing systems. In the past, lack of on-line sensors prevented direct control of product quality parameters. Instead, correlations between process outputs and product attributes were developed. Therefore, previously proposed control schemes aimed to control product quality indirectly through secondary process variables (Schonauer, 1995).

Quality evaluation is an important operation performed routinely in the food industry. This task is increasingly being automated as computer technology develops and as machine-vision technology becomes a useful on-line quantification technique. Improved quality usually brings higher revenues and, perhaps more important, higher consumer satisfaction.

Raw materials used in food processes are mainly of biological origin, and their compositional and physical nature can vary considerably. Such variations can introduce significant, unmeasurable disturbances to the process that make manual control unreliable. Besides the complexity caused by raw material variability, food-processing systems also exhibit longer dead time and are typically non–minimum-phase systems. For systems with significant time delays, improved control performance over PID controllers has been achieved with the Smith predictor (Seborg et al, 1989) and the minimum-variance controller scheme (Isermann, 1989) provided that the time delay is estimated accurately. Incorrect estimates of delays can cause poor performance for these control systems (Clarke et al, 1987a).

A new class of controllers, MPCs, is available. One form of MPC, the generalized predictive controller (GPC), has been found to overcome the limitations associated with the minimum-variance and Smith predictor control schemes (Clarke et al, 1987a, 1987b). Schonauer and Moreira (1995) and Brescia and Moreira (1997) successfully employed a GPC with an ARX model to control a twin-screw food extruder and a continuous fryer, respectively. Haarsma (1994) applied dynamic matrix control (DMC) to a continuous frying process, achieving good results. More recently, artificial neural network control has been developed to control a frying process (Huang et al, 1998).

Clarke et al (1987a, 1987b) commented that GPC coupled with an on-line identification routine is capable of stable control of processes with variable parameters and variable dead times and with model orders that change, provided input/output data are sufficiently rich. This type of controller is called adaptive. In food-processing systems, the operating environment can drift to the extent that raw material properties (particle size, moisture content, composition, etc) vary and machines wear or deteriorate due to long-term use. Constantly changing dynamics, variable dead times, and model parameters that change depending on the direction and size of control movement are typical in many food processes (Schonauer, 1995). These typical disturbances are not easily modeled because on-line sensors to measure them are not readily available. Therefore, fixed controllers may fall short in compensating for such disturbances over time, making adaptive controllers more attractive for food-processing systems. Adaptive control has been successfully applied to continuous-flow grain dryers (Nybrant, 1986; Moreira & Bakker-Arkema, 1990) and a food extruder (Schonauer & Moreira, 1995).

One of the difficulties of applying adaptive control to food-processing systems is that in addition to normal wear of machine parts, abrupt failures caused by actuator and sensors malfunctions and failures arising from the process itself are common. When these faults occur, they have to be detected very fast and be distinguished from wear so that the process can continue to run safely. In conventional adaptive control, the process model parameters are recursively estimated and used to recompute the controller parameters at every sampling instance. However, estimator transients due to disturbances and parameter tracking are passed to the controller. Closed-loop performance monitoring (Gustafsson & Graebe, 1998) is a recent technique that can be used to update controllers only when the system has changed and the parameters converge with sufficient confidence.

Future areas of research include better identification of the process dynamics using a combination of black box modeling and fundamental modeling (gray box modeling). The combination of system identification and physical modeling may give more information on how the unknown and uncertain parts of the process should be designed. This gives the opportunity to achieve more knowledge about the functionality and physical understanding of the process. As a result, the controller will be more efficient and the final product quality improved.

Today, with better and faster computers and development of novel on-line sensors, advanced control strategies will be able to cope with most of the complexity of food-processing systems, such as nonlinearities, time delay, uncertainties, and variations in process variables. Advances in the area of statistical process control, fuzzy logic, and neural network controls will play an important role in the future of automation in the food-processing industries.

FUTURE TRENDS

According to Willians (1998), the development of PLCs was very important in the control of discrete or sequencing processes like those encountered in the automobile industry. Distributed control systems (DCSs) were essential in the application of sophisticated continuous control for chemical industries. The food industry, like many others, requires a mixture of both sequencing and continuous controls to efficiently control its batch processes (recipe handling, canning, bottling, packaging, etc) and continuous processes (baking, frying, fermentation, distillations, etc) over the entire plant. However, the difficulty arises when trying to make those two systems communicate with each other. PLCs are based on relay logic and DCSs on analog control systems. When digital control was introduced, there was not enough processing power to provide both very high I/O and processing speed and complex calculations for continuous control in one system.

Today, in addition to PLCs and DCSs, personal computers are being used to run a control program that executes logic sequences and control loops without any special control hardware. These systems are called soft logic or PC-based control. However, they have their limitations for process control, since PCs are engineered basically to provide low cost and reliability to suit office based data-processing primary applications.

With the progress in processing power of microchips, roughly doubling every 18 to 24 months, DCSs are now available with PLC features and PLCs with all the capabilities of a DSC. These new PLC-based hybrids are being called PCSs or process control systems.

Another important area that is changing fast is software development. Programming tools need to be very flexible to deal with the complex combination of continuous and batch processes in the food industry. Today, a single programming tool can be used to complete automation of most processes: for example, function block for process applications, sequential function charts for batch sequencing, ladder logic for machine interlocks, and so on, thus reducing costs in configuration and maintenance.

Advancing equally fast is the power of supervisory control and data acquisition software (SCADA) running on PCs under Windows NT. With the new generation of PC-SCADA products being more flexible, it will be simple to use office products such as Access, Excel, and Word to schedule batches, store and download recipes, archive batch records, present management information, and so on.

In the area of communication, a new product called DataFlow (part of the merging process information management, PIM) will provide support for communication with PLC and control system products. DataFlow will exchange information between many systems, supply presentation aids capabilities, support numerous file formats and data types, enable pictures, sound, and movies clips to accompany information, and provide a help screen.

CONCLUSION

This chapter presents an overview of control of food-processing systems. Automatic control of food-processing systems is necessary to maintain final product quality, improve process efficiency, and reduce waste. Limited understanding of the physicochemical interactions occurring during processing, large and unpredictable variation on raw material composition and characteristics, and long dead time make it difficult to design proper controls for food-processing systems.

The combination of empirical modeling and fundamental modeling will significantly improve the understanding of food-processing system dynamics. With better and faster computers and the development of new on-line smart sensors, advanced control techniques will be easily implemented to deal with most of the complexity of food-processing systems. Greater attention should be given to advancement of control techniques such as closed-loop performance monitoring, statistical process control, fuzzy logic, and neural network controls.

Sensors and measuring devices have improved significantly in the last decade. Today, smart sensors can measure, self-compensate for disturbances, and even amplify signals. Advances in machine-vision technology will contribute to better characterization of product properties on line.

Another important area of continuing improvement is digital computer control. LC hybrids, produced by the integration of PLCs and DLCs, will be able to efficiently handle the complex combination of batch and continuous controls required in the food industry. New programming tools are also being developed to improve control applications and communications

---

REFERENCES

Brescia, L. & Moreira, R.G. (1997). Modeling and control of a continuous frying process: II. Control development. *Trans Chem Eng Inst* 75(C), 12–20.

Clarke, D.W., Mohtadi, C. & Tuffs, P.S. (1987a). Generalized predictive control: part I. The basic algorithm. *Automatica* 23, 137–148.

Clarke, D.W., Mohtadi, C. & Tuffs, P.S. (1987b). Generalized predictive control: part II. Extensions and interpretations. *Automatica* 23, 149–160.

Gustafsson, F. & Graebe, S.F. (1998). Closed-loop performance monitoring in the presence of system changes and disturbances. *Automatica* 34, 1311–1326.

Haarsma, G. (1994). Project Report: Development of a Dynamic Matrix Controller for a Frying Process. Netherlands: Dept of Agricultural Engineering and Physics, Wagneningen University.

Huang, Y., Whittaker, E.D. & Lacey, R. (1998). Internal model control for a continuous snack food frying process using neural networks. *Trans ASAE*. 41(5):1519–1525.

Institute or Food Technologists Research Committee. (1993). America's food research needs: into the 21st century. *Food Technol* 47(3S), 1S–40S.

Isermann, R. (1989). *Digital Control Systems. Vol 2. Stochastic Control, Multivariable Control, Adaptive Control, Applications*. New York, NY: Springer-Verlag.

McGrath, M.J., O'Connor, J.F. & Cummins, S. (1998). Implementing a process control strategy for the food processing industry. *J Food Eng* 35, 313–321.

Moreira, R.G. & Bakker-Arkema, F.W. (1990). Feedforward/feedback adaptive control for commercial cross-flow grain driers. *J Agric Eng Res* 45, 107–116.

Morris, E. (1998). The state of food manufacturing. *Food Eng*, September, 66–82.

Nybrant, T. (1986). *Modeling and Control of Grain Driers*. Uppsala, Sweden: Dept of Technology, Uppsala University. Rep UPTC 8625.

Ogata, K. (1970). *Modern Control Engineering*. Englewood Cliffs, NJ: Prentice Hall.

Prett, D.M. & Garcia, C.E. (1988). *Fundamental Process Control*. Boston, MA: Butterworth-Heinemann.

Schonauer, S. (1995). Product quality adaptive control system for a food extender. Doctorate dissertation, Department of Agricultural Engineering, Texas A & M University. College Station, TX.

Schonauer, S. & Moreira, R.G. (1995). Development of a fixed-GPC controller for a food extruder based on product quality attributes. Part I: control development, implementation and analysis. *Trans Inst Chem Eng* 73(C4), 200–210.

Seborg, D.E., Edgar, T.F. & Mellichamp, D.A. (1989). *Process Dynamics and Control*. New York, NY: John Wiley.

Skjöldebrand, C. (1991). Moving from art to science in food processing. *Food Technol Int Eur* 115–118.

Takahashi, Y., Rabins, M.J. & Auslaner, D.M. (1972). *Control and Dynamic Systems*. Menlo Park, CA: Addison-Wesley.

Willians, A. (1998). Process control: the way ahead. *Food Proc*, (suppl), October, 2–26.

CHAPTER 2

Food Processing Systems

Many industrial food processes are truly multivariable, meaning that there are strong interactions between process inputs and outputs; such processes often have long delays (dead time), have input constraints, show non–minimum-phase behavior, and have time-varying dynamic behavior. In addition, due to the large variability of raw materials and the complex physicochemical reactions that occur during processing, food-processing systems are characterized by nonlinearities and uncertainties that cannot be handled by classical linear feedback, making the design of a proper controller a difficult task.

Examples of industrial control system applications include pasteurizers, brew kettles, baking ovens, food extruders, continuous deep-fat fryers, heat exchangers, boilers, evaporators, cookers, dryers, and fermentors in processes such as coffee roasting, sugar refining, spray drying, snack manufacturing, aseptic processing, brewing, waste water treatment, and thermal processing.

From the control point of view, a system is composed of controlled variables (operating/process parameters), manipulated variables (process output), and disturbances. In a control scenario, the number of manipulated (independent) variables must be at least as large as the number of controlled (dependent) variables. The controlled variables should measure product quality directly or strongly affect it. The manipulated variables should have a large effect on the controlled variables (large gain), should rapidly affect the controlled variables (minimum delay, small time constant), and should affect the controlled variables directly rather than indirectly (Schonauer & Moreira, 1995a). Chapter 3 presents the background needed to better understand the fundamentals of process control.

This chapter presents a series of examples of food-processing systems, with emphasis on process description, list of inputs and outputs used for control, and dynamic characteristics.

FOOD EXTRUSION PROCESS

A food extruder is a high-temperature, short-time (HTST) bioreactor that transforms a variety of raw ingredients into intermediate and finished products, including pet foods, snack foods, breakfast cereals, pasta products, and texturized protein foodstuffs. It is a unique process since it accepts relatively dry materials, adds liquids to plasticize the raw

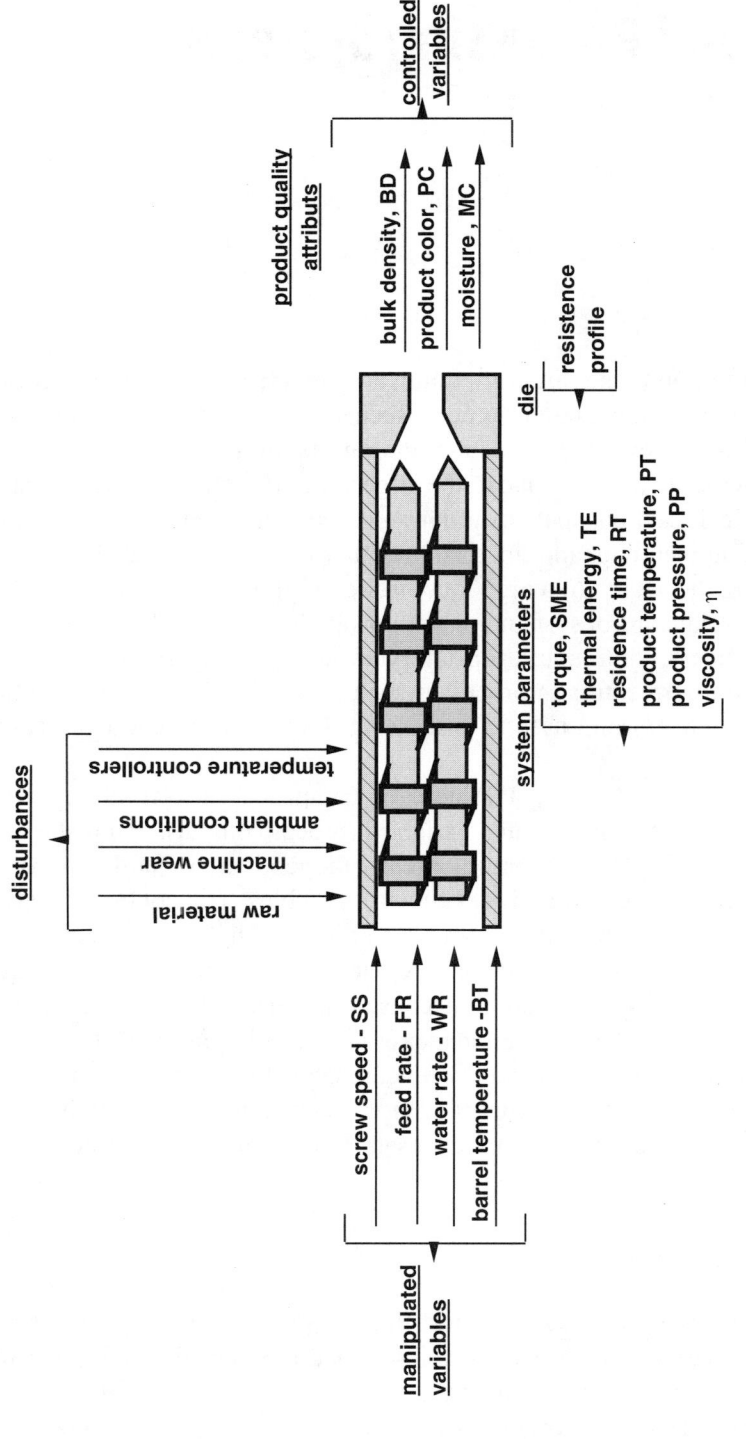

product quality = f(die, system parameters)
system parameters = f(disturbances, operating variables, raw material, start-up)

Figure 2–1 Twin-screw food extrusion process: operating variables, disturbances, system parameters, and product quality attributes

material, gelatinizes starch, denatures protein, and inactivates enzymes before expanding (texturizing) the finished product. It may also be used as a thermal process to eliminate undesirable flavors, to inactivate growth inhibitors, and to modify starch (eg, by cross-linking or substitution) (Schonauer, 1995). The extrusion cooking may be characterized as an energy- and space-efficient technique for producing a variety of products depending on the selected raw materials, process conditions, and mechanical configuration (Harper, 1981).

The cooking extruder combines several unit operations—mixing, cooking, kneading, shear, cooling, and/or final shaping/forming. The combination of operations is possible because of a multitude of controllable variables such as feed rate, total moisture in barrel, screw speed, barrel temperature, screw profile, and die configuration. Figure 2–1 illustrates the critical processing parameters and dependent variables for a twin-screw food extruder.

The major functional ingredient in expanded products is starch. During the extrusion process, starch is plasticized with water and subjected to a specific mechanical and thermal energy treatment. The final product quality is determined by the physicochemical changes that occur in the starch subjected to high-shear, low-moisture, and high-temperature conditions. At the exit of the die, the product is expanded (texturized) as a result of internal steam pressure buildup in the extrudate.

Food extruders can be classified as single screw or twin screw. Single-screw extruders dominate the food industry, but the application of twin-screw extruders is growing because they offer greater flexibility in controlling both product and process parameters. The single-screw extruder exhibits inherent instability because it is a frictionally driven process (Harper, 1981). The twin-screw extruder is inherently more stable. Corotating twin-screw extruder technology offers improved ingredient conveying, enhanced mixing, and more uniform residence time distribution. These characteristics provide improved process stability and/or greater product flexibility.

Schonauer (1995) presented an excellent detailed analysis of the state of art in food extrusion control, discussed the main operating principles, and gave a thorough analysis of the main system variables. The reader is also referred to Schonauer and Moreira (1995a, 1995b) for some aspects of extrusion modeling and control.

Importance of Food Extrusion Automatic Control

Food extrusion processes are multiple-input, multiple-output (MIMO) systems involving strong relationships among mass, energy, and momentum transfer as well as complex physicochemical transformations such as gelatinization and dextrinization of starch and denaturization of proteins.

Advantages of automating cooking extruders include the achievement of constant product quality, automatic start-up and shutdown, process optimization, and more effective information management. The ideal control of a food extrusion process system must aim at controlling the product characteristics by keeping conditions constant at the die despite disturbances to the process such as extruder wear and raw material variability.

System Variables for Food Extrusion Process Control

Extrusion process parameters include raw materials, feed rate, in-barrel moisture content, screw speed, barrel temperature profile, and screw/die configuration. They are considered to be independent operating variables, with the exception of the screw/die configuration, which is typically fixed for a specific application. Changes in the operating parameters will cause changes in dependent process variables such as die product temperature, die pressure, and viscosity, as well as product quality attributes (PQAs), through changes in specific mechanical energy (SME) and thermal energy inputs during the product residence in the extruder.

Food Extrusion Fixed Control Variables

In addition to the controlled, manipulated, and disturbance variables, a food extrusion process also has fixed control variables. These variables are defined as those that have a significant effect on process variables or product attributes but are not manipulated or changed during processing. Typical examples of extrusion fixed control variables are the screw geometry and die size and shape (Schonauer, 1995). The modular aspect of the twin-screw extruder screws allows a variable screw configuration, but it can only be changed after shutdown.

Food Extrusion Manipulated Variables

Extruder variables that have a significant effect on process variables or product attributes and can be readily changed or manipulated during processing are considered manipulative variables for control purposes.

Examples of manipulated variables in a food extrusion process include screw speed, SS, feed rate, FR, and water rate added, WR (or moisture in the barrel, MIB). Moisture in the barrel is the water added in relation to the feed rate that changes the raw material properties. Changes in water added cause changes in SME, pressure, and strain (shear × time) applied to the extrudates, resulting in product differences. Moisture disturbances generally occur from raw material moisture content differences or feed rate fluctuations.

Screw speed, SS, has always been a predominant variable for controlling single-screw extruders because it causes rapid changes in die pressure. Feed rate, FR, is automatically tied to the screw speed in these scenarios due to the flood feeding mechanism (Schonauer, 1995). SS and FR are typically independent in twin-screw extrusion processes. Use of screw speed (or rpm) as a variable for control purposes can affect the residence time of the product in the process. Although feed rate has an impact on finished product quality, most production facilities would want to maximize feed rate (throughput) at all times, so using feed rate in control scenarios must be approached cautiously (Schonauer & Moreira, 1995a).

Food Extrusion Disturbances

Disturbances variables can be classified as internal or external. An *internal* disturbance originates within the system, while an *external* disturbance originates outside the system. In a food extruder, start-up conditions, screw wear, and raw material variability are some examples of disturbances that can affect the process and thus the final product quality.

Several researchers have documented differences in operating equilibrium points caused by the start-up differences or the direction of variable change. Nejman et al (1992) showed that not every state is reproducible in a food extrusion process and that reproducibility is strongly dependent on the perturbation history of the extrusion variables. This nonlinearity confounds the process and emphasizes the need for intelligent control systems. A proper designed control system should be able to prevent major upsets that would cause metastable operating points.

Another important disturbance in food extrusion systems is screw and barrel wear (Schonauer, 1995). Wear causes the process reaction to change over time. As the screw diameter decreases (or barrel diameter increases), the process efficiency decreases. Screw profiles in a food extruder are designed with restrictions, mixing paddles, and leakage clearances to assure mixing and adequate dissipative energy (SME) for the cooking process. As the process wears, these clearances increase, changing the process balance. Shear elements (usually bilobal paddles) lose effectiveness as the metal wear increases the shear passages. Thus, energy transfer in paddle areas decreases with wear. The net effect of wear is a composite between effects of screw, barrel, and bilobal paddle wear. Appropriate design of screw/paddle sections could be self-compensating (Schonauer, 1995).

Another source of disturbances is the wear of the die. As a restrictive element, the die undergoes wear that relieves the restriction. Again, this changes the balance of the process, resulting in different pressures and shear input (Schonauer, 1995).

Since raw materials in food extruders are of biological origin, they often vary in composition. Changes in raw material affect viscosity and therefore screw fill, which creates instantaneous changes in outputs.

A twin-screw food extruder process generally shows fluctuations in temperature, pressure, and torque even when operating in nominal steady-state conditions. The origins may be multiple, including temperature control through switching of heating and cooling phases, dependence of the pressure transducer response on temperature, pressure effects due to screw rotation, and surging of material related to screw design (Schonauer, 1995).

Food Extrusion Controllable Variables

Controlled variables are the parameters of the process that indicate product quality or the operating condition of the process. In a food extrusion process, the controlled variables are the product quality attributes, such as density, color, and moisture content.

Temperature, feed rate, pressure, moisture content, and composition of the dough affect the appearance of the product. The skillful operator will manipulate water added, screw speed, barrel temperature, and possibly feed rate to maintain quality within specifications. The final cooking zone of the extruder barrel is that area where homogeneity is achieved and texturizing (expansion) occurs. If process conditions (temperature and pressure) are maintained constant at the die, product quality will also be constant. Temperature and pressure typically increase rapidly, and the resulting material viscosity is such that the extrudate will be expelled from the extruder die to yield the desired final product texture, density, color, and functional properties.

Several investigators have noted the difficulty with direct on-line measurement of finished PQAs. This remains one of the limitations for direct control of quality. Control of finished product quality can rarely be achieved by regulation of pressure due to the many extrusion variables that influence pressure, including feed rate, temperature, screw speed, and die fouling.

Typical PQAs measured off-line include color of extrudate, bulk density, expansion (diameter, lineal, ratio), texture (breaking strength), water solubility index (WSI), water absorption index (WAI), gelatinization, dextrinization, sensory attributes, dimensional parameters (diameter and length), and surface texture (Schonauer, 1995).

If implemented on line, quality measurement can replace one to two control personnel and eliminate a large portion of off-specification product. Finished-product quality can be monitored and adjusted through use of machine-vision sensors that measure dimensional and surface parameters of the product (Schonauer & Moreira, 1995a).

Food Extrusion Dynamics Analysis

This work is described in more detail by Schonauer and Moreira (1995a, 1995b). The food extruder used was a corotating, intermeshing twin-screw extruder (MPF50, APV Baker; Grand Rapids, MI). Figure 2–2 shows a diagram of the extruder with the process variables commonly monitored in a food extrusion process.

The barrel and the screw configuration were typical of a puffed–snack food production extruder. The barrel length was 15:1 L/D. The MPF-50 was a fully intermeshing corotating design with a 50.8-mm (2-in) diameter screw, powered by a 25-hp DC motor. The barrel was of the clamshell style, with five zones of independent temperature control. Heat was supplied by electrical cartridge heaters. Cooling was provided by chilled water parallel to the heater circuits. The heating and cooling control schemes were time-proportional PID schemes and on/off controllers, respectively. The raw material was degermed yellow cornmeal (moisture content = 12.5% w.b.; protein = 7.6% d.b., carbohydrate = 87.3% d.b.).

Data Collection and Analysis

Experimental data were collected at 0.5-second intervals (2 Hz) by two data acquisition systems, Genesis (Iconics; A.) and In-Touch (WonderWare; CA), through a 486 PC. The steady-state data were analyzed on a Sun (Sun; CA) workstation using Matlab (Mathworks; MA) software.

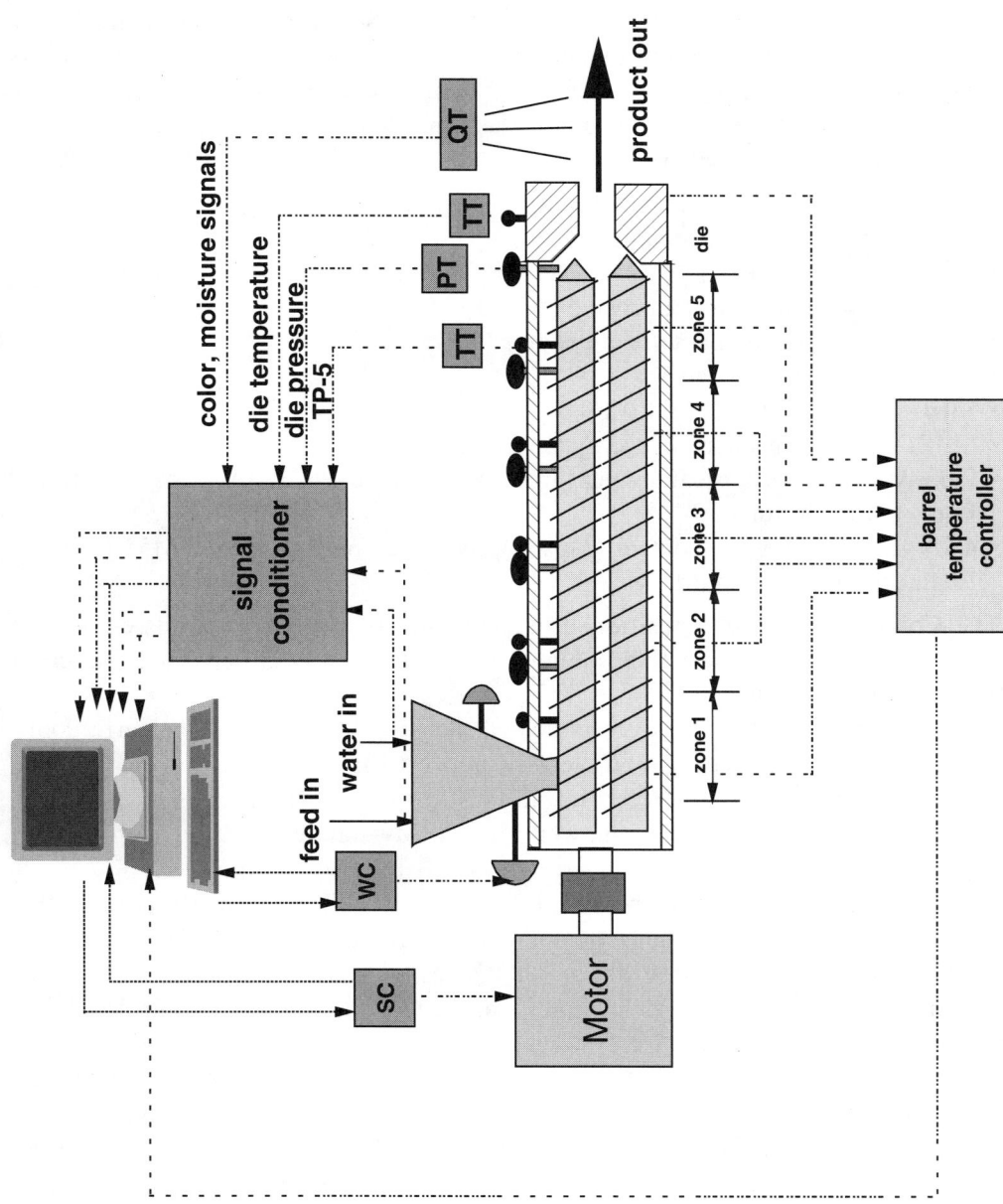

Figure 2–2 Twin-screw food extrusion process and control system

Instrumentation

Six zones of product temperature were monitored (five barrel zones and one die neck) to describe the product temperature profile. J-type thermocouples located in the Sensotron (Hunting Beach, CA) pressure transducers were inserted flush with the barrel surface. Pressure was measured at each of the same zones using Sensotron pressure transducers. The extruder was instrumented to provide heating and cooling information. Flow meters were installed in each of the cooling-zone flow streams, and thermocouples monitored the inlet and outlet temperatures. Barrel heater status (on/off) was also monitored. Examples of logged variables are given below:

- FR—yellow cornmeal feed rate (kg/h)
- WR—process-water-added rate (kg/h)
- SS—screw speed (rpm)
- BT5—barrel temperature in zone 5 (°C)
- PTDIE—product temperature at the die
- TOR (%)—motor % load
- MIB—moisture in barrel (% w.b.)
- SME—specific mechanical energy (kW/h/kg)
- CB—yellow/blue indicator, measured on line with a Colorex sensor (Infrared Engineering; Waltham, MA)
- CM—collect moisture content (% w.b.), measured on line with MM55 (Infrared Engineering; Waltham, MA)

The Colorex sensor was positioned over the bed of collects (extrudates) on the extruder takeout conveyor at a fixed height. The Colorex was calibrated using color standards prior to each run.

Dynamic analysis of the process indicated that two PQA variables clearly dominated—CB and CM. They exhibited clear response to independent process variable changes with good steady-state correlation to bulk density (BD) and sensory attributes. One process variable, PTDIE, also showed high steady-state correlation with BD but showed slow response dynamics.

Process dynamics were driven by WR, BT5, FR, and SS in that order of priority. WR change or FR change (WR constant) yielded similar responses due to MIB similarities. MIB in actuality was the primary driver for the response. BT5 was important in the response of PTDIE, but excessive delay times made it less important as a control variable. SS responded quickly, but gains were minuscule. FR will be used, but practicality will require it to be fixed. When FR is held constant, use of WR mirrors MIB, making the choice of using WR or MIB inconsequential.

Relative gain array analysis confirmed a multivariable rather than multiloop system because high interaction indicated that decouplers would be required for multiloop control. The gains indicated a nonlinear system over the response range explored.

Another important aspect in designing a control system for food processes is to consider the process constraints. It may be necessary to impose limits on the input to maintain a stable operating regime, protect equipment, or prevent process shutdown. In the case of the food extrusion process considered above, the constraints imposed in the inputs were (Schonauer & Moreira, 1995b)

- FR: 36.4 to 72.7 kg/h
- WR: 2.3 to 5.5 kg/h
- SS: 430 to 500 rpm
- MIB: 17% to 22% w.b.

CONTINUOUS DEEP-FAT FRYING PROCESS

Deep-fat frying is one of the oldest processes of food preparation. For decades, consumers have desired deep-fat fried products, such as potato chips, french fries, doughnuts, extruded snacks, fish sticks, and the traditional fried chicken products, because of their unique flavor-texture combination.

The frying technology is important to many sectors of the food industry: suppliers of oils and ingredients, food service operators, the food industries, and manufacturers of frying equipment. The amounts of food fried and oils used at both industrial and commercial levels are vast. The United States produces over 2.5 million metric tons (MMT) (5 x 10^9 lb) of snack foods per year, the majority of which are fried (Snack Food Association, 1997). More than 500,000 institutional and commercial restaurants in the United States use approximately 1 MMT (2 x 10^9 lb) of frying fats and oils annually (O'Brien, 1993).

Deep-fat frying is defined as a process of cooking and drying through contact with hot oil that involves simultaneous heat and mass transfer. The quality of the products from deep-fat frying depends not only on the frying conditions but also on the type of oils and foods used during the process. Oils play a dual role in the preparation of fried foods because they serve as a heat transfer medium between the food and the fryer and also contribute to the food's texture and flavor characteristics.

In deep-fat frying of foods, the temperature of the heated oil, the frying time, and the fryer type (batch or continuous) are factors that affect the process. The chemical composition of the frying oil, the physical and physicochemical constants, and the presence of additives and contaminants also influence the frying process. Additives or contaminants can have a marked effect on the palatability, digestibility, and metabolic utilization of a fried food. The food weight/frying oil volume and surface area/volume ratios determine how much fat penetrates the food.

A deep-fat fryer consists of a chamber where heated oil and a food product are placed. The size of the fryer may range from small oil baths (used in food services) to large continuous industrial baths. In a batch fryer, the product is placed in small net cages that are lowered into the oil bath. In a continuous fryer, a conveyer belt transports the product through the bath. The product is often pushed through the bath by means of screens and/or paddles.

The speed and efficiency of the frying process depend on the temperature and quality of the frying oil. The temperature of the oil is usually between 150 and 190°C. Oil turnover time (mass of used oil/oil usage rate) is generally around 10 hours. It is important to understand what happens to the temperature, moisture content, and oil content of the product during the frying process to determine safe temperatures and turnover times of the frying oil for a given fryer type (Moreira et al, 1999).

A continuous fryer system consists of at least five independent sets of equipment: (1) the kettle or tank containing the frying oil; (2) a heating unit with a control system for generating thermal energy; (3) a conveying system for moving the product into, through, and out of the frying process; (4) a fat system, which pumps and filters the frying oil; and (5) an exhaust system for removing the hot vapors emerging from the product.

The output capacity of continuous fryers varies from 250 to 25,000 kg/h for french fries, 100 to 2000 kg/h for tortilla chips, and 100 to 2500 kg/h for potato chips. A continuous frying system used for frying nuts can have a capacity from 500 to 3000 kg/h.

Continuous Fryer Control Variables

Examples of manipulated variables in a continuous frying process include oil temperature, OT, submerger speed, SS, and takeout conveyor, TC. Change in the oil temperature will generally result in a slow response. Change in the submerger speed affects the residence time of the product in the fryer. The takeout conveyor regulates the temperature change in the product after it has been removed from the oil bath.

The most important controlled variables for a continuous fryer are those related to the PQAs. Color, oil content, and moisture content are some examples of these variables.

Examples of disturbances that can affect the frying process include ambient conditions (temperature), oil quality, and raw material variability.

Continuous Fryer Dynamics Analysis

Figure 2–3 presents a schematic of a multizone continuous fryer used in the study by Brescia and Moreira (1997). This is a "split-apart" type of fryer. The hood and the product conveyor system can be raised for cleaning and inspection. The fryer oil capacity is about 327.5 kg. The oil was "turned over" periodically (approximately every 10 hours) to keep a constant oil quality and level.

The frying oil is heated by an external heat exchanger. Recirculated oil from the heat exchanger is introduced to the kettle at several points along the fryer. At the end of the fryer kettle, the oil is withdrawn from the container for heating and circulation.

The raw material is dispensed from a conveyor belt and dropped into the hot oil and conveyed along the cooking zone (a distance of 5.59 m) by means of a series of rotating paddles (not shown), which dunk, separate, agitate, and control the advance of the product as it is fried.

Oil flows from the frying zones through exit pipes, while fresh oil and/or recycled oil is introduced through inlet pipes. After the products pass the free zone, they will contact a series of flighted submerger conveyors (total length of 4.27 m), which hold the product below the surface of the oil while controlling their advance through the fryer. The fried products are then removed from the kettle by means of a takeout conveyor (2.7 m long)

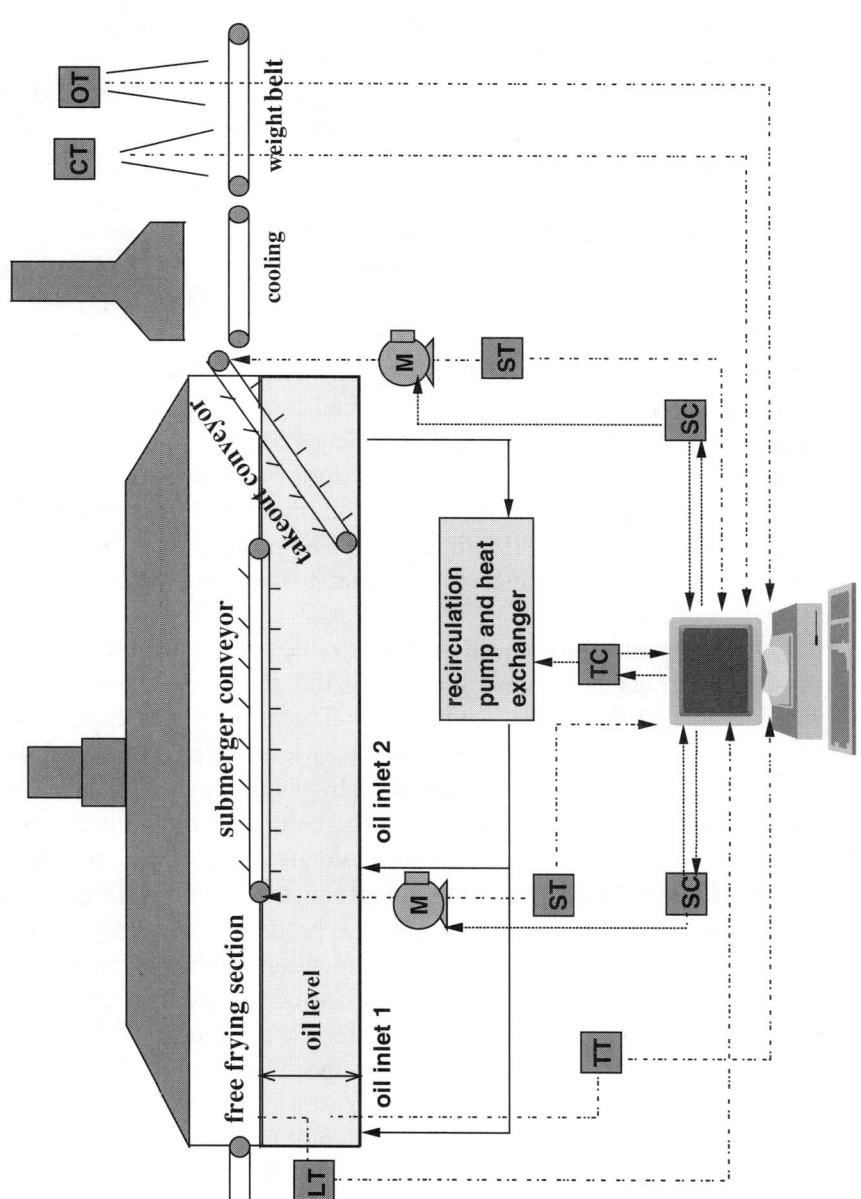

Figure 2–3 Continuous deep-fat frying process and control system

positioned at a 30° angle, which deposits the products onto a cooling conveyor. The cooling conveyor is of open-weave mesh construction and is located within a chamber that contains an opening to an exhaust fan. Drainage of oil from the chips along the takeout conveyor is enhanced by heat, which may be provided by flow of fryer exhaust gases and/or overhead heaters. After this section, a weight belt is in place, where the sensors for PQA measurements are located. The quality attributes measured on line are color, moisture content, and oil content. The same sensors described earlier in the "Food Extrusion Dynamics Analysis" section were used to measure the PQAs in this process.

The oil temperature typically ranges from 188°C to 193°C, with 190°C being the optimum temperature. The speed of the conveyors is expressed as a percentage of the maximum attainable speed. The optimum operating conditions of the submerger and takeout conveyors are 67% and 70%, respectively. Optimum operating conditions result in a residence time of 105.27 seconds up to the color sensor and 115.37 seconds up to the moisture-oil sensor.

Dynamic analysis of the process showed that color of the product decreased when the temperature of the oil was increased. The color turned to a darker yellow as a result of a browning reaction because the product was more cooked. The moisture content of the product also responded inversely to an oil temperature positive change. Increasing the temperature caused more water to evaporate from the product during frying. The oil content increased due to an increase in oil temperature. The oil occupied the spaces left by the water during frying. Therefore, increasing water loss during frying resulted in increased final oil content of the product.

Increasing the submerger speed increased the color value and moisture content while decreasing the oil content. This is because increasing the submerger conveyor speed decreased the residence time of the product in the fryer. The product was exposed to the hot oil for a shorter period, so the color turned lighter as less water evaporated during frying. As a result of less cooking, less oil was absorbed by the product.

Increasing the takeout conveyor speed increased the color, left the moisture content the same, and increased the oil content. Reducing the residence time of the product in the takeout conveyor allowed less oil to drip off from the product surface, resulting in higher oil content during cooling. In the takeout conveyor, the product was not in the oil but was still under the fryer hood, thus maintaining its temperature higher than the ambient temperature. The decrease in temperature of the oil also caused the viscosity at the surface of the product to increase, thus making it more difficult for the oil to drip off the product's surface. Therefore, an increase in the takeout conveyor speed resulted in more oil in the final product. Color changes in this case could be the result of density reduction of the product as oil content increased, thus resulting in lighter color (more yellow).

Step tests revealed that the color and the oil content should be used as the controlled variables in the control scenario, since they show the largest and fastest response to input changes. All the analyzed variables turned out to be highly correlated. An important correlation was the one between color and moisture content, indicating that it would be inefficient to try to control color and moisture content at the same time. The high interaction among control loops suggested the use of a multivariable control scheme.

The fastest response was obtained with the submerger speed and the slowest one with the oil temperature. Oil content reacted very fast to changes in submerger speed and takeout speed, but slowly to oil temperature changes.

The input constraints for the process described above were defined as the following (Brescia & Moreira, 1997):

- OT—oil temperature: 188 to 193°C
- SS—submerger speed: 50% to 75%
- TS—takeout conveyor speed: 30% to 90%

HIGH-CAPACITY CONTINUOUS-FLOW GRAIN-DRYING PROCESS

The control of continuous-flow dryers consists of controlling the speed of the unload augers and/or the temperature of the drying air. The basic control objective is to minimize overdrying/underdrying of the grain moisture content while maximizing dryer capacity under acceptable conditions of grain quality deterioration and energy consumption.

Adequate control of the dryer is essential to reduce the quality damage because of overdrying and underdrying of the grain. Overdrying causes breakage and stress cracks in grains. Overdrying of grain is also costly, mainly because the price of grain is based on a specific moisture content. For example, corn is priced in the United States at 15.5% moisture content. The weight or shrinkage loss of 25,400 MT by 1.0% (ie, to 14.5%) results in shrinkage losses of 297 MT and costs in the United States of $17,544 when corn sells at $59.1/MT and of $29,240 when corn sells at $98.4/MT.

In addition to the loss in weight, overdrying is costly because of the extra energy required to dry the grain beyond the moisture content at which it is priced by the market. A 1.0% overdrying requires 87.8 MJ/MT in energy and costs $10,570 per 25,400 MT, assuming that the energy cost is $5.00/1055 MJ.

Underdrying of grain is more serious, since wet spots and spoilage may result. If considerable mixing takes place soon after drying, some underdrying is not serious, as moisture equalization will occur.

Continuous-flow grain dryers are usually manually controlled. Manual control of a drying system is a complicated task. It requires extensive experience on the part of the dryer operator not to overdry or underdry the grain.

The development of automatic controller for grain-drying processes has encountered considerable difficulties. The two main reasons are (1) the complexity of the process (nonlinear and time varying) and (2) the lack of reliable commercial on-line moisture meters. Developments in digital computers, microelectronics, and control theory have contributed greatly to the implementation of control systems in the drying processing industry.

Various high-temperature dryers are used in the grain industry. The recommended air temperature and airflow rates vary among the dryer types. The basic off-farm continuous-flow dryer designs are (1) cross-flow, (2) concurrent flow, (3) counterflow, (4) mixed flow, (5) rotary, (6) fluidized bed, and (7) spouted bed (Brooker et al, 1992).

The four major continuous-flow grain-drying systems are cross-flow, concurrent flow, counterflow, and mixed flow. They are classified by the relative direction of the grain and air movement through the dryer. In cross-flow dryers, the air and grain move in perpendicular directions; in concurrent-flow dryers, they move in parallel directions; and in counterflow dryers, they move in opposite directions. The flow of the air and the grain in mixed-flow dryers is a combination of cross-flow, concurrent flow, and counterflow.

Each continuous-flow grain dryer is supplied with a cooling section in which the hot, dried grain is cooled to within 5°C of the ambient dry-bulb temperature. During the cooling process, 0.5% to 1% of moisture is removed, depending mainly on the residence time of the grain in the cooler (Brooker et al, 1992). The relative directions of the grain and the air in the drying and cooling sections are the same in the major dryer types except in the concurrent-flow dryers (in which the grain is cooled by counterflow).

Continuous-Flow Grain Dryer Control Variables

The input variables that affect the moisture content of the grain in a continuous-flow dryer are

- The grain residence time
- The drying air temperature
- The air flow rate

Theoretically, any of these variables can be used to control the process. However, in practice, continuous-flow grain dryers are manually controlled by changing the residence time and keeping the air flow rate and the air temperature constant.

The air flow rate is maintained at a level that is a desired compromise between capacity and specific energy consumption (higher air flow rate gives higher capacity but higher energy consumption, too). The air temperature is generally kept constant at a maximum value, indirectly determined by the maximum allowed grain temperature. The residence time is determined by the grain flow rate or the discharge rate, which is determined by the speed of the electric motor of the discharge auger.

In continuous-flow grain dryers, the main dynamics are created by the changing state of the grain as it flows through the dryer. In the ideal case, this flow is homogeneous and the grain velocity is equal in all parts of the dryer. However, in practice, the process dynamics are affected by such flow disturbances as shrinkage and mixing of the grain during drying.

Inhomogeneities of the grain (eg, initial moisture content and drying characteristics) cause rapid fluctuations in process outputs. These fluctuations are considered process noise with statistical properties that are dependent on the variability of grain characteristics.

Another source of disturbance to the process is the discharge mechanism characteristics. These are normally dependent on the grain properties (moisture content, grain shape). The actual discharge rate may vary in spite of a constant motor speed.

Nybrant (1986) showed that the dynamics of continuous-flow grain dryers (cross-flow and concurrent flow) were nonlinear. Keeping the control input proportional to the inverse of the discharge rate made the process model linear. In the case of a concurrent-flow grain dryer, Nybrant (1986) found that the process showed non–minimum-phase behavior (the process model had a zero on the outside of the unit circle).

Continuous-Flow Grain Dryer Instrumentation

An automatic controller has been implemented and tested on a cross-flow grain dryer. The work is described in detail by Liu (1998). The physical input to the process is the discharge rate, determined by the reference voltage to the discharge motor servo. The main output is the outlet moisture content. The main disturbance is caused by changes in grain inlet moisture content.

The dryer has a cylindrical shape, with the drying air flowing through the grain column from the inside to the outside. The dryer consists of a drying section, cooling section, unload system, fan, and burner (Figure 2–4). The upper part of the tower consists of the drying section, with a grain turn-flow located midway in the drying column; the lower part of the tower is the cooling section, in which ambient air is used to cool the grain before it leaves the dryer. The air exiting the cooling section is directed to the inlet of the burner and is recycled.

A sampling box was added to the outlet of the dryer to facilitate installation of a moisture sensor. The sampling box keeps the probe of the sensor surrounded by a constant flow of grain.

The control system consists of moisture and temperature sensors, the unload mechanism of the dryer, and the computer plus software. The computer communicates with the sensors and the drive motor through a data acquisition card. One digital I/O is used to control the auto-control contact, three analog inputs are used by the sensors (two for moisture and one for temperature), and one analog output is for the motor control of the unload auger.

The rpm of the unload auger is proportional to a 0- to 5-V input to the controller of the unload motor. The maximum speed of the auger at 5 V of input is 293 rpm and results in the maximum discharge capacity of 60.2 tonne/h (51.2 tonne of dry matter per hour).

Two moisture sensors (TRIME-GW, MESA Systems Co.; Framingham, MA) were employed to measure the inlet and outlet grain moisture contents. The moisture sensor is based on time-domain reflectometry technology; two metal probes are emerged in the grain column; a high-frequency pulse is generated and propagated along the probes; and at the end of the probes, the pulse is reflected back to its source. The moisture content of grain is related to the transit time of the pulse in the probes. The inlet sensor was placed at the top of the dryer, and the outlet sensor in the sampling box.

An 8-m-long PT100 resistance temperature detector (RTD) was installed in the drying-air plenum to measure the average drying-air temperature. A G418 RTD conditioner was used to convert the temperature signal to 0 to 5 V of output.

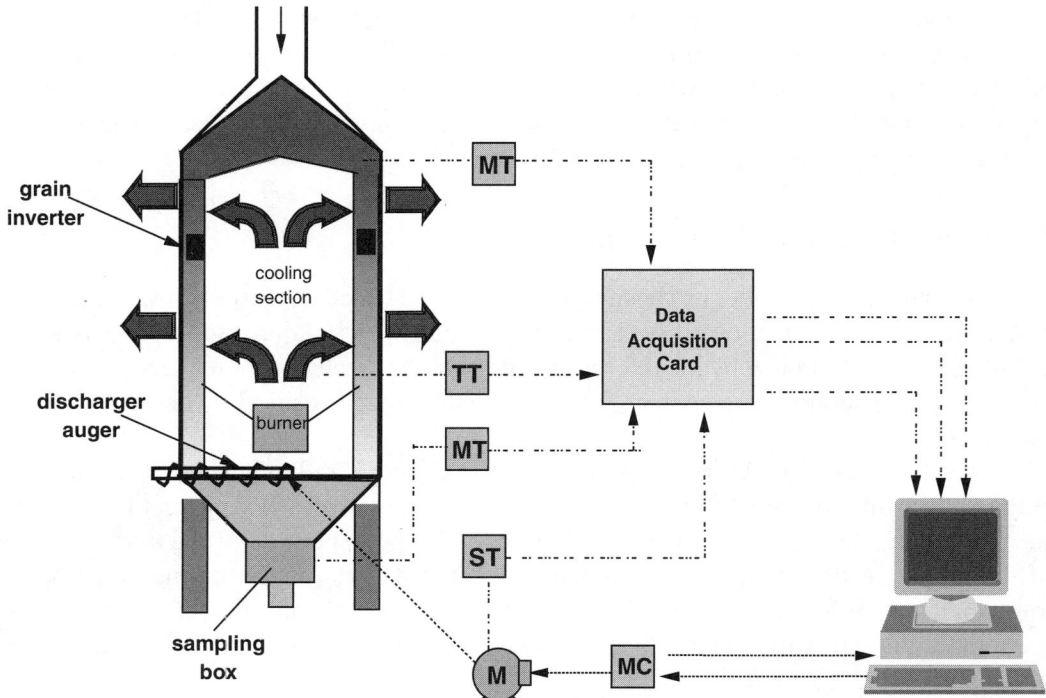

Figure 2–4 Cross-flow grain-drying process and control system

MULTI-EFFECT EVAPORATORS

The control of evaporation processes is important in the food industry to reduce the irregularity of final product quality and improve energy use.

Evaporation can be defined as the concentration of a solution by boiling the solvent. The food industry uses this technique for manufacturing many products, including tomato paste, orange juice concentrate, milk concentrate, and sugar.

The simplest evaporator system, known as a single-effect evaporator, consists of one heat exchanger, one liquid pump, and one vapor/removal section. Single-effect evaporation may be acceptable for some systems but is generally energy inefficient in large systems. The combination of several single-effect evaporators into a configuration known as a multi-effect evaporator improve the process efficiency. In a multi-effect evaporation system, the vapors produced in one effect are used as the heating medium in downstream effects (Perry & Green, 1997).

In a multi-effect evaporator, the steam is used only in the first effect. In a triple-effect evaporator, for example, the partially concentrated product leaving the first effect is introduced as a feed into the second effect. After additional concentration, the product from the second effect becomes the feed for the third effect. The product then leaves the third-effect evaporator at the desired concentration. The feed-flow system arrangements used in the industry are backward, forward, and parallel (Perry & Green, 1997).

Evaporators can be classified as circulating and noncirculating types. Circulating evaporators can be of natural or forced-circulation types. In natural circulation evaporators, the basic moving force in the heat exchanger is the difference in densities that exist within the system. Forced-circulation evaporators use a recirculating pump to maintain the desired liquor rate on the heat exchanger and thus a good heat transfer rate (Perry & Green, 1997).

Circulating evaporators are either short-tube vertical (STV) evaporators or long-tube vertical (LTV) evaporators. STV evaporators have a heat exchanger (calandria) placed inside the cylindrical evaporator body. In an LTV evaporator, the heat exchanger can be integrally mounted in the evaporator body or connected externally.

Noncirculating evaporators are divided into two types: rising film and falling film. In a rising-film evaporator, a low-viscosity liquid is allowed to boil inside a long vertical tube. A falling-film evaporator has a thin liquid film moving downward under gravity on the inside of the vertical tubes.

Control Variables in an Evaporator System

The most important control objectives for an evaporator are to provide high-energy efficiency, precise composition control, and maximum throughput with adequate safety and environmental protection. In most evaporators, the main goal is to accurately control the product composition at a low energy consumption per pound of feed material (Nisenfeld, 1985).

In an evaporator, the most common manipulated variables include the liquor from each effect, the steam flow to the first effect, condenser-cooling medium flow, pressure, and feed rate. Typical controlled variables are the product composition and product inventory (Ritter & Andre, 1970).

Level measurement and control are particularly important in evaporator systems. Level is a measure of the accumulation of liquid in the evaporator and is directly related to the residence time of the liquor in the system. In a natural circulation system the liquid head on the heat exchanger is an important factor in establishing the recirculation rate. In forced-circulation systems, level must be controlled to ensure the proper submergence (Nisenfeld, 1985).

In designing a control scheme for any evaporator, or any food-processing system, the physical limitations of equipment must also be taken into consideration. Equipment limitations define the constraints within which a continuous control scheme operates. If the operating point of the evaporator is near the equipment constraints, override control techniques may be used to maintain operations within the constraints. Typical constraints in an evaporator system are condenser capacity, vacuum-system capacity, heat-exchanger capacity, and internal or external stream-flow capacities (Nisenfeld, 1985).

Disturbances that can affect the evaporator system operation typically enter through the feed as rate, composition, or temperature changes. In the heat exchangers, disturbances to the evaporator operation may result from changes in heating medium flow, quality, or pressure. Condenser operations can also be a source of disturbances through changes in the coolant flow, temperature, or a system pressure, all of which affect the energy outflow from an evaporator. Environmental conditions also can disturb the evaporator performance by affecting the condenser duty or the heat loss through the evaporator walls. Fouling in the heat exchanger can also affect the evaporator performance. It generally appears as a constraint function in an evaporator control system (Nisenfeld, 1985).

Evaporator Control System

A double-effect vacuum evaporator was chosen by Ritter and Andre (1970) to study the design of a conventional control configuration (Figure 2–5) to concentrate 81 kg/h of a sugar solution from 5% to 15% w.b. The process consisted essentially of an STV first effect, an LTV, forced-circulation second effect, a cyclone separator in the external circulation path around the second effect, and a water-cooled total condenser (not shown) that was maintained at the desired operating pressure by means of a steam ejector and pressure controller. Both saturated steam and aqueous feed solution containing 3 weight percent sugar were introduced into the first effect, while the concentrated product containing 9 weight percent sugar was withdrawn from the circulating solution in the second effect.

Three manipulated (S_i, B_1, and B_2) and three controlled (C_2, W_1, and W_2) variables were selected in this study. The second-effect product composition, C_2, was controlled through manipulation of the steam rate, S_i, while the first-effect, W_1, and second-effect, W_2, inventories were controlled through manipulation of the first-effect product rate, B_1, and the second-effect product rate, B_2, respectively.

The study demonstrated that product composition is relatively insensitive to variations in product flow rate. The first-effect product composition was less sensitive to load disturbances (feed rate and concentration) than the second-effect product composition. It was concluded that the critical composition was most sensitive to fluctuations in the evaporator heat load and should thus be controlled through direct manipulation of that variable. Inventory control is thus readily achieved by manipulating the appropriate flow rate from each effect, without undue influence on the product composition.

Although pressure has a great influence in product composition, the pressure of the second effect was ignored as a possible controlled variable. To reduce steam use, pressure was maintained at the lowest possible value in this study.

BREW KETTLE OPERATIONS

The production of beer consists of converting barley (or another type of grain), water, and hops into a fermentable liquid. The brewing process has four steps: malting, mashing, brewing, and fermenting. Once the beer is produced, it is filtered, pasteurized, and aged before being packed. The decisions that each brewer makes during the process, and then in the fermentation process itself, result in an array of beer varieties (Hardwick, 1995).

In the brewing industry, the quality of the final product is very important. Improper control of the boiling of wort in the brew kettle, for example, can result in a final product

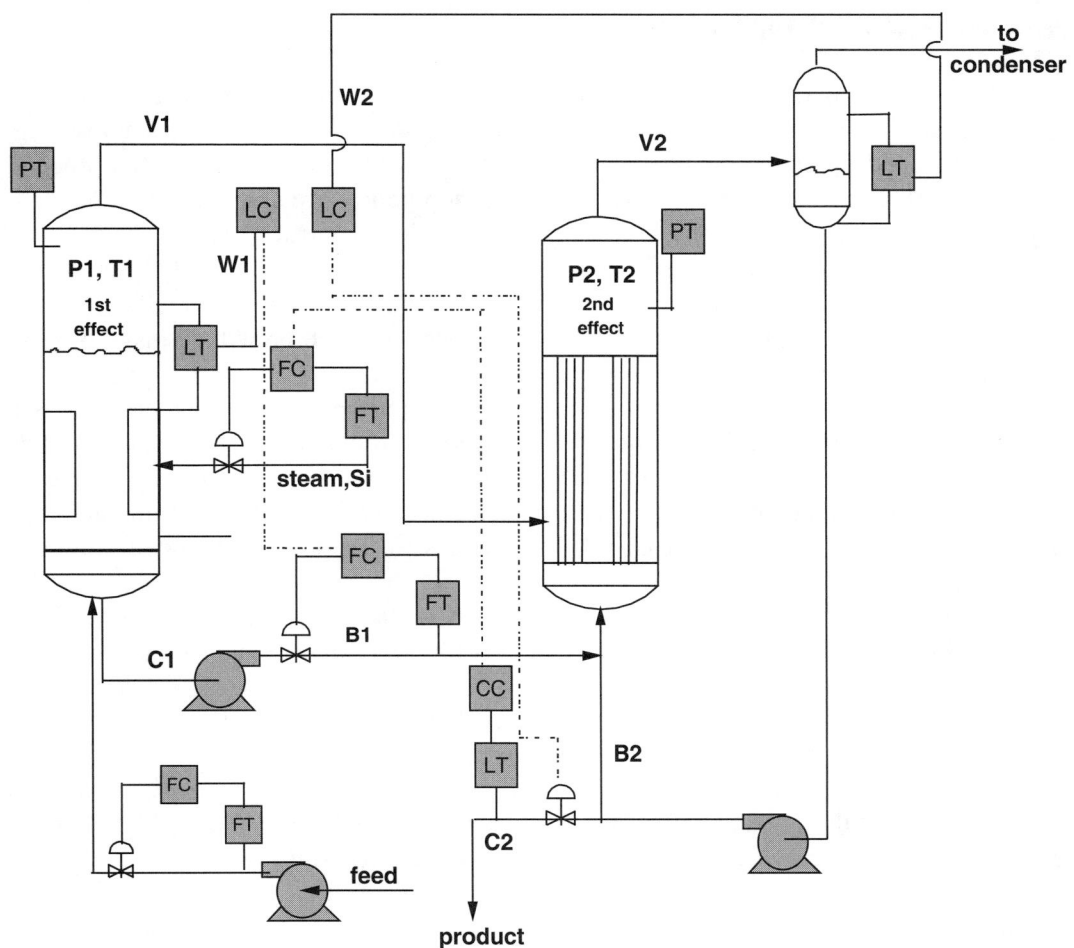

Figure 2–5 A double-effect evaporator process and control system

with insipid color, excessive bitterness, hazy appearance, and inconsistent taste from bottle to bottle (Lewis, 1995).

For product quality and consistency, boiling should start as soon as the kettle is filled and continue for 2 to 3 hours through the addition of hops and runoff phase. Plant productivity is maximized by using large brew kettles. The process efficiency is improved by maximizing heat transfer and evaporation rate. In addition, minimizing the fluctuations on the steam draw reduces the peak loading and energy loss through vents (Lewis, 1995).

The control of the brew kettle process is very difficult due to the following factors: the nonlinear characteristics of the boiling process; the critical changes in the boiling characteristics as the hops are added; and the pressure differences caused by either ambient conditions or venting fans switching on and off. Another important factor is that a boilover represents a serious safety hazard (Hardwick, 1995).

Brew Kettle Control Application

A stainless steel kettle with a bottom and two side steam jackets and a central percolator was controlled using BrainWave model predictive control software (Food Engineering, 1998) developed by Universal Dynamics Technologies (Vancouver, BC). In the process, wort enters through the bottom of the kettle, and hop pellets are added in three batches. Figure 2–6 illustrates a schematic of the system with all the control variables.

The main problem with the brewing process is process repeatability. Due to the nonlinear characteristics of the boiling liquid, it is difficult to maintain repeatable setpoints at the brew kettle. In addition, changes in the formulas for each brew type make process control a difficult task.

The difference between the nonboiling and the boiling liquid was identified as a suitable parameter to control. The control objective was to optimize evaporation and thus maintain the product quality by controlling the process to a certain boil height for each phase of the boil. Reducing boil height variations filled the kettle more completely, thus increasing the plant capacity. Measured disturbances to the process included hop flow rate and wort flow rate.

The manipulated variable selected for the system was the steam flow rate, and the controlled variable was the liquid boil height. Feedforward signals included data from the hop blower and wort flow meter.

The application of this control strategy resulted in an improved evaporation rate, reduced steam pressures at peak loading, and improved hop utilization.

This example illustrates the application of one of several commercial available software products designed to control industrial processes that are nonlinear, higher order, dead-time varying, or involving changing time constants and gains.

PASTEURIZATION SYSTEMS

In the food industry, pasteurization is one of the most used heating processes to destroy pathogenic microorganisms. It is widely used for liquid foods such as milk, ice cream, juices, and beer. The process is designed to destroy and inactivate harmful microorganisms in each particle of the product by using a defined time/temperature treatment while minimizing organoleptic and nutritional properties damage of the product.

The pasteurization process consists of four main stages:

1. preheating of incoming product
2. final heating to pasteurization temperature using hot water or steam
3. holding the product for a fixed time to achieve the desired microbiological effect
4. cooling of product using water/refrigerant

Other ancillaries can also be included, depending on the process requirements, such as separation, deaeration, and homogenization. The main requirements for the microbiologically safe operation of the pasteurization process are (1) selection and maintenance of

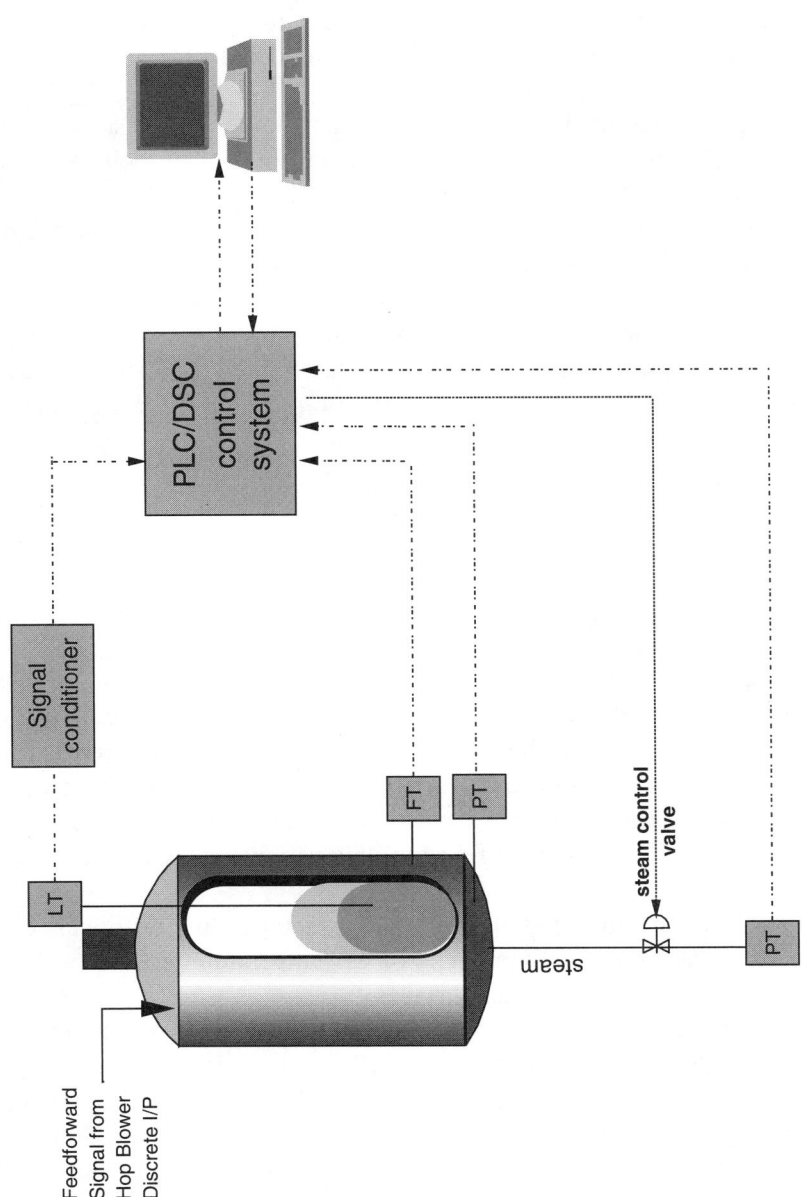

Figure 2–6 Brew kettle process and control system

appropriate pasteurization conditions and (2) prevention of product recontamination downstream of the pasteurizer (Hasting, 1992).

The key components of a pasteurization process are (1) heat exchanger, (2) flow control, (3) holding tube, (4) flow diversion, and (5) process controller.

The main function of the heat exchanger is to heat and cool the product to the desired temperatures. The most common types used in the food industry are plate, tube, and scraped surface.

The flow control is designed to provide a constant flow to the pasteurizer under all normal operating conditions. Some examples of flow control devices include a mechanical device, a positive timing pump, and a flow control valve/flow meter.

In the holding tube, each particle is guaranteed to receive the desired heating treatment. The product is held at least a minimum time at the pasteurization temperature. Holding tubes are available in tubes and plates with heat exchangers. Product viscosity and flow rate are the key parameters in holding-tube design.

Flow diversion systems are used to prevent unpasteurized product from leaving the holding tube. Examples of flow diversion devices are pneumatic valves. The diversion system should be located at the end of the holding tube.

The process controller's main function is to monitor and control critical temperatures and activate the flow diversion system in case the product came out from the holding tube at a lower temperature.

Pasteurization Control Variables

Generally, the main objective of a pasteurization control system is to ensure that the desired quality standards (temperature and nutritional) are achieved, regardless of disturbances to the process such as holding-tube fouling, heat exchanger wear, and raw material variability. The main manipulated variables are the flow rates through the timing pump (to the heater), the flow diversion device (from the holding tube), and the steam valve (to the water heater). An example of a controlled variable is the product temperature from the holding tube.

The quality control and ability to maintain desired temperature profile are dependent on the response speed, accuracy, and sensitivity of the temperature probe as well as on the reliability and consistency of maintenance services.

Hasting (1992) mentioned that it is very important that all the pasteurizer equipment downstream of the holding tube be easy to clean (CIP), bacteria tight, able to be disinfected prior to using, and free of dead spaces. Contamination in the pasteurizer is of great concern. The product can be contaminated by the unpasteurized product due to perforation of heat exchanger tubes/plates, backmixing during diversion, or valve leakage and by the fluid media due to perforation of the heat exchanger tubes/plates.

Milk Pasteurization Process Control

According to the US Public Health Service (1989, 1993), the ordinance to produce Grade A pasteurized milk (PMO) involves the process of heating every particle of milk in properly designed and operated equipment to one of the seven temperature and holding-time combinations. Before 1988, only analog controllers were allowed to control the pasteurization process. It was only in 1988 that the Milk Safety Branch of the US Food and Drug Administration (FDA) authorized the use of microprocessor-based control systems for the pasteurization of dairy products. However, a separate analog controller is still required to monitor and control temperatures and flow-rates of these processes.

Negiz et al (1998) developed a controller for a high-temperature, short-time (HTST) pasteurization system. The system used was a pilot-scale continuous milk pasteurizer with a capacity of 400 gpm. The HTST system consists of a balance tank, a regeneration unit, a booster pump, a timing pump, a heating unit, a holding tube, and a flow diversion device (Figure 2–7). The raw milk is transferred from the balance tank to the regeneration tank (plate heat exchanger) with a booster pump. In the regeneration unit, the raw milk passes on one side, and hot pasteurized milk coming from the holding tube passes on the other side, so that the raw milk is initially heated by recovering some of the heat that is available from the pasteurized milk. A timing pump then sends the initially heated raw milk to the heating unit (a plate heat exchanger), where it is heated to a temperature above the required pasteurization temperature. The hot milk then flows through a holding tube, where it is held for a specified length of time. A flow diversion device, located at the end of the holding tube, directs the hot milk to be cooled (if its temperature is the desired one) or sends it back to be reprocessed (if its temperature is lower than the legal limit). The timing pump is used to provide a fixed flow rate as required by the PMO. The required milk pasteurization conditions in an HTST system are temperatures between 71 and 75°C and holding-tube residence time between 15 and 40 seconds.

The process has several temperature and flow rate sensors, automatic control valves, and variable speed pumps (Figure 2–7). The control system consists of on-line data acquisition system and temperature and flow control loops. The signals recorded by the data acquisition system include the flow rate measurements, product, hot water inlet and outlet temperatures from the heating unit, differential pressure, and flow diversion valve position. The temperature is measured with a T-type thermocouple, the pressure differential inside the regenerator with a differential pressure switch, and the flow rate with a magnetic flow meter.

The control signals generated by the control system algorithm are current signals to the pneumatic steam valve (to control the temperature) and to the variable speed pump (to control the flow rate).

The control system was based on the lethality equivalent of 72°C or above 18 seconds and was designed to provide accurate control of the HTST system. The multivariable controller uses the product total lethality to determine temperature and flow rate setpoints.

CONCLUSION

Examples of food-processing systems and controls were presented in this chapter. Interest in automation and computerized control of food-processing systems has increased recently, and a number of applications can be found in the literature. With advances in control theory, better and reliable on-line sensors, and faster and less expensive new

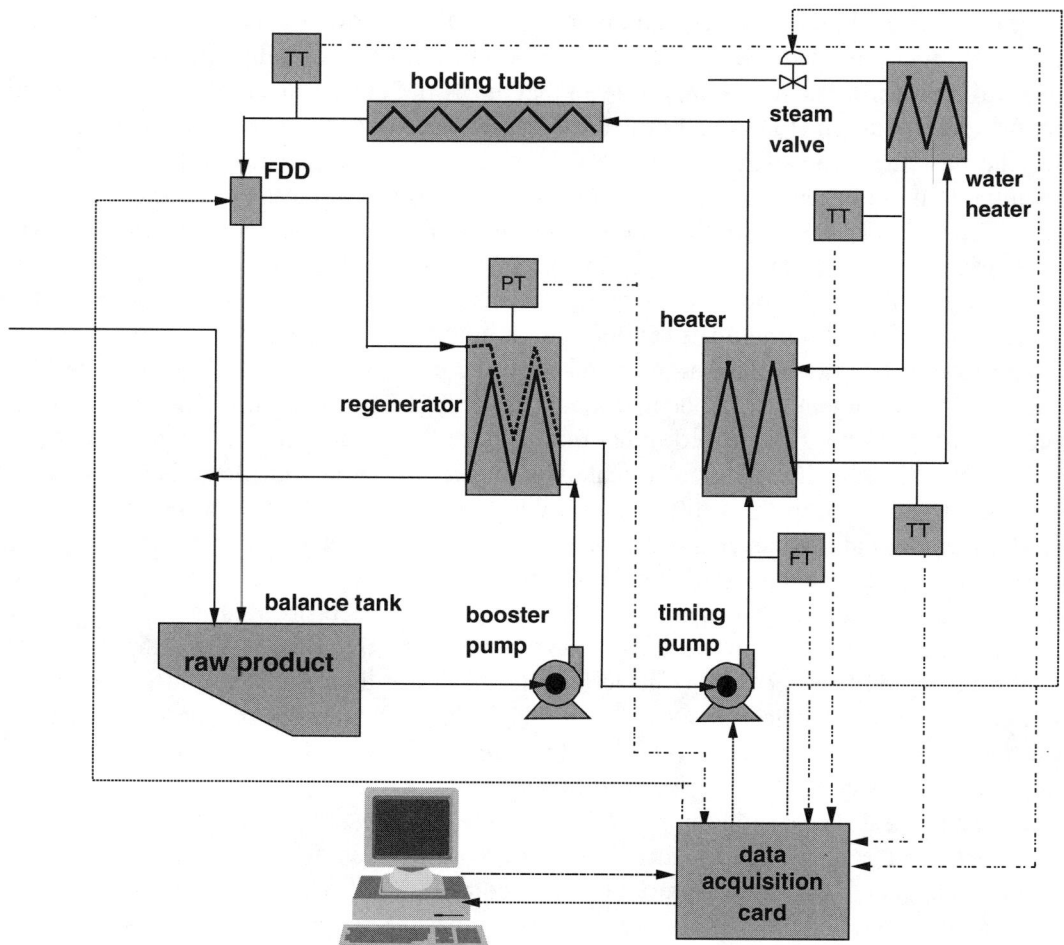

Figure 2–7 HTST Pasteurization Process and Control System

computers, today it is possible to implement control systems in even the very complex processing plants found in the food industries.

REFERENCES

Brescia, L. & Moreira, R.G. (1997). Modeling and control of a continuous frying process: I. Dynamic analysis and system identification. *Trans Chem Eng Inst* 75(c), 3–11.

Brooker, D.B., Bakker-Arkema, F.W. & Hall, C.W. (1992). *Drying and Storage of Grains and Oilseeds.* New York, NY: Van Nostrand Reinhold.

Hardwick, W.A. (1995). *Handbook of Brewing.* New York, NY: M. Dekker.

Harper, J.M. (1981). *Extrusion of Foods.* Vols 1 & 2. Boca Raton, FL: CRC Press.

Hasting, A.P.M. (1992). Practical considerations in the design, operation and control of food pasteurization processes. *Food Control* 3(1), 27–32.

Lewis, M.J. (1995). *Brewing.* New York, NY: Chapman & Hall.

Liu, Q. (1998). Stochastic modeling and automatic grain dryers: optimizing grain quality. East Lansing, MI: Michigan State University. PhD dissertation.

Moreira, R.G., Castell-Perez, M.E. & Barrufet, M.A. (1999). *Deep-Fat Frying: Fundamental and Applications*. Gaithersburg, MD: Aspen Publishers.

Negiz, A., Ramanauskas, P., Cinar, A., Schlesser, J.E. & Armstrong, D.J. (1998). Modeling, monitoring and control strategies for high temperature short time pasteurization systems. 2. Lethality-based control. *Food Control* 9(1), 17–28.

Nejman, N., Shakourzadeh, K. and Bouvier, J. (1992). State reproducibility and relative gain analysis in twin-screw extrusion cooking. In *Food Extrusion Science and Technology*. pp 519–526. Edited by J.L. Kokini, C. Ho, & M. Karwe. New York, NY: Marcel Dekker.

Nisenfeld, A.E. (1985). *Industrial Evaporators: Principles of Operation and Control*. International Society of Measurement and Control (ISA). NC.

Nybrant, T. (1986). *Modeling and Control of Grain Dryers*. Uppsala, Sweden: Institute of Technology. Uppsala University. Rep UPTEC 8625 R.

O'Brien, R. (1993). Foodservice use of fat and oils. *INFORM* 4, 913–921.

Perry, R.H & Green, D.W. (1997). *Perry's Chemical Engineers' Handbook*. 7th ed. New York, NY: McGraw-Hill.

Ritter, R.A. & Andre, H. (1970). Evaporator control system design. *Can J Chem Eng* 48, 606–701.

Schonauer, S.L. (1995). Product quality adaptive control system for a food extruder. College Station, TX: Texas A&M University. PhD dissertation.

Schonauer, S.L. & Moreira, R.G. (1995a). Development of a fixed-GPC controller for a food extruder based on PQA. Part I: system identification. *Trans Inst Chem Eng* 73(c), 189–199.

Schonauer, S.L. & Moreira, R.G. (1995b). Development of a fixed-GPC controller for a food extruder based on PQA. Part II: control development, implementation and analysis. *Trans Inst Chem Eng* 73(c), 200–210.

Snack Food Association. (1997). *State of the Snack Food Industry Report*. Alexandria, VA: Snack Food Association.

US Public Health Service. (1989). Grade A: Pasteurized Milk Ordinance. Washington, DC: US Government Printing Office.

US Public Health Service. (1993). Grade A: Pasteurized Milk Ordinance. Washington, DC: US Government Printing Office.

CHAPTER 3

Background

Automatic control has become an integral part of modern food manufacturing today. It is essential, for example, in controlling product quality attributes such as color, bulk density, moisture, and oil content as well as pressure, temperature, viscosity, humidity, and flow rate in food processing.

The area of automatic control has evolved from James Watt's centrifugal governor to control the speed of a steam engine in the eighteenth century to the most recent developments in modern control theory such as adaptive, fuzzy logic, and closed-loop monitoring.

There is constant advancement in theory and practice of automatic control, but the principles on which process control is based are not changing. New developments help extend our understanding of the basic principles. In addition, there is continuous improvement of software and hardware available from vendors. Finally, there is variation in the application and practice of process control theory from one industry to another, and industries tend to learn from each other.

This chapter will introduce the basic concepts of process control for those readers who have little contact or experience with this subject matter.

DEFINITIONS OF BASIC CONCEPTS

Control theory is a branch of the general subject of systems theory. Mathematically speaking, systems theory is the study of the interactions and behavior of a group of elements organized toward a goal or set of goals. Systems theory is more concerned with mathematical properties, while control theory deals with physical applications. So a control system is considered to be any system that exists for the purpose of regulating and controlling a flow—for example, of energy, information, or money (Brogan, 1982).

Control theory can be divided into two categories: classical and modern. Most classical control techniques were developed for linear constant coefficient systems with one input and one output. Laplace transform and transfer functions are the language of classical techniques. Modern control theory can cope with more complex plants, such as multiple-input, multiple-output (MIMO) systems, nonlinearities, and time-variant coefficients. The most recent developments in modern control theory include optimal, adaptive, supervisory, and fuzzy logic control.

Few real food-processing systems are exactly linear over their whole operating range, and few of them have parameter values that are precisely constant forever. But many food systems approximately satisfy these conditions over a sufficiently narrow operating range. This chapter reviews the classical methods that are applicable to linear, constant coefficient systems.

The terminology necessary to describe control systems will be defined in this section.

- ***Process:*** from the production point of view, a progressively continuing operation that consists of controlled actions systematically directed toward a desired product. Food extrusion is a process that consists of feeding, mixing, cooking, and forming. Figure 3–1 illustrates a food extrusion process with all important variables needed for process control.
- ***Variables of a Control System***
    1. *Controlled Variables:* independent variables of a process; the parameters of the process that indicate product quality or the operating condition of the process. They are also called output, factors (variables) that are caused by a system and are used as measures of performance for the given system. In a food extrusion process, the controlled variables are the product quality attributes, such as density, color, and moisture content.
    2. *Manipulated Variables:* dependent variables of the process used to cause a change in the process. They are also called control input and are defined as factors (variables) that are used to modify the system behavior. Therefore, inputs are variables that cause or stimulate a change in system behavior. Extruder variables that have a significant effect on process variables or product attributes and can be readily changed or manipulated during processing are considered manipulated variables for control purposes. These manipulated variables include the screw speed, barrel temperature, feed rate, and water rate.

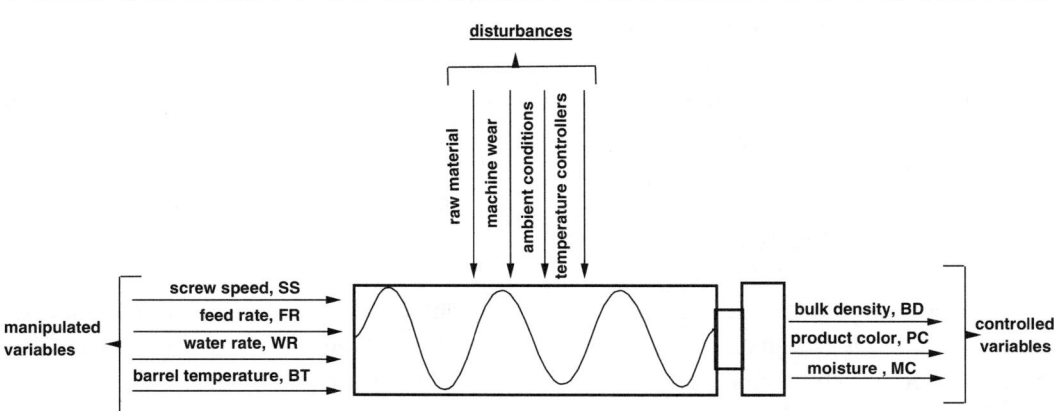

Figure 3–1 Manipulated, controlled, and disturbance variables for a food extrusion process

3. *Disturbances:* signals that tend to adversely affect the value of the output of a system; they can be classified as internal or external. An *internal* disturbance is a disturbance that originates within the system, while an *external* disturbance originates outside the system. In a food extruder, screw wear is an internal disturbance, and raw material variability is an external disturbance to the system in control.

4. *Loads:* all variables affecting a controlled variable, other than the one being manipulated. Both loads and manipulated variables may influence a controlled variable from either the inlet or outlet side of the process.

5. *Deterministic Disturbances:* variables that can be described exactly in analytical form.

6. *Stochastic Disturbances:* signals that cannot be exactly described or predicted.

- *Dynamics:* time-dependent behavior of a process.
- *System:* a set of interconnected elements organized toward a goal or a set of goals. Systems can be physical, biological, economic, and so on.
- *Exogenous Input:* an input factor (or variable) that is determined by factors completely independent of or external to the system.
- *Desired Output (or Setpoint):* a system output variable that is a means of satisfying an objective (or goal) for that system.
- *Fixed-Structure System:* a system whose structure is fixed in time (time invariant): that is, the way it responds to inputs does not change over time. In a fixed-structure system, system elements, their interconnections, and the parameters that describe them do not change with time.
- *Time-Varying System:* a system in which the system structure does change with time.
- *Adaptive Control System:* a system that has the ability to self-adjust or self-modify under unpredictable changes in the environment conditions or structure.
- *SISO:* single-input, single-output system.
- *MIMO:* multiple-input, multiple-output system.
- *Feedback Control:* an operation that, in the presence of a disturbance, tends to reduce the difference between the output of a system and the reference input (setpoint) and that does so on the basis of this difference.
- *Servomechanisms:* a feedback control system in which the output is some mechanical position, velocity, or acceleration.

THE CONTROL PROBLEM

The control problem can be stated as follows: determine the one value of the manipulated variable that establishes a balance among all the influences (loads/disturbances) on the controlled variable and keeps this variable constant at the desired value.

The control problem can be described by the relationship among the controlled, manipulated, and load variables. The manipulated variable and loads may either increase or decrease the value of the controlled variable, depending on the process. Changes in the controlled variable reflect the balances between loads and the manipulated variable.

Consider, for example, a single–controlled variable process such as the heat exchanger presented in Figure 3–2. To maintain the temperature of the product (hot water) in this process, the control system manipulates the position of the steam valve (thus changing the steam flow rate). However, the water temperature can also be affected by changes in the water flow rate, the inlet temperature, the steam enthalpy, the degree of fouling in the heat exchanger, and the ambient temperature. So the process needs to be controlled regardless of changes in these disturbances or load variables.

An increase in the steam-valve opening, steam enthalpy, inlet temperature, and ambient temperature tends to raise the water temperature, while flow rate and heat exchanger fauling cause a reduction in the water temperature. The temperature responds to the net effect of these influences. If all the loads remained constant, the steam valve could be adjusted until the water temperature was constant at the setpoint (desired value).

In reality, these load variables do not remain constant during the process operation. For example, variations in inlet temperature and flow rate upset the product temperature, requiring a different steam-valve position to maintain the water temperature to the desired

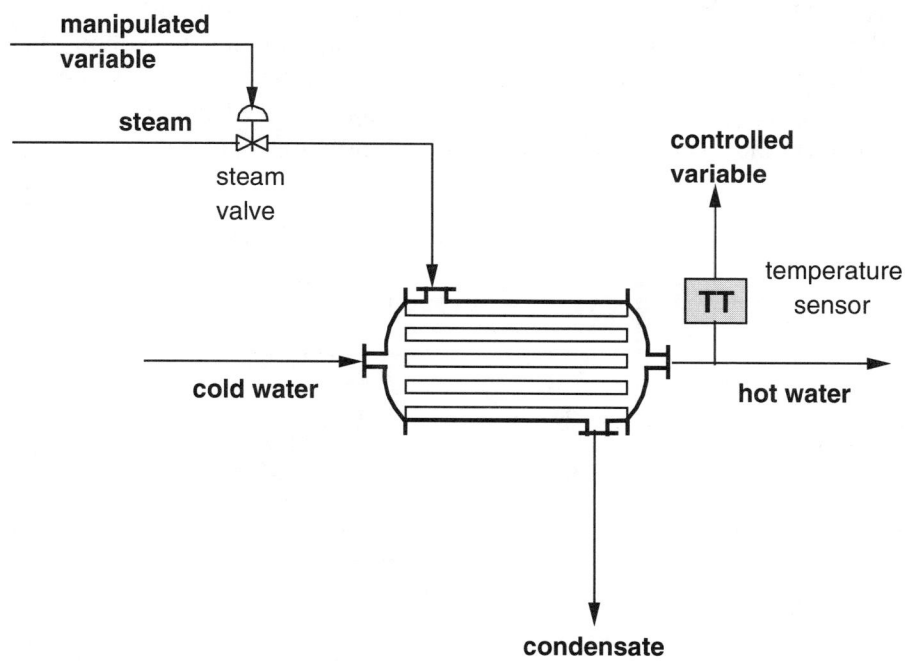

manipulated variable = steam valve
controlled variable = hot water temperature

Figure 3–2 A simple process represented by a steam-heated heat exchanger

value. This can be accomplished by an automatic control system. The objective of the control system is to determine and continuously update the valve position as the load conditions change.

It is important to note that any control system solves this same basic problem, independently of how complex the control system may be. The control problem can be solved in only two ways: by using feedback or feedforward strategies.

The steps required to design a control system are described in Figure 3–3. Once the controlled, manipulated, and load variables have been identified, the next step is to mathematically describe the process and signals (disturbances, reference variables, initial values). The models can be exact or approximate depending on the design method and the application. Generally, models of the signals can only be estimated by approximating the real behavior to step changes, for example, although they are rare in practice. However, modern process computers can be applied to obtain more exact models of deterministic and stochastic signals.

LAPLACE TRANSFORM

Differential equations play an important role in modeling dynamic continuous-time systems (Chapter 4). While there are classical methods for solving differential equations, the Laplace transform approach has become a popular method for solving differential equations with constant coefficients (describing linear time-invariant dynamic systems). There are essentially three steps involved in the Laplace transform approach to analysis: (1) to transform the system differential equation into an algebraic equation; (2) to solve

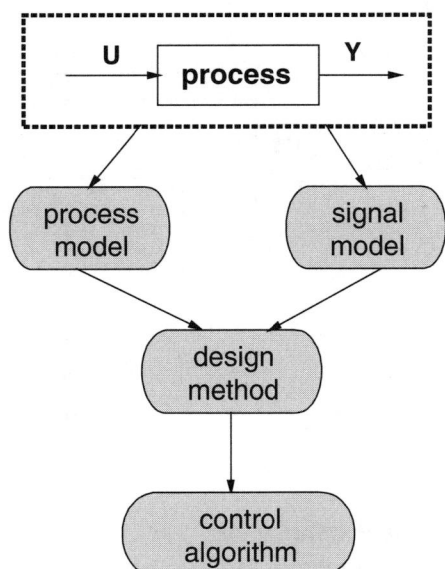

Figure 3–3 The flow diagram of a control algorithm design

the algebraic equation for the transformed system output $Y(s) = \mathcal{L}[y(t)]$; and (3) to take an inverse transform $(\mathcal{L}^{-1})$ to determine the system behavior through time, $y(t) = \mathcal{L}^{-1}[Y(s)]$.

The Laplace transform of a function of time is given by:

$$\mathcal{L}[f(t)] = F(s) = \int_0^\infty f(t)e^{-st}\,dt$$

(1)

The symbol $\mathcal{L}[\ ]$ denotes the Laplace transformation of the function included in the brackets. The result of the transformation is the function $F(s)$, and a complex variable s is defined as

$$s = \sigma + j\omega$$

(2)

where σ is the real, ω the imaginary part of s, and $j = \sqrt{-1}$. Appendix 3–A presents some fundamentals on the Laplace transform and examples of applications.

BLOCK DIAGRAM AND TRANSFER FUNCTIONS

A control system generally consists of a number of components. A *block diagram* is used to demonstrate the function performed by each component. The block diagram can be defined as a pictorial representation of the functions performed by each component and the signals flow. In a block diagram, all system variables are linked to each other through blocks that symbolize the mathematical operation on the input signal to the block that produces the output. The transfer functions of the components is entered in the corresponding blocks, which are connected by arrows to indicate the direction of the flow signal. Figure 3–4 illustrates a block diagram of a closed-loop control system for a continuous-flow grain dryer. The arrowheads pointing toward the blocks indicate the input, and the arrowheads leading away from the blocks represent the output. These arrows are referred to as signals.

Blocks can be connected in series only if the output of one block is not affected by the following block. Any number of cascaded blocks can be replaced by a single block, the transfer function that is the product of the individual transfer functions. A more complex block diagram involving many feedback loops can be simplified by a step-by-step rearrangement, using rules of block diagram algebra. Figure 3–5 shows some of these rules.

Transfer functions are used to characterize the input-output relationships of linear time-invariant systems in terms of the system parameters and are a property of the system itself, independent of the input or driving function. The transfer function $G(s)$ is defined as the ratio of the output signal $Y(s)$, divided by the input signal $U(s)$, given that all system initial conditions are zero. It includes the units necessary to relate the input to the output, but it does not provide any information regarding the physical structure of the system.

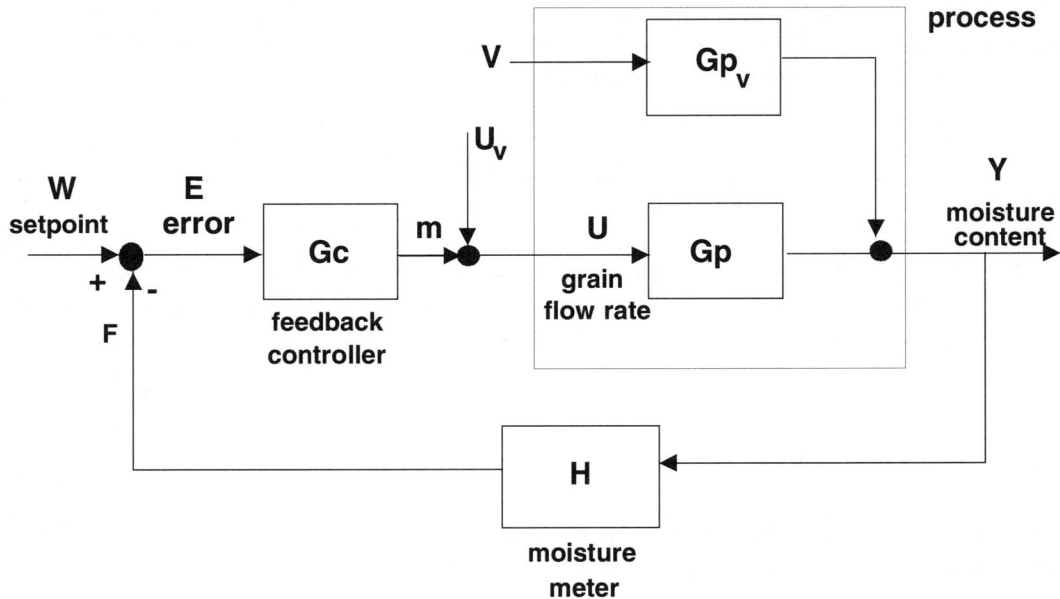

Figure 3–4 Closed-loop control of a single-input, single-output continuous-flow grain dryer

CLOSED-LOOP AND FEEDBACK CONTROL SYSTEMS

Closed-loop control systems are feedback control systems. Therefore, the output signal in a closed-loop system has a direct effect upon the control action. On the basis of the difference (acting error) between the actual value and the setpoint (desired value) of the controlled variable (output), the feedback controller calculates the changes that need to be made in the manipulated variable (input) to bring the output of the system to the setpoint thus reducing the acting error.

To illustrate the concept of feedback control, Figure 3–6 shows a manual control of a SISO continuous-flow grain-drying process. On the outlet side of the dryer, there is an indicator (moisture meter) to provide operator information on the actual value of the grain outlet moisture content (controlled variable). The operator wants to keep the grain outlet moisture closed to the desired value. If the operator watches the moisture meter and finds that the moisture is higher than the setpoint, he or she will reduce the grain flow rate by adjusting (closing) the auger rpm. The setpoint, of course, is in the operator's mind, and the operator makes all of the control decisions. It is quite possible for the moisture content to become lower so that the operator will have to repeat the operation in the opposite direction until the desired value is reached and kept constant during the operation.

If the operator is replaced by an automatic controller, as shown in Figure 3–7, the system becomes an automatic feedback control system. The actual outlet grain moisture content (the process output), measured by the moisture meter, is compared to the setpoint value, and an acting error signal is generated. The output moisture content is converted

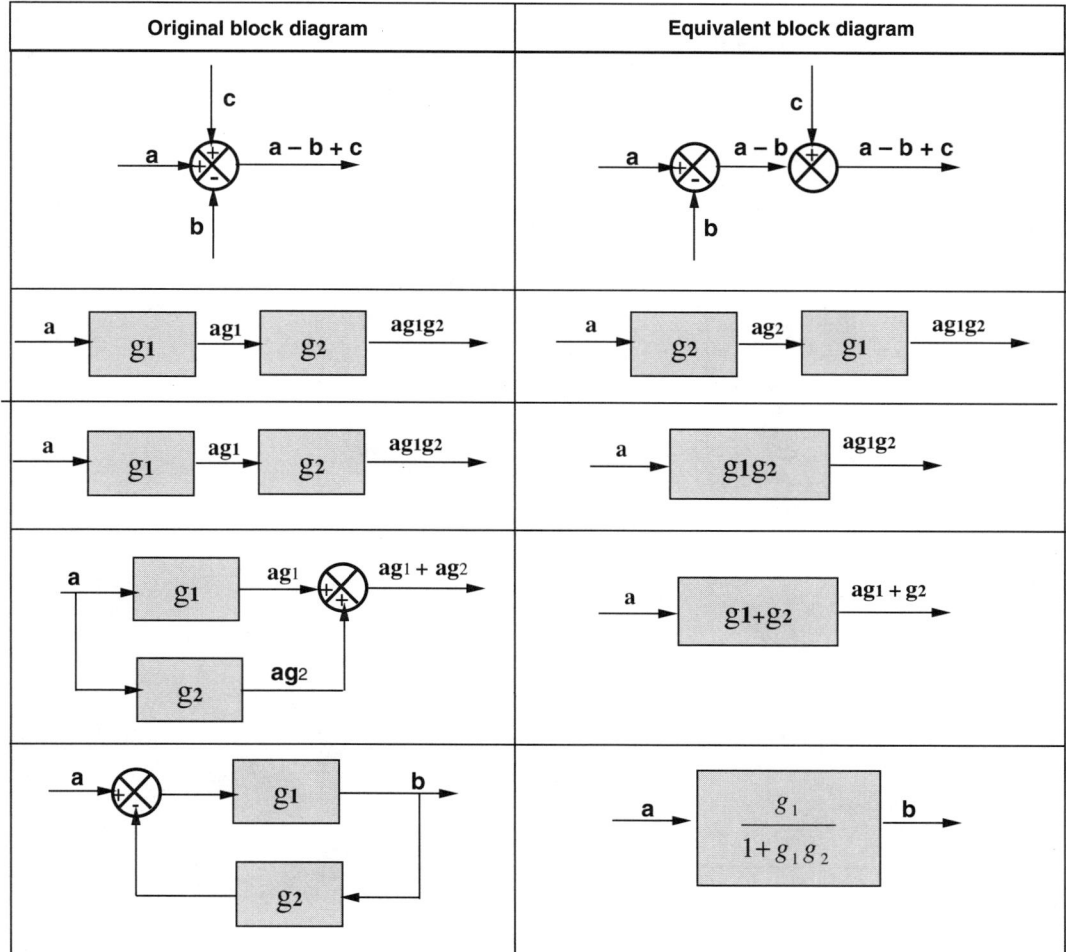

Original block diagram	Equivalent block diagram

Figure 3–5 Rules of block diagram algebra

to the same units of the setpoint by a transducer. On the basis of the acting error value, the controller calculates the changes that need to be made in the auger rpm to reduce and then eliminate that error.

Thus, manual and automatic controls operate similarly. In this case, the operator's eyes correspond to the acting error device, the brain corresponds to the automatic controller, and the operator's muscles represent the actuator.

The derivation of the closed-loop transfer function from the block diagram in Figure 3–4 can be easily performed by applying the block diagram algebra described in Figure 3–5. The setpoint or reference (W) is the input to the control system, the manipulated variable (grain flow rate) is the input to the process, and the controlled variable (moisture content) is the output. For simplicity, let the disturbances U_v in the motor rpm (auger)

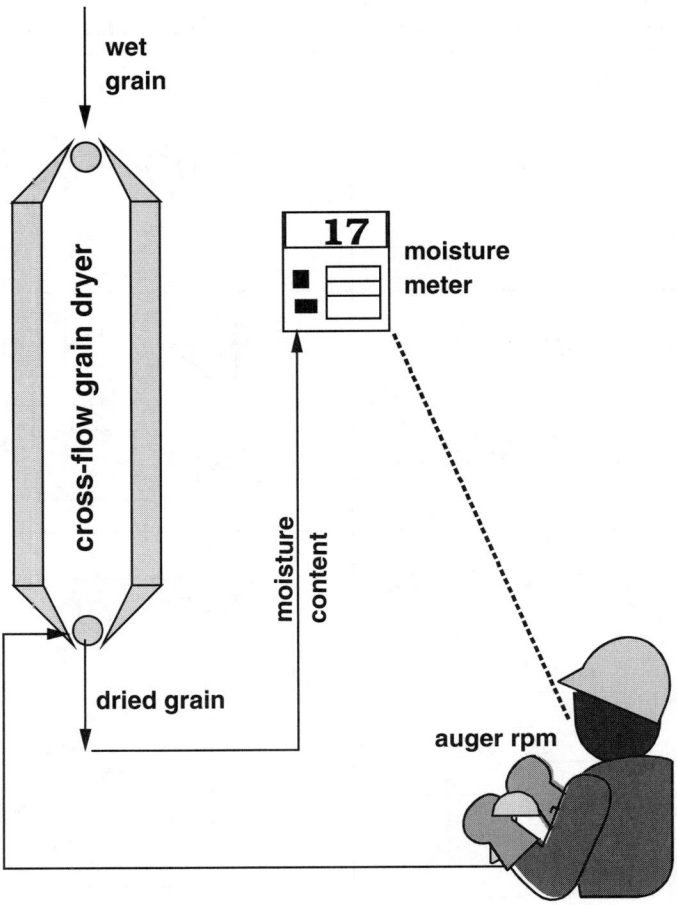

Figure 3–6 Manual control of a single-input, single-output continuous-flow grain dryer

and V in the process be zero. Then the transfer function of the closed-loop control system can be derived as

Error = setpoint - controlled variable (measured)

$$E = W - F$$

(3)

Controlled variable = controller transfer function × process transfer function × error

$$Y = G_c G_p E = GE$$

(4)

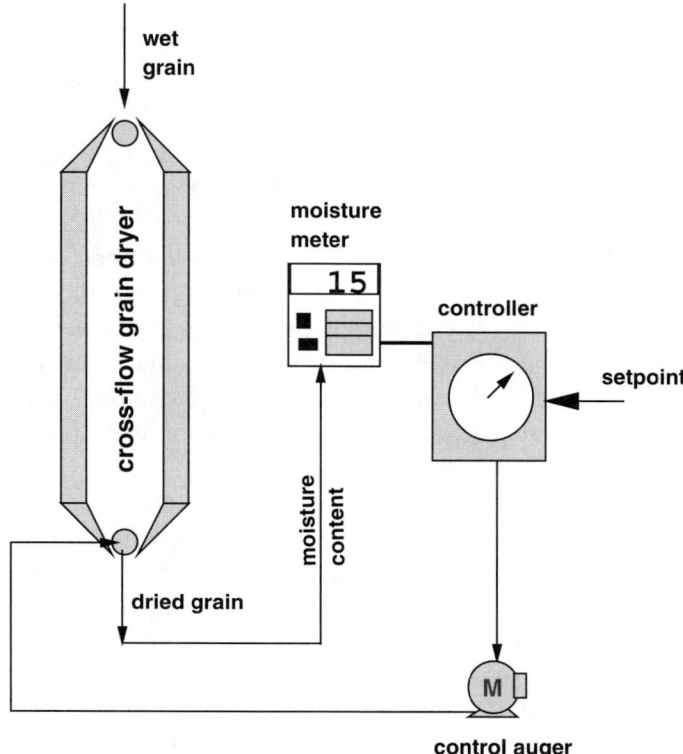

Figure 3–7 Automatic control of a single-input, single-output continuous-flow grain dryer

Controlled variable (measured) = controller variable × measuring device transfer function

$$F = YH$$
$$Y = (W - F)G$$
$$Y = (W - YH)G$$
$$Y + YGH = WG$$
$$Y(1 + GH) = WG$$
$$\frac{Y}{W} = \frac{G}{1 + GH}$$

(5)

Equation 5 is the transfer function of the feedback control system of a SISO continuous-flow grain dryer. The system transfer function (G) contains all system components (augers, amplifiers, motors, generators, etc). The feedback path transfer function (H) is generally a device that converts the controlled variable into a suitable signal for input to the error detector.

The Feedback Controller

The first feedback mechanisms were mechanically connected directly to the process and the manipulated variable. Later, pneumatic and electronic controllers were developed. Today, the state of the art is distributed control through digital systems. Regardless of the hardware used, the basic function of the feedback controller is the same: to solve the control problem.

All feedback controllers must have certain common elements (for SISO systems) (Figure 3–8): (1) two inputs—the measurement signal of the transmitter and the reference value (setpoint)—and (2) one output. Within the controller, the measured and setpoint signals are subtracted and an error is generated (if this difference is not zero). This error is the input to the control algorithm that generates the output. The controller also includes a manual control signal generator that can be activated by the operator in case of emergencies or during start-up.

In simple loops, the output signal from the control algorithm will directly position a valve, while in more complex schemes, this signal will be an input to another instrument. Typically, the controller will have an associated operator interface displaying, for example, setpoints, measurement, current output, and automatic/manual status.

All feedback control loops share the same characteristics: open loop versus closed loop, positive versus negative feedback, and oscillation.

Figure 3–9 illustrates a schematic and the block diagram of the feedback temperature control loop installed in a heat exchange system. The control objective is to maintain the hot water temperature at the setpoint despite variations in the inlet cold water temperature and flow rate. This is done by manipulating the steam flow rate. A feedback control scheme would entail measuring the hot water temperature, comparing it to the setpoint, and then adjusting the steam flow rate.

Open Loop versus Closed Loop

Once a feedback controller is installed in a system and placed on automatic, a *closed loop* is created. The controller output affects the measurement, and vice versa. If this effect is broken in either direction, the loop is said to be *open*, and feedback control no longer exists. Examples of events that can open the feedback loop include placing the controller in manual, failure of the sensor or transmitter, saturation of the controller output at 0% or 100% of scale, or failure of the valve actuator. The first thing to check when a control loop does not seem to be operating properly is whether the loop is closed.

Positive versus Negative Feedback

Feedback can be positive or negative, and the difference is crucial to the controller performance. Every feedback will have a means of changing the controller action, which defines the direction of the controller response to a change in measurement. *Increase-increase*, or direct action, causes the controller to increase its output in response to an increasing measurement signal. *Increase-decrease*, or reverse action, causes the controller to decrease its output when the measurement signal increases.

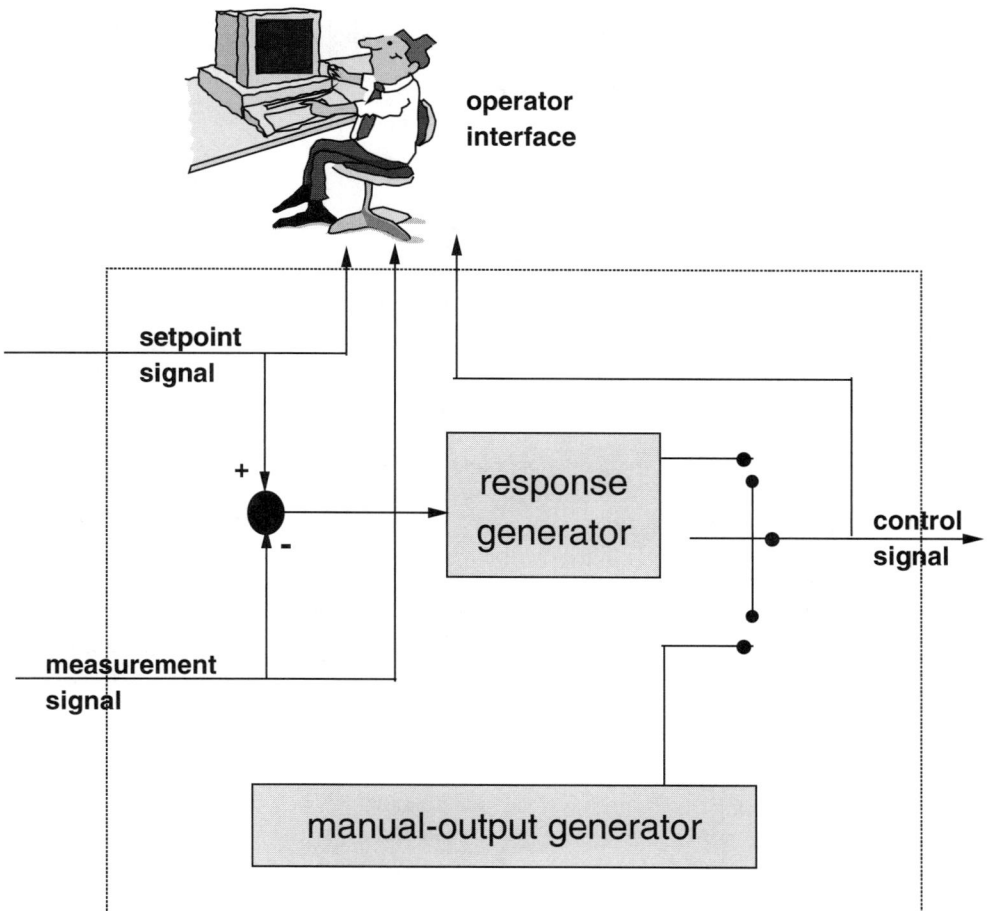

Figure 3–8 Basic components of a feedback controller

Figure 3–10a illustrates the behavior of an output-temperature control loop installed on the heat exchanger. The steam valve is set air-to-open (ie, fail closed). This means that an increasing control signal will open the valve to increase steam flow. The controller action is set to increase-increase, which is incorrect. Manual control can bring the system to the setpoint, but as soon as the controller is placed on automatic, the loop will become unstable. Any small disturbance that increases the temperature will also cause an increase in controller output. The controller will then open the valve, causing the temperature to increase further and the valve to continue to open. The result is a temperature drop. If the temperature is dropped due to a small disturbance, the controller will close the valve, and the temperature will fall even more. In turn, this will cause the valve to close even more.

For this feedback loop to be successful, it must have negative feedback. Figure 3–10b shows the same feedback loop, except that the controller has been set to increase-decrease

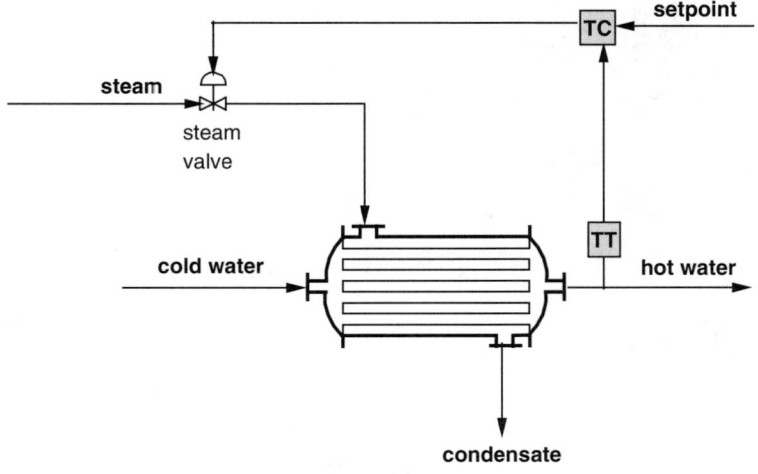

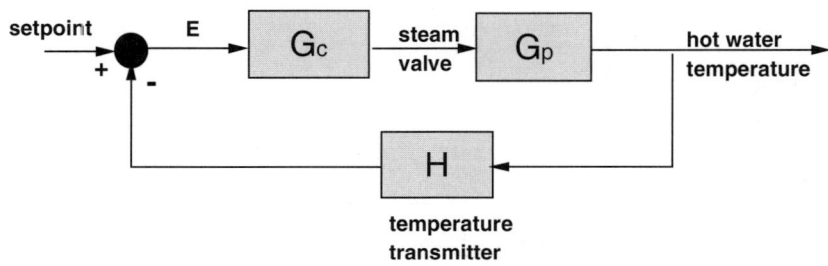

Figure 3–9 Schematic of a feedback control of a heat exchanger

---

action. The controller then responds to an increase in temperature by closing the valve. A decrease in temperature causes the controller to open the valve. These responses drive the measurement back toward the setpoint.

Therefore, selection of the proper control action is fundamental for correct performance of the system. The correct choice for feedback action will depend on the system.

Oscillation

Oscillation is the natural response of a feedback control loop to an upset. Take, for example, the closed-loop control system of the heat exchanger shown in Figure 3–9. If the water temperature starts to move away from the setpoint, an error is produced, causing the controller to change the manipulated variable (adjust the steam valve). However, the water temperature does not respond immediately due to the lags and/or dead time within the process. It takes some time for the controller to return the water temperature to the setpoint. During this period of time, the measurement reverses itself and so does the controller, causing both the controlled and manipulated variables to oscillate. The characteristics of

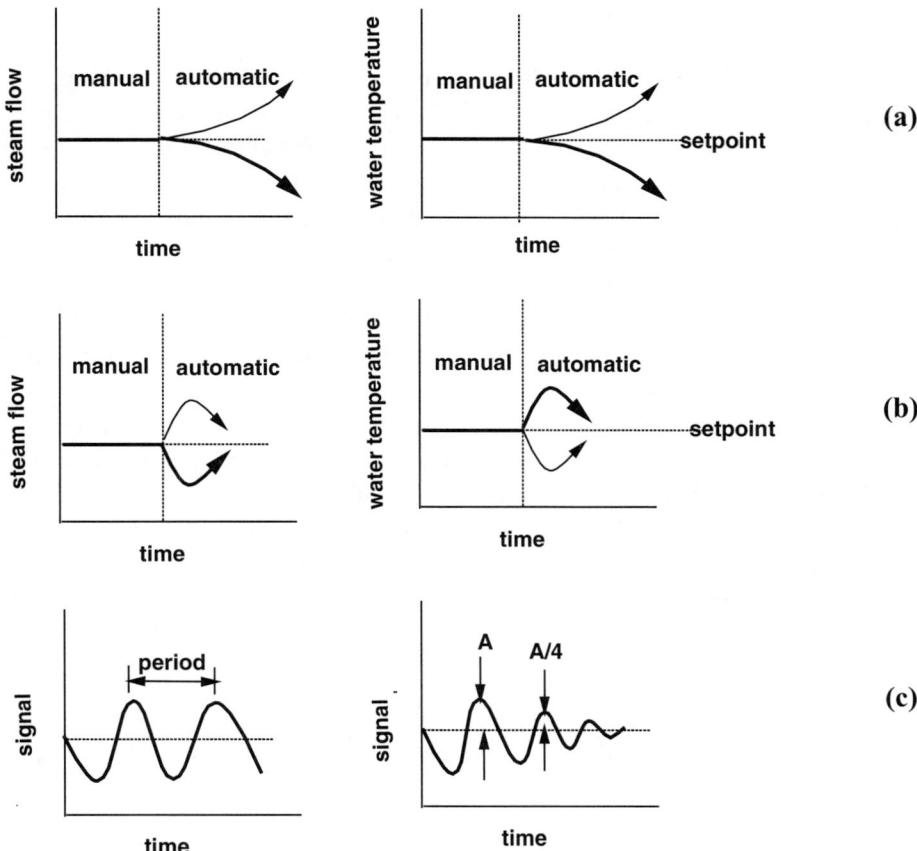

Figure 3–10 Control action for closed-loop feedback control, including (a) Positive feedback; (b) negative feedback; (c) oscillatory signals

this oscillation will depend on the adjustments to the controller responses. For a good control, the oscillation in the measurement signal should steadily decay and return to the setpoint thus reestablishing a balance among the load and manipulated variables. Figure 3–10c shows a typical oscillation with a period of a cycle measured as the time between two consecutive positive or negative peaks. Another oscillation shown is the one that is decaying to a constant signal. A popular criterion used to evaluate control performance is the quarter-amplitude decay (Figure 3–10c). A loop showing quarter-amplitude decay will stabilize faster, indicating that the controller is good.

Process Characteristics

Lags in the process can affect the performance of feedback control loops. The cause and characteristics of these lags need to be evaluated before control modes can be selected. Lags include dead time and capacity element.

Dead time, also called transportation lag, is defined as the time delay between changes in the control input and the beginning of its effect on the control output. Dead time represents an interval in which the controller has no information about the effect of a control action already taken. Dead time is present in most food-processing systems. An example of a process that has dead time response is shown in Figure 3–11. A hopper auger deposits material on a moving belt, and a weight transmitter measures the amount of material being transported. A step change in the control signal (opening the valve) will cause an increase in the material flow rate. The weight sensor will detect this change only after the material has traveled from the hopper to the sensor—that is, after a delay or dead time. Although dead time does not slow down the process, the longer the delay the more difficult it will be to control the system. Therefore, dead time should be reduced by properly locating transmitters.

A *capacity element* is defined as the part of the processing system where material or energy is accumulated. Consider, for example, the liquid-level system shown in Figure 3–12 representing a single capacity element (material storage). The inlet flow is manipulated to affect the liquid level, and the outlet flow is the load variable. The level is maintained constant as long as the inlet and outlet flows are equal. A step change in the control signal (valve position) will cause a difference in inlet and outlet flows, resulting in

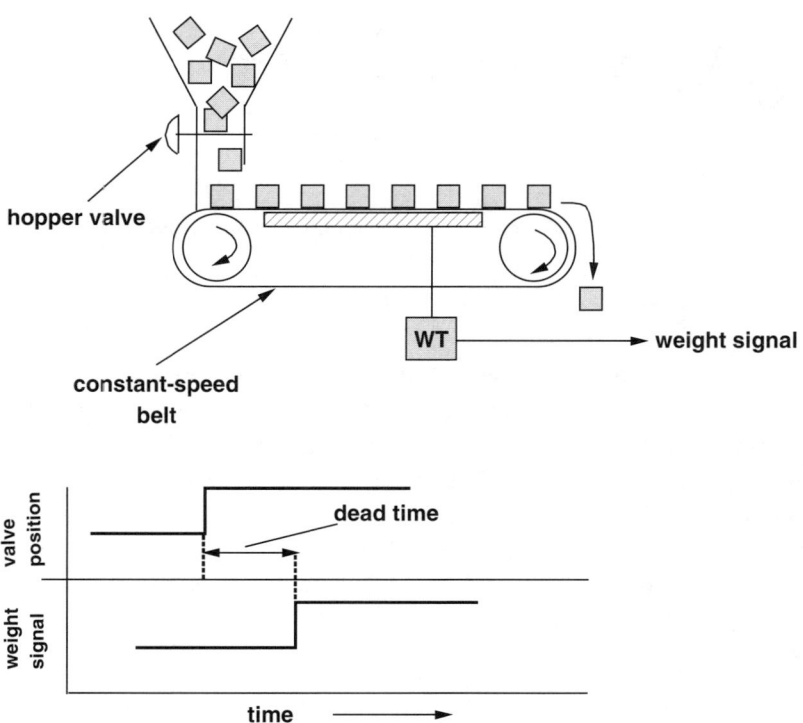

Figure 3–11 Constant-speed weighing-belt system

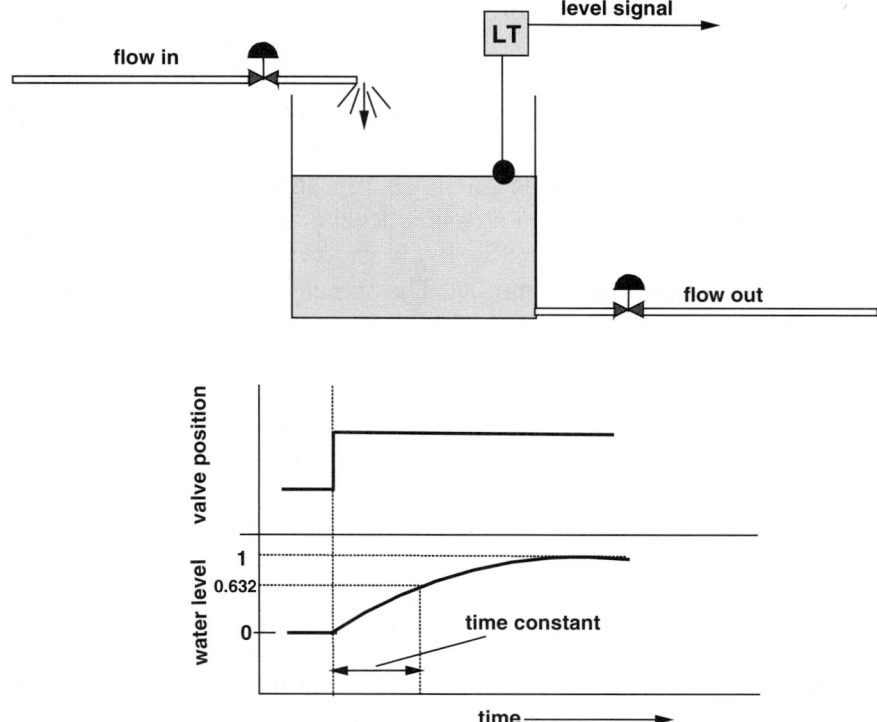

Figure 3–12 Liquid-level system

an increase in the liquid level. As the level increases, pressure gradually increases across the drain valve, thus raising the outlet flow. As a result, the two flows are brought back in balance. However, the response to the change in the process input is not instantaneous; it will follow an exponential decay from its initial value to its final value. The size of a capacity element is measured by its time constant—that is, the time required to complete 63.2% of the final response. This residence time is roughly the residence time of the capacity element: in this case, the tank capacity divided by the throughput. This percentage is the same for all first-order processes. Therefore, the time constant of a first-order process indicates how fast the process responds to sudden input changes.

Feedback Control Modes

The four basic feedback control modes used in the industry are on-off, proportional (P), integral (I), and derivative (D) control.

On-Off

On-off (also called two-position) controllers are the simplest and most inexpensive form of feedback control loops. They are used in heating systems and domestic refrigerators. On-off controllers are also commonly used in storage facilities to aerate or dry (at low temperature) batches of cereals, potatoes, or other agricultural products.

Consider, for example, the simplified system shown in Figure 3–13a, where a feed-back control loop is used to aerate a grain bin. For simplification, it is assumed that the ambient temperature is below the setpoint temperature and that the relative humidity is high enough so no moisture is removed and no condensation occurs. The control objective is to maintain the grain temperature at a desired level (7°C) by turning the fan on or off. The thermocouple measurement (in millivolts) is sent to a temperature transducer, where it is amplified and then sent to the controller. The output signal from the controller is sent to a silicon-controller rectifier (SCR), which converts this signal to a form compatible with the electrical fan.

The on-off control function has only two outputs:

$$\text{output} = \begin{cases} on & \text{if error } \geq 0 \\ off & \text{if error } < 0 \end{cases}$$

(6)

In the example, the controller turns the fan on when the grain temperature is above the setpoint. However, lags and/or dead times in the system cause the temperature to rise before reversing and moving toward the setpoint. When the temperature falls below the setpoint, the controller stops the fan and the cycle repeats again (Figure 3–13b). The main disadvantages of the on-off controllers are continuous cycling of the controlled variable and excessive wear of the control element (Seborg et al, 1989).

Proportional Control

In proportional control, the controller output is algebraically proportional to the error input signal to the controller:

$$m(t) = K_p e(t)$$

(7)

where $m(t)$ is the controller output and K_p the controller gain. Equation 7 is called the control algorithm. Essentially, the proportional controller is an amplifier with an adjustable gain. Proportional controllers are simple, are the easiest to tune, provide good stability, have a very rapid response, and are relatively stable. The major disadvantage of proportional-only controllers is their inability to eliminate the steady-state errors that occur after the setpoint change—that is, they exhibit offset or steady-state error. Offset will occur with proportional-only control regardless of the value of K_p. Figure 3–14 shows the offset response of a proportional controller in a system similar to the SISO heat exchange system to a step change in the setpoint. The plot also shows the case for no control, which is called the open loop, or the normal dynamic response of the process itself. A block diagram of a proportional controller is shown in Figure 3–15.

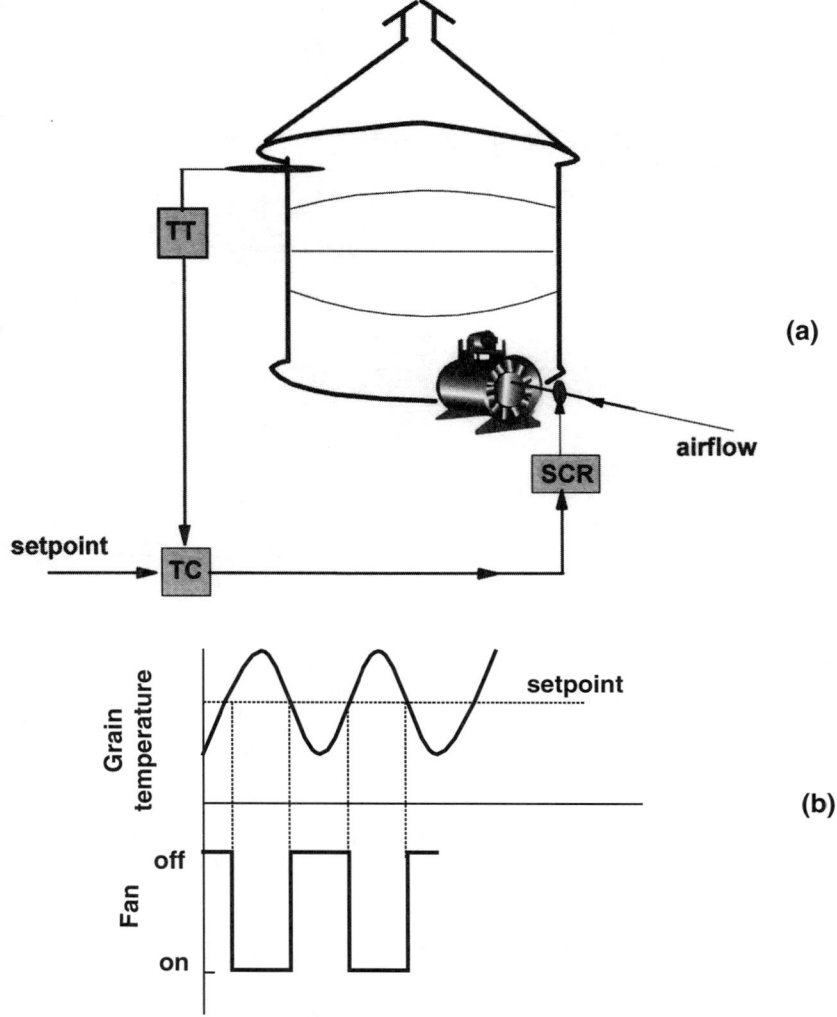

Figure 3–13 On-off feedback controller for an aeration of agricultural products during storage. (a) The process; (b) two-position controller.

Integral Control

Integral control, also called reset control, is really an integration of the input error signal. Integral control action is based on the principle that the controller output $m(t)$ is proportional to both the size and the duration of the actuating error signal $e(t)$:

$$\frac{dm(t)}{dt} = K_i e(t)$$

(8)

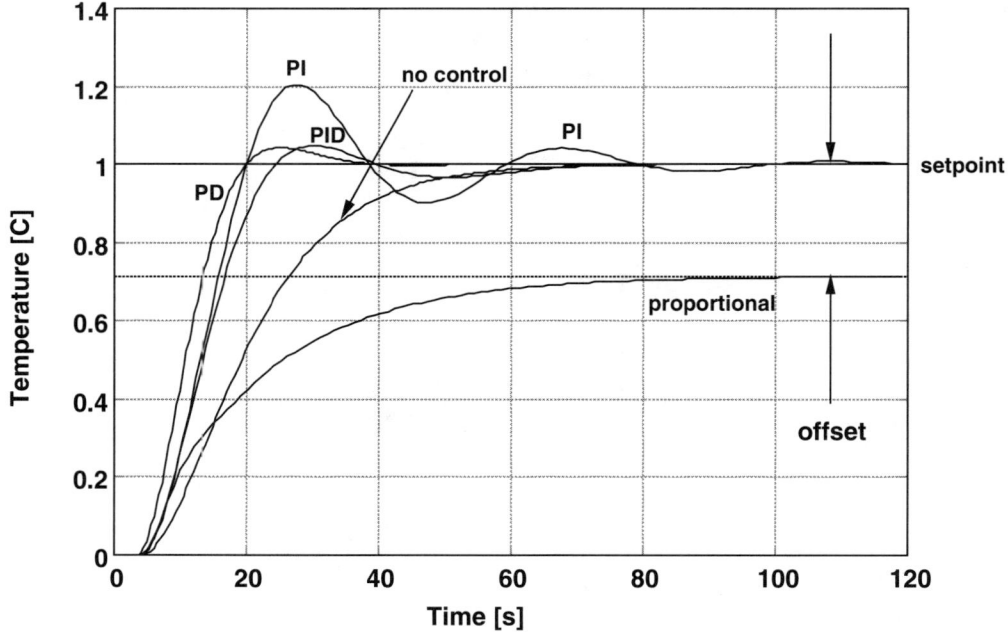

Figure 3–14 Step responses of the heat exchanger with feedback control

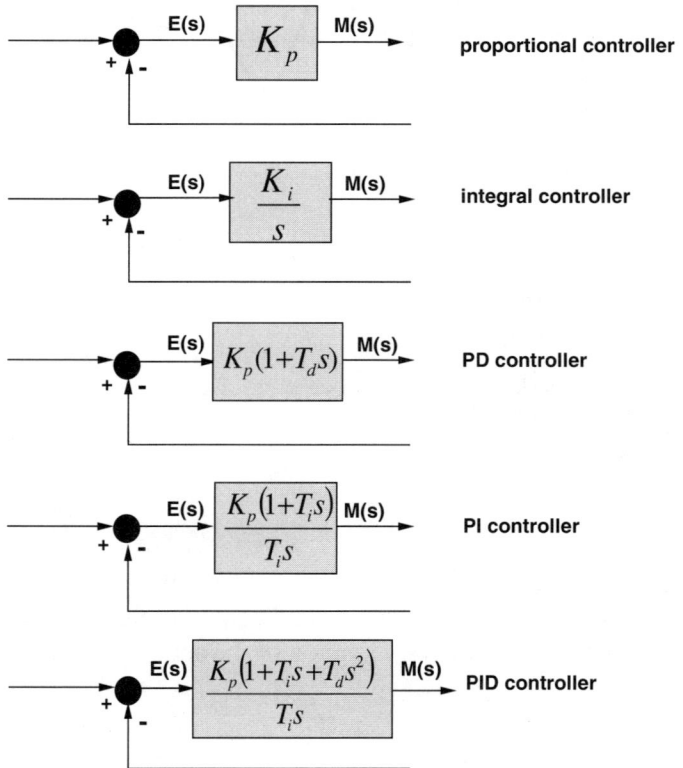

Figure 3–15 Block diagrams of different feedback control modes

or

$$m(t) = K_i \int_0^t e(t)dt$$

(9)

where K_i is an adjustable constant. The transfer function of the integral controller (Figure 3–15) is

$$\frac{M(s)}{E(s)} = \frac{K_i}{s}$$

(10)

(Note that Equation 10 is the Laplace transform of Equation 9.) If the error is doubled over a previous value, $m(t)$ is moved twice as fast. When the controlled variable is at the setpoint ($e(t)=0$), the value of $m(t)$ remains stationary. So integral mode causes the controller output to change as long as the error $e(t)$ is different from zero. When a sustained error occurs, the integral error becomes very large, and the controller output eventually saturates. Integral control action may also lead to oscillatory response. In practice, integral controllers are seldom used alone, since little integral action occurs until an error signal persists for some time.

Proportional-plus-Integral (PI) Control

In practice, integral control is usually combined with proportional control. The basic control action for a PI controller is defined by the following equation:

$$m(t) = K_p e(t) + \frac{K_p}{T_i} \int_0^t e(t)dt$$

(11)

and the transfer function (Figure 3–15) of the PI controller is

$$\frac{M(s)}{E(s)} = K_p \left(1 + \frac{1}{T_i s} \right)$$

(12)

where T_i is the integral time. The integral time adjusts the integral control, while the proportional constant K_p affects both the proportional and integral parts of the controller. The inverse of the control time is called *reset rate*, which is the number of times per minute that the proportional part of the control action is duplicated.

The advantage of PI controllers is fast response and no offset. A disadvantage is a decrease in stability due to the presence of the integral mode. Figure 3–14 illustrates the PI control response of the heat exchanger system of Figure 3–9 to a unit step change in the setpoint. It shows how the integral action eliminates offset but causes the loop to oscillate (this behavior is called reset windup). Reset windups are common in batch processes, during the start-up, or after larger setpoint changes when sustained errors are likely. Sustained error can saturate the control output, leading to overshoot when the measurement approaches the setpoint. In this case, switches may be added to the integral circuit to prevent the controller from reducing the error signal to zero. Today, commercial controllers are available that provide *antireset windup* by halting the integral control action whenever the controller output saturates.

Tuning a PI controller is also more difficult than tuning a proportional-only controller, since there are two adjustable constants that need to be tuned separately but each depends on the other.

Derivative Control

Derivative action (also called rate control action) is based solely on the rate of change of the error signal. An ideal derivative action can be described as

$$m(t) = K_d \frac{de(t)}{dt}$$

(13)

where K_d is an adjustable constant. The transfer function of the integral controller is

$$\frac{M(s)}{E(s)} = K_d s$$

(14)

Equation 13 shows that as long as the error is constant—that is, $de/dt = 0$—the controller output $m(t)$ remains stationary. Therefore, derivative control is usually found in combination with proportional control.

Proportional-plus-Derivative (PD) Control

The control action of a proportional-plus-derivative controller is defined as

$$m(t) = K_p e(t) + K_p T_d \frac{de(t)}{dt}$$

(15)

and the transfer function (Figure 3–15) is

$$\frac{M(s)}{E(s)} = K_p (1 + T_d s)$$

(16)

where T_d, the derivative time, is the time interval by which the rate action advances the effect of the proportional control action. The derivative mode moves the controller output as a function of the rate of change of the controlled variable, which adds phase lead to the controller, increasing the speed of response. Figure 3–14 shows the PD control response of the heat exchanger system to a step change in the setpoint. In this case the PD controller improves the dynamic response of the controlled variable by decreasing the time it takes the process to reach steady state.

If the controller output measurements tend to be noisy (contain high-frequency or random fluctuations), as in the case of flow control loops, derivative action is seldom used unless the measurement is filtered.

Proportional-plus-Derivative-plus-Integral (PID) Control

The derivative control action can be combined with proportional and integral actions to form the three-mode or PID controller. This combined action has the advantages of each of the three individual control actions. The equation of the PID controller is

$$m(t) = K_p e(t) + K_p T_d \frac{de(t)}{dt} + \frac{K_p}{T_i} \int_0^t e(t)\,dt$$

(17)

and the transfer function (Figure 3–15) is

$$\frac{M(s)}{E(s)} = K_p \left[1 + T_d s + \frac{1}{T_i s} \right]$$

(18)

PID controllers give rapid responses and do not exhibit offset, but it is very difficult to tune them. The interaction among the control modes makes it difficult to adjust the three parameters independently in standard PID controllers (pneumatic and electronic). They often require extensive and continuous adjustment to stay properly tuned. Digital PID controllers, on the other hand, are easier to tune, with no interaction among the modes (Seborg et al, 1989).

PID controllers offer very fine control when good tuning is implemented. Figure 3–14 shows the PID control response of the heat exchanger system to a step change in the setpoint. The response was faster and less oscillatory than the PI controller but slower than the PD controller for the system studied.

In conclusion, the typical responses of a feedback control system to a step change in load variables are as follows: if no feedback control is used (open loop, no control), the process reaches the new steady state slowly; proportional control speeds up the process response and reduces offset; adding an integral action eliminates offsets, but the response is more oscillatory; adding a derivative action reduces both the degree of oscillation and the response time.

Tuning Methods

Once the control system is installed, the controller settings need to be tuned so that the control system can perform satisfactorily. The techniques used for adjusting controller settings are classified into open-loop and closed-loop methods. One of the best techniques for tuning PID controllers is the Ziegler-Nichols method, known as the *ultimate method* or *loop tuning*. This method requires the determination of the ultimate gain (sensitivity) and ultimate period for the loop. The ultimate gain (K_u) is the maximum allowable value of the gain (for proportional-only controllers) for which the closed loop is stable. The period of the sustained oscillation is the ultimate period (P_u). The PID controller settings are then calculated from K_u and P_u using the Ziegler-Nichols tuning relations shown in Table 3–1. The Ziegler-Nichols relations were empirically developed to give a decay ratio of ¼ (see Figure 3–10c).

Although widely used in the industry (Shrinskey, 1988), the Ziegler-Nichols tuning method is time consuming and should be used with caution. The values shown in Table 3–1 should be used as first estimates. If a process model or frequency response data are available, transient response and frequency response criteria can be used to determine controller settings and sometimes are superior to the Ziegler-Nichols setting method. However, if process information is not very accurate or incomplete, plant setting may still be required to fine-tune the controller.

OPEN-LOOP AND FEEDFORWARD CONTROL SYSTEMS

Open-loop control systems are control systems in which the output has no effect on the control action. In an open-loop control system, the output is neither measured nor fed back for comparison with the input. An open-loop control system is shown in Figure 3–16. An example of such system is the bread machine, where mixing, raising, kneading, and raising again operate on a time basis. The bread machine does not measure the output that is the final product, the bread quality (color, volume, etc). Open-loop control systems are used when the relationship between the input and output is known and when internal or external disturbances are not present in the process.

Feedforward control is defined as the control of measurable disturbances by compensation for them before they materialize. Feedforward control is designed to prevent errors from occurring as opposed to feedback control, which is designed to eliminate errors. On the basis of the values of the measuring disturbances, feedforward controllers calculate the required values of the manipulated variables to maintain the controlled variable at the setpoint.

Figure 3–17 illustrates an example of a feedforward control for a SISO system. If an external disturbance, v, of a process system can be measured before it acts on the output

Table 3–1 The Ziegler-Nichols Controller Settings

Settings/Controllers	P	PI	PD	PID
K	0.5 K_u	0.45 K_u	0.6 K_u	0.6 K_u
T_i	---	P_u/1.2	--	0.5 P_u
T_d	---	--	P_u/8	P_u/8

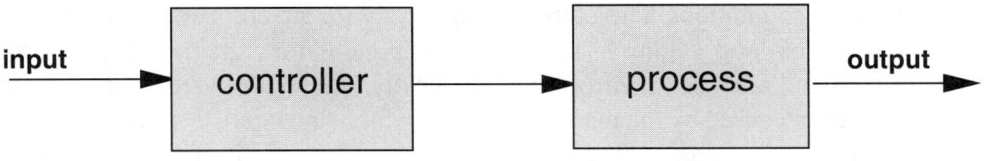

Figure 3–16 Open-loop control system

variable, y, then feedforward control can be used to improve the control performance with respect to this disturbance. Immediately after a change in the disturbance v, the process input, u, is manipulated by a feedforward control, G_{ff}, which does not wait until the disturbance has affected the control variable, y, as a feedback control does. The main disadvantages of feedforward controllers are:

- The disturbance must be measured on line. This may not be feasible for many applications, mainly in the food industries, where on-line sensors to measure raw material properties such as particle size distribution, composition, and rheological properties are not yet available.
- An accurate mathematical description of the process should be available so we can know how changes in the disturbances and manipulated variables affect the controlled variables.

Ideal feedforward control may be not physically realizable, but practical approximations are common and provide good control performance.

Feedforward Design Using Energy and Material Balances

Pure feedforward control is rarely found. In practical applications, feedforward control is normally combined with feedback control. Feedforward control can minimize the transient error, but it will not cancel the effects of unmeasurable disturbances under

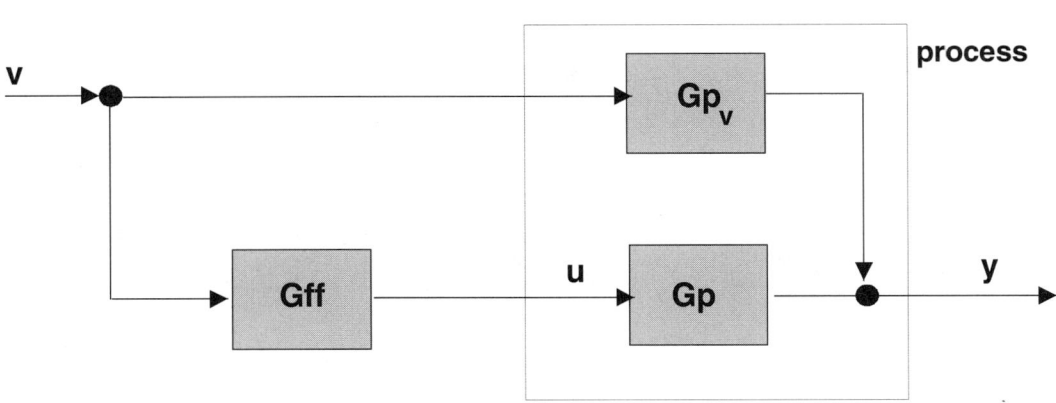

Figure 3–17 Feedforward control of a single-input, single-output system

normal operating conditions. Therefore, it is necessary for a feedforward control system to include a feedback loop. Figure 3–18 shows a continuous-flow grain-drying process with a feedforward and feedback control loops. Basically, the feedforward control minimizes the transient error caused by the measurable disturbance (inlet moisture content), while the feedback control compensates for any imperfections in the functioning of the feedforward control and provides for corrections for unmeasurable disturbances.

In a continuous-flow grain dryer, a feedforward controller measures continuously the main disturbance variable (ie, the inlet moisture content). Subsequently, if the manipulated

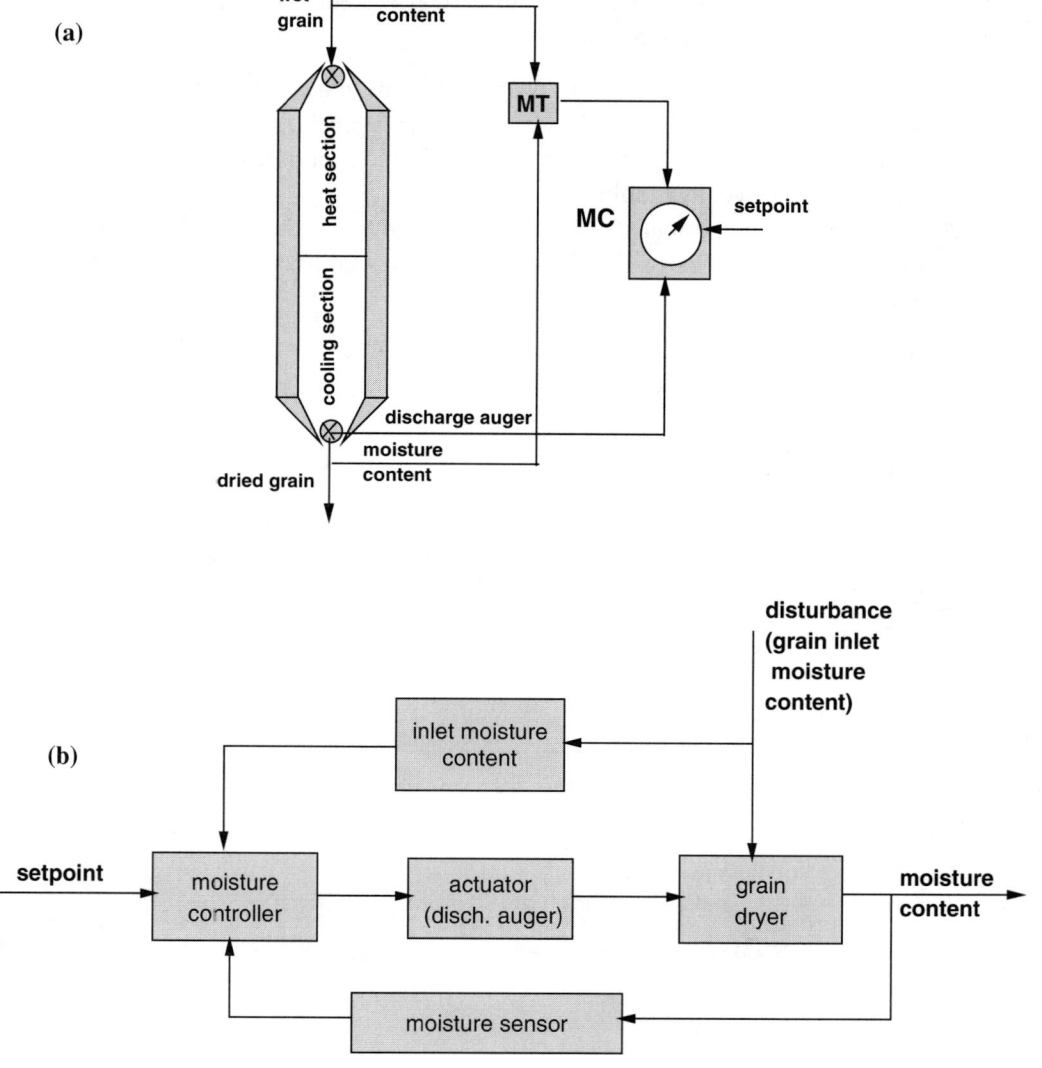

Figure 3–18 Continuous-flow grain dryer. (a) Moisture control system; (b) block diagram.

variable is the discharge auger rpm, a process control computes the residence time for which the grain should remain in the dryer. If a correction of the residence time is required due to a change in the inlet moisture content, corrective action is taken at the moment the grain enters the drying section. Thus, no time delay is encountered with feedforward control, in contrast to feedback control, which has an inherent 0.5- o 2.5-hour dead time in a continuous-flow grain dryer.

Generally, feedforward control calculations are based on materials and energy balances. To illustrate this design method, consider the feedforward control system and continuous-flow grain dryer shown in Figure 3–18. In this case, the feedforward algorithm is based on a material balance in the water vapor and in the air. In a typical cross-flow dryer (see Figure 3–19), the rate of moisture evaporated from the grain in the dryer per unit of dryer depth is (Platt et al, 1992)

$$\text{moisture removed / depth} = \dot{m}_a H(W_{out} - W_{in})$$

(19)

where $\dot{m}_a$ is the airflow rate, W the air humidity ratio, and H the dryer length. To reduce the grain moisture to the desired value, the grain flow rate needs to be calculated as

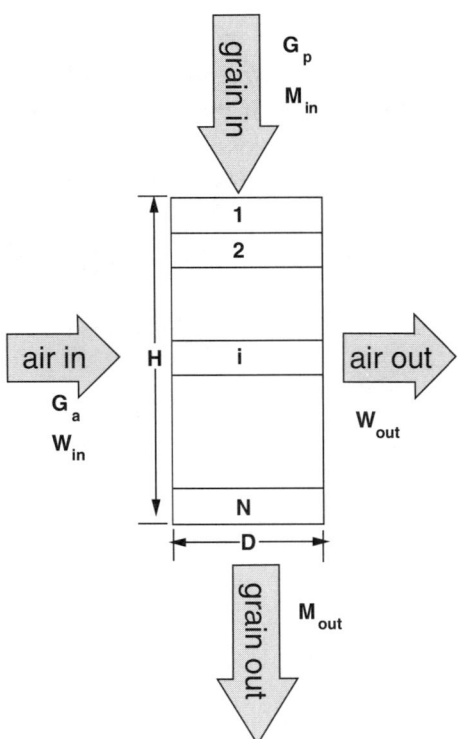

Figure 3–19 Schematic of a cross-flow grain dryer

$$\dot{m}_p = \frac{\dot{m}_a H(W_{out} - W_{in})}{D(M_{in} - M_{set})}$$

(20)

where $\dot{m}_p$ is the grain flow rate, D is the dryer width, and M and M_{set} are the grain moisture content and the moisture setpoint, respectively. The residence time of the grain in the dryer can be calculated as

$$T_r(t) = \frac{\rho(1-\varepsilon)H}{\dot{m}_p(t)} = \frac{\rho(1-\varepsilon)D[M'_{in}(t) - M_{set}]}{\dot{m}_a(t)[W_{out}(t) - W_{in}(t)]}$$

(21)

where ε is the grain bed porosity and M' the arithmetic average of the initial moisture content of all grain currently in the dryer (Moreira, 1989). Typically, the inlet grain moisture content (M_{in}) is not constant, so the manipulation of the grain flow rate according to Equation 21 takes into consideration changes in the input air humidity as well as input grain moisture. Discretizing and converting Equation 21 to a dimensionless form results in

$$T_r^*(t) = \frac{C[M'^*_{in}(k) - M^*_{set}]}{\dot{m}_a^*(k)[W^*_{out}(k) - W^*_{in}(k)]}$$

(22)

where $C = [\dot{m}_p(0)DM_{in}(0)]/[\dot{m}_a(0)HW_{in}(0)]$ and $M^*_{in}(i,k)$ is the average dimensionless grain moisture that occupies the ith row of the dryer at the kth time interval ($t^* = t^*_k$). To calculate $M^*_{in}(1,k)$, the average dimensionless inlet moisture of the grain entering the dryer during the $k - 1^{st}$ time interval ($k \geq 2$), requires an on-line measurement of M_{in}, so

$$M^*_{in}(1,k) = \frac{\int_0^{t_k^*} M^*_{in} dt^* - \int_0^{t_{k-1}^*} M^*_{in} dt^*}{T_r^*(k-1)/N}$$

(23)

where $M^*_{in}(1,1) = 1$ and $M^*_{in}(N,k-1) = M^*_{in}(i,k-N)$ for $k > N$ and $M^*_{in}(N,k-1) = 1$ for $k \leq N$. The values of $M^*_{in}(N,k-1)$ are found by shifting the data along the rows of the dryer at the end of each time interval:

$$M^*_{in}(i+1,k=1) = M^*_{in}(i,k) \qquad \text{for } 1 \leq k < N$$

(24)

where $M^*_{in}(1,k)$ is given by Equation 23. The residence time, $T_r^*(k)$, is calculated for each time interval using Equation 22 and the grain flow rate $= 1/T_r^*(k)$ adjusted accordingly.

Figure 3–20 shows the closed-loop response to a 20% step change in the inlet moisture content. The objective of the controller is to maintain the outlet moisture content close to the setpoint ($M^*_{out}(0) = 0.91$), regardless of disturbances in the loads (in this case the inlet grain moisture content), starting at $t^* = 0.2$. With the feedforward controller, the system is

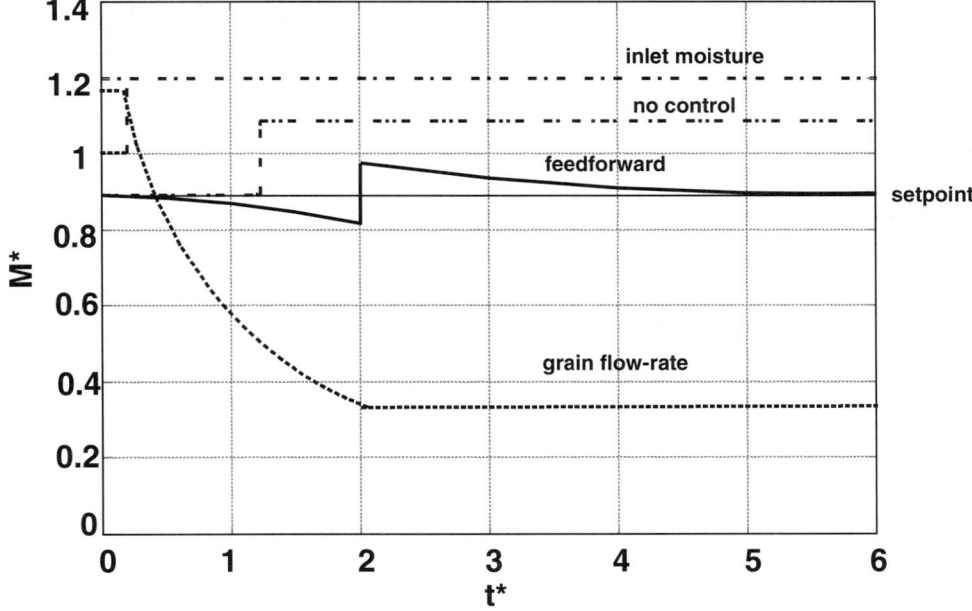

Figure 3–20 Cross-flow drying-process response to a 20% step change in the grain inlet moisture content.

maintained at the setpoint. There is a transient period before the system reaches the desired output. The setpoint is achieved only after two dryer volumes of grain have exited the dryer when the inlet moisture content stops changing (Platt et al, 1992). This is because the outlet moisture content of each grain layer depends on the initial moisture contents of all grain layers already in the dryers.

Feedforward Design Using Dynamic Models

For a feedforward and feedback control combination, a method for determining the transfer functions for the feedforward and feedback controllers will be presented. Consider the system shown in Figure 3–21, where the process transfer function $G_p(s)$ and the disturbance transfer function $G_v(s)$ are known. The output $Y(s)$ is given by

$$Y(s) = G_c(s)G_p(s)E(s) + G_v(s)V(s)$$

(25)

where

$$E(s) = W(s) - Y(s) + G_{ff}(s)V(s)$$

(26)

where $G_{ff}(s)$ is the feedforward transfer function. Substituting Equation 26 into Equation 25 results in

$$Y(s) = G_c(s)G_p(s)[W(s) - Y(s)] + [G_c(s)G_p(s)G_{ff}(s) + G_v]V(s)$$

(27)

The closed-loop transfer function for load changes can then be derived as

$$\frac{y(s)}{V(s)} = \frac{G_v(s) + G_c(s)G_p(s)G_{ff}(s)}{1 + G_c(s)G_p(s)}$$

(28)

For a constant setpoint $[W(s) = 0]$, the objective of the control system is to maintain $Y(s) = 0$ regardless of changes in the disturbance variable $[V(s)]$, so from Equation 28 it can be seen that the effect of $V(s)$ can be eliminated if

$$G_v(s) + G_c(s)G_p(s)G_{ff}(s) = 0$$

(29)

or

$$G_{ff}(s) = \frac{G_v(s)}{G_c(s)G_p(s)}$$

(30)

Once the controller transfer function $G_c(s)$ is properly designed, the feedforward transfer function $G_{ff}(s)$ can be obtained from Equation 30 so that the closed-loop control system can provide the desired performance.

For an application of feedforward-feedback dynamic control design, consider the simplified block diagram of a food extruder system. What is desired is a feedforward control scheme to maintain the die pressure, DP, at the setpoint despite disturbances in feed moisture content (Figure 3–22). This example is based on the work of Moreira et al (1990).

The system dynamics are shown in Figure 3–23. The feed rate response is characterized by a 22-second delay and has a shape of second-order function. It shows that an increase in feed rate (keeping everything else constant) causes an immediate increase in pressure due to the increase in the number of filled flights followed by a pressure reduction due to an increase in temperature (due to energy dissipation) and thus viscosity of the melt in the die. So the die-pressure response to a step up in the feed rate can be characterized by (Figure 3–23)

$$G_p(s) = \frac{DP(s)}{FR(s)} = \frac{22e^{-22t}}{4489s^2 + 49.58s + 1}$$

(31)

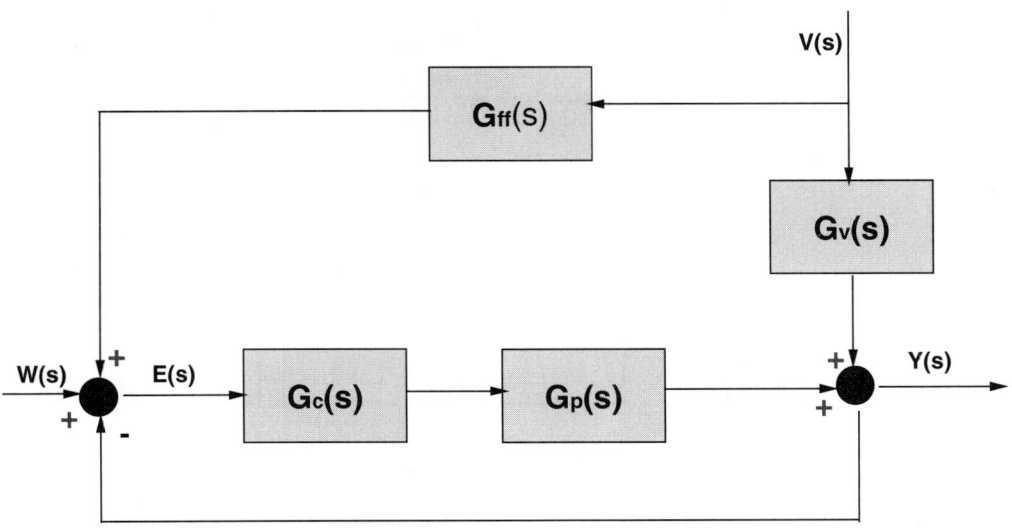

Figure 3–21 Feedforward-feedback control system

The feed moisture also shows a 22-second delay, and the response is a combination of first-order function with lead-lag. The response shows a brief increase and subsequent exponential decay in the die pressure. This response indicates that the die pressure is very sensitive to changes in moisture. Decreasing the feed moisture content results in an increase in the extrudate viscosity and then die-pressure. The die pressure response to a step down in the feed moisture can be described as (Figure 3–23)

$$G_p(s) = \frac{DP(s)}{FM(s)} = \frac{(2417.7s + 20.40)e^{-22t}}{1950s^2 + 95s + 1}$$

(32)

The step-down change in the screw speed shows a rapid pressure response (no time delay). The results indicate that an increase in screw speed instantly increases the specific mechanical energy (SME). Increase in screw speed adds dissipative energy to the system, causing an increase in the melt temperature and thus viscosity reduction. This response can be characterized by a first-order function (Figure 3–23)

$$G_p(s) = \frac{DP(s)}{SS(s)} = \frac{30}{36s + 1}$$

(33)

This is a simplified description of the extrusion process. In reality, the process is nonlinear and multivariable. A more in-depth analysis of the system will be given in Chapter 6. The objective here is just to illustrate how to design a feedforward control in a food-processing system in which variations in raw material composition (such as water content) are present.

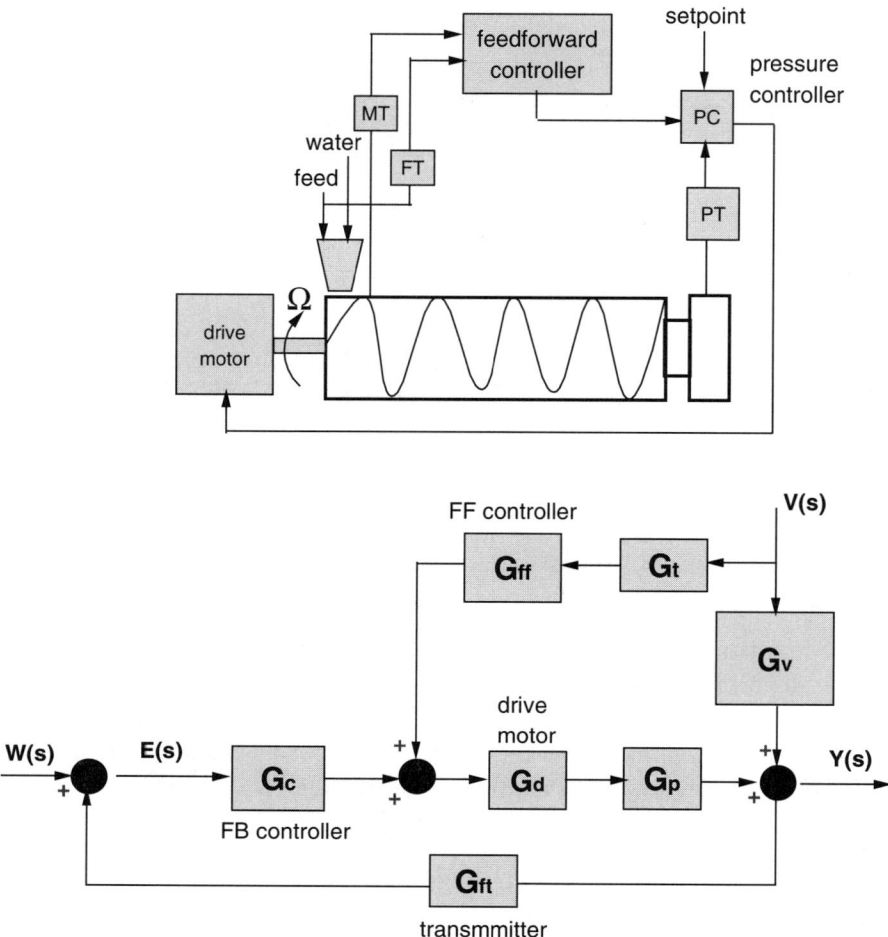

Figure 3–22 Feedforward control with feedback control loop for die pressure of a food extrusion process

Generally, when operating a food extruder, it is economically efficient to run the process at its maximum throughput, thereby keeping the feed flow rate at its maximum value. So this leaves us with two inputs. Since the screw speed showed no delay in the die pressure response to a unit-step input, it could be selected as the input variable and the feed moisture variations can be selected as the major disturbance.

In the block diagram of Figure 3–23, the load transmitter, the feedback transmitter, and the drive motor are assumed to have negligible dynamics, so that $G_t(s) = K_t \ G_{ft}(s)$, and $G_d(s) = K_d$. By using block diagram algebra (see Figure 3–5), the closed-loop transfer function for disturbance changes can be described as

$$\frac{Y(s)}{V(s)} = \frac{G_v + G_t G_{ff} G_d G}{1 + G_c G_d G_p G_{ft}}$$

(34)

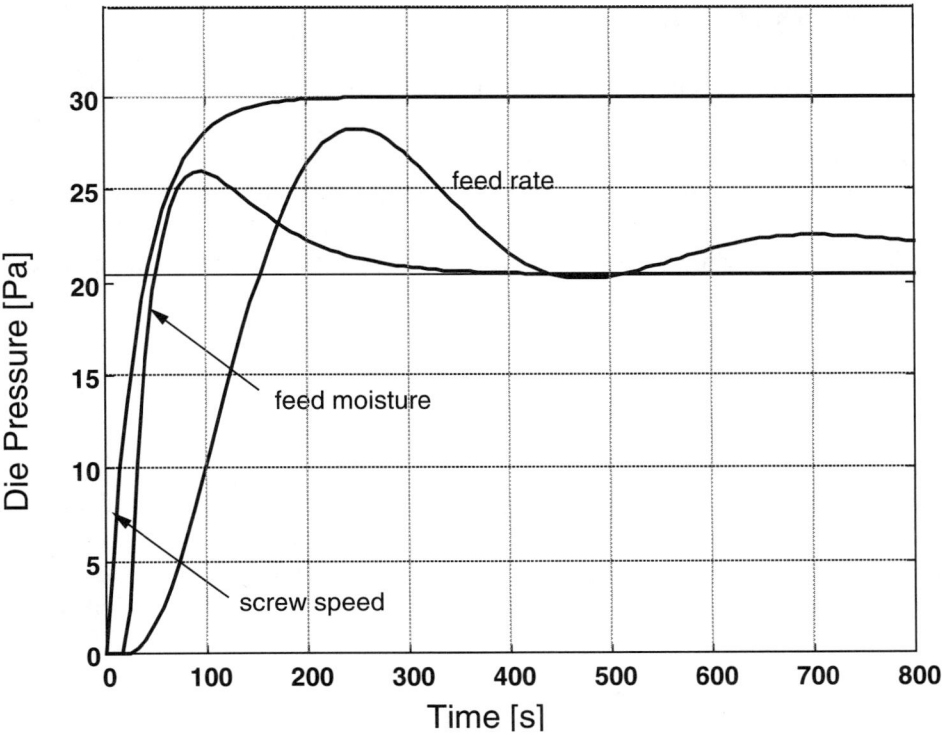

Figure 3–23 Unit-step response for a simplified food extrusion process

So, at constant setpoint [$W(s) = 0$], we want $y(s) = 0$ despite changes in the feed moisture content [$V(s) \neq 0$]. From Equation 34, this condition is satisfied if

$$G_v + G_t G_{ff} G_d G_p = 0$$

(35)

Thus, the ideal feedforward controller can be calculated as

$$G_{ff}(s) = \frac{G_v(s)}{G_t(s)G_d(s)G_p(s)}$$

(36)

In our example, $G_v(s)$ is described by Equation 32 without the time delay e^{-22t} and $G_p(s)$ by Equation 33. From Equation 36, the ideal feedforward controller is

$$G_{ff}(s) = -\left(\frac{80.59s + 0.68}{1950s^2 + 95s + 1}\right)\left(\frac{36s + 1}{K_t K_d}\right)$$

(37)

Dynamic Compensation: Lead-Lag

Lead-lag compensation can provide good approximations to ideal feedforward controllers. Lead compensation improves the speed of response and reduces overshoot but gives little improvement in steady-state performance. Lag compensation gives large improvement in steady-state performance, but the response is slower. So, for improvement in both transient and steady-state responses, lead and lag networks are used simultaneously in a single lead-lag network. A feedforward controller consisting of lead-lag compensation will have the following characteristics

$$G_{ff} = K_{ff} \frac{(T_1 s + 1)}{(T_2 s + 1)}$$

(38)

where K_{ff}, T_1 and T_2, are parameters that depend on the system with $T_2 > 0$ for stability. Consider, for example, the case where the noise transfer function relating extrudate moisture content and feed moisture, $G_v(s)$, and the process transfer function relating extrudate moisture and screw speed, $G_p(s)$, of the extrusion process are described as

$$G_v(s) = \frac{MC(s)}{FM(s)} = \frac{K_v}{58s + 1}, \quad G_p(s) = \frac{MC(s)}{SS(s)} = \frac{K_p}{63s + 1}$$

(39)

where K_v and K_p are the noise and process gains, respectively. From Equation 26, the feedforward controller is described as

$$G_{ff} = -\left(\frac{K_v}{K_t K_d K_p} \right)\left(\frac{63s + 1}{58s + 1} \right)$$

(40)

Equation 40 shows that the feedforward controller is a lead-lag compensator.

Like feedback controllers, feedforward controllers require tuning after the controller has been installed in the process. For a lead-lag feedforward controller, the tuning consists of selecting and adjusting the controller parameters, which can be done in three steps, as described by Seborg et al (1989).

CLOSED-CONTROL SYSTEM DESIGN FOR LINEAR SISO SYSTEMS

In the classical control problem, usually there are three control design objectives:
1. System stability under all system operating conditions
2. Good steady-state error performance
3. Good system dynamics or transient response

Design for Steady-State Error Performance

The steady-state error is defined as

$$e_{ss} = \lim_{t \to \infty} e(t)$$

(41)

where $e(t)$ is the error defined in Figure 3–4. Steady-state error is always defined with respect to a particular input, such as step or ramp. Consider the system shown in Figure 3–4 with $V(s)$ and $U(s)$ equal to zero. The closed-loop transfer function is

$$\frac{Y(s)}{W(s)} = \frac{G_c(s)G_p(s)}{1 + H(s)G_c(s)G_p(s)}$$

(42)

The transfer function between $E(s)$ and $W(s)$ is

$$\frac{E(s)}{W(s)} = \frac{1}{1 + H(s)G_c(s)G_p(s)}$$

(43)

so $E(s)$ is equal to

$$E(s) = \frac{W(s)}{1 + H(s)G_c(s)G_p(s)}$$

(44)

Applying the final value theorem, FVT (see Appendix 3–A) to Equation 44, we can find the steady-state performance of a stable system. So the steady-state error is

$$e_{ss} = \lim_{t \to 0} e(t) = \lim_{s \to 0} sE(s) = \lim_{s \to 0} \frac{sW(s)}{1 + H(s)G_c(s)G_p(s)}$$

(45)

Equation 45 is the basis for design for steady-state error performance. As an example of application of steady-state error, consider the following transfer functions:

$$G_p(s) = \frac{15}{s(s+5)(s+1)}; \quad G_c(s) = K_p; \quad H(s) = 1$$

(46)

We can determine the steady-state error for the system for a step input $W(s) = M/s$:

$$E(s) = \frac{M(s+5)(s+1)}{s(s+5)(s+1) + 15K_p}$$

(47)

and from the FVT we can get the steady-state error as

$$e_{ss} = \lim_{s \to 0} \left[\frac{(s+5)(s+1)}{s(s+5)(s+1)+15K_p} \right] = 0$$

(48)

Note that the FVT does exist for *some* values of K_p. If we apply the Routh test (Appendix 3–B) to the characteristic equation of $E(s)$, $e(t)$ is bounded for $0 < K_p < 8$.

We can determine now the steady-state error of the above system if the input is a ramp function, $W(s) = M/s^2$:

$$E(s) = \frac{M(s=5)(s+1)}{s\left[s(s+5)(s+1)+15K_p \right]}$$

(49)

$$e_{ss} = \lim_{s \to 0} sE(s) = \frac{M}{K_p}$$

(50)

For stability, $e_{ss} > M/8$. This example shows that while the steady-state error for the step input is zero, e_{ss} for the ramp input is nonzero and inversely proportional to K_p.

Design for Dynamic Performance

In general terms, the objective of control system design for dynamic performance is a system that adjusts quickly and well to input changes. Several common measures of system dynamic performance are defined with respect to a step change in the system input, W. Systems that respond well to step-input changes often react well in a dynamic sense to other types of input changes. Figure 3–24 shows the dynamic performance measures defined with respect to a unit-step input:

- T_r = the "rise time," defined as the time required for the output to go from zero to 90% of the input value
- T_s = the "settling time," defined as the time required for the output to reach and stay within a range about the final value size specified by the absolute percentage of the final value (usually 5% or 2%)
- M_t = the maximum percent "overshoot," defined as the maximum peak value of the output curve measured from unit; it indicates the relative stability of the system

The dynamic performance of a linear system is determined by the location of the system poles in the s-plane. So an approach used to design control systems involves choosing $G_c(s)$ so that the resulting pole locations will give desired values for the dynamic performance measures described above.

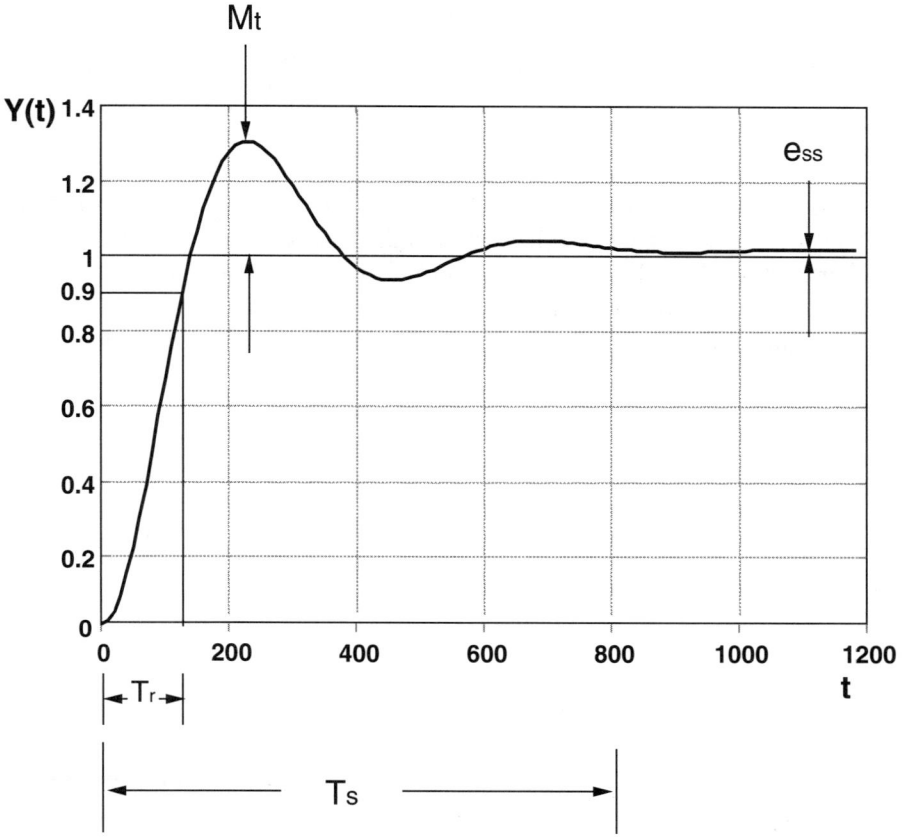

Figure 3–24 Dynamic performance measures for a unit-step input

Dominant System Poles

These are the poles that are nearest the origin without zeros in close proximity. Properly locating the dominant system poles can simplify the design process considerably.

Consider the following system subject to a unit-step input:

$$G(s) = \frac{Y(s)}{U(s)} = \frac{5}{(s+1)(s+5)}$$

(51)

So

$$Y(s) = \frac{5}{s(s+1)(s+5)}$$

(52)

and

$$y(t) = \left[1 + \frac{5e^{-t} - e^{-5t}}{4}\right], \quad t \geq 0$$

(53)

From Equation 51, we can see that the pole that tends to be more dominant is at $s = -1$. This can be better visualized in Equation 53. The term associated with $s = -1$, $5e^{-t}$, has the larger coefficient, 5. In addition, the term e^{-t} persists longer than e^{-5t} due to the pole $s = -5$. This response is plotted in Figure 3–25.

Now consider the case of the dominant pole approximation that includes only the dominant pole:

$$Y(s) = \frac{1}{s(s+1)}$$

(54)

and

$$y(t) = 1 - e - t, \quad t \geq 0$$

(55)

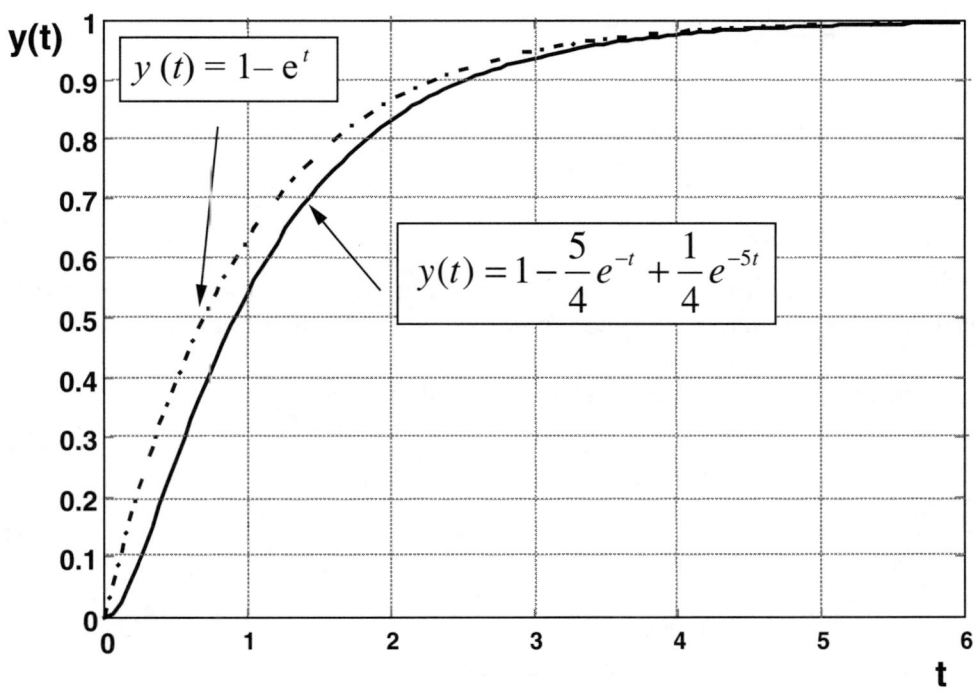

Figure 3–25 Dominant pole approximation

The response is shown in Figure 3–25. Note that the larger approximation error occurs for smaller values of t, where the effect of the pole at $s = -5$ is more significant.

Response Due to Dominant Complex Pole Pair

In control system design, it is common to make a complex pole pair dominant. Consider the following transfer function:

$$\frac{Y(s)}{W(s)} = \frac{a^2 + b^2}{(s - a + jb)} = \frac{a^2 + b^2}{(s + a)^2 + b^2} = \frac{\omega_n^2}{s^2 + 2\zeta\omega_n s + \omega_n^2}$$

(56)

The complex poles are at $s = -1 \pm jb$ and the natural frequency $\omega_n = (a^2 + b^2)^{1/2}$ and the damping ratio $\zeta = a/\omega_n$. Figure 3–26 illustrates this relationship geometrically.

The variables ζ and ω_n relate to the polar coordinates of the complex pole pair. The radial distance from the origin of the s-plane to either complex pole is ω_n. The cosine of the angle θ shown is ζ.

The unit-step response of a system with this complex pole pair is shown in Figure 3–27. From Equation 56, the time domain response to a unit-step input is

$$y(t) = 1 - e^{-\zeta\omega_n t}\left(\cos\omega_2 t + \frac{\zeta}{\sqrt{1 - \zeta^2}}\sin\omega_r t\right)$$

(57)

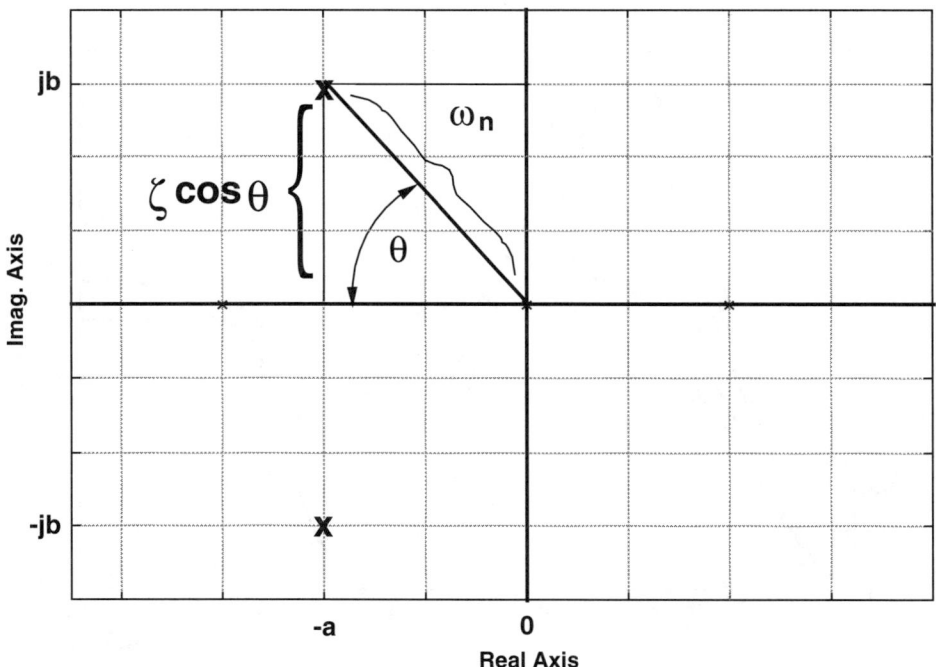

Figure 3–26 Geometric interpretation of ζ and ω_n

where $\omega_r = \omega_n\sqrt{1-\zeta^2}$. From Figure 3–27, the following points are taken from the step response:

- The parameter ζ determines the system overshoot. The following equation relates overshoot to the ζ value:

$$M_t = e^{-\left(\frac{\zeta\pi}{\sqrt{1-\zeta^2}}\right)}$$

(58)

- For a given ζ, the parameter ω_n determines the speed of the system response. It can be seen that the rise time T_r is inversely proportional to ω_n for a given ζ value. An increase in ω_n "speeds up" the response of $y(t)$ proportionally. Figure 3–28 illustrates the effect of ω_n on $y(t)$. The values used to generate Figure 3–28 were ω_n equal to 1 or 2 and ζ equal to 0.5. Note that the two responses have the same overshoot because the poles have the same ζ and θ. However, the response due to poles at $s_a = -1 \pm 1.75j$ is twice as rapid as that due to poles at $s_b = -0.5 \pm 0.75j$. This is because ω_n is twice the value for s_b as for s_a.

From the above discussion it is possible to describe locations for a complex pole pair that will result in desired values for overshoot and rise time for the system of Equation 56 with a step input.

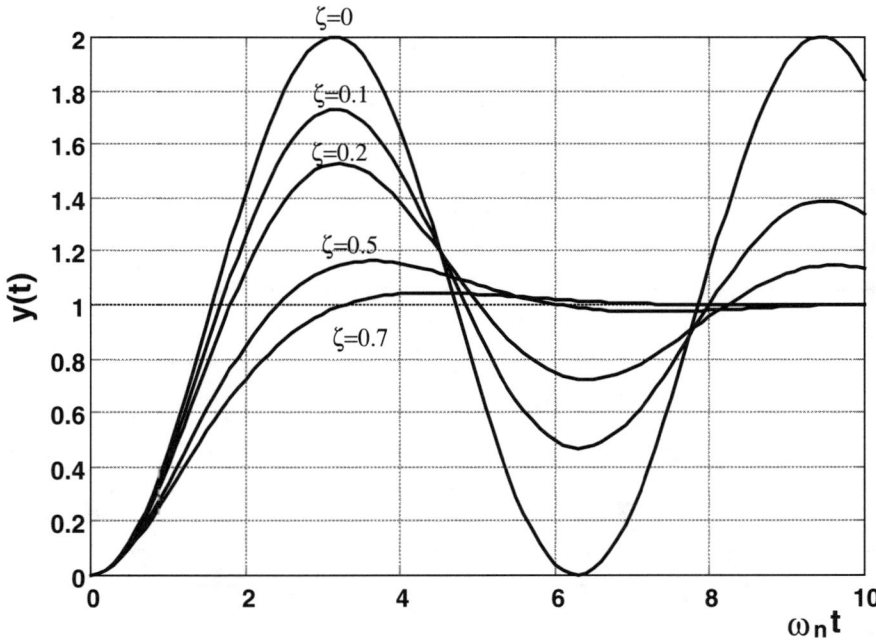

Figure 3–27 Step response due to a complex pole pair

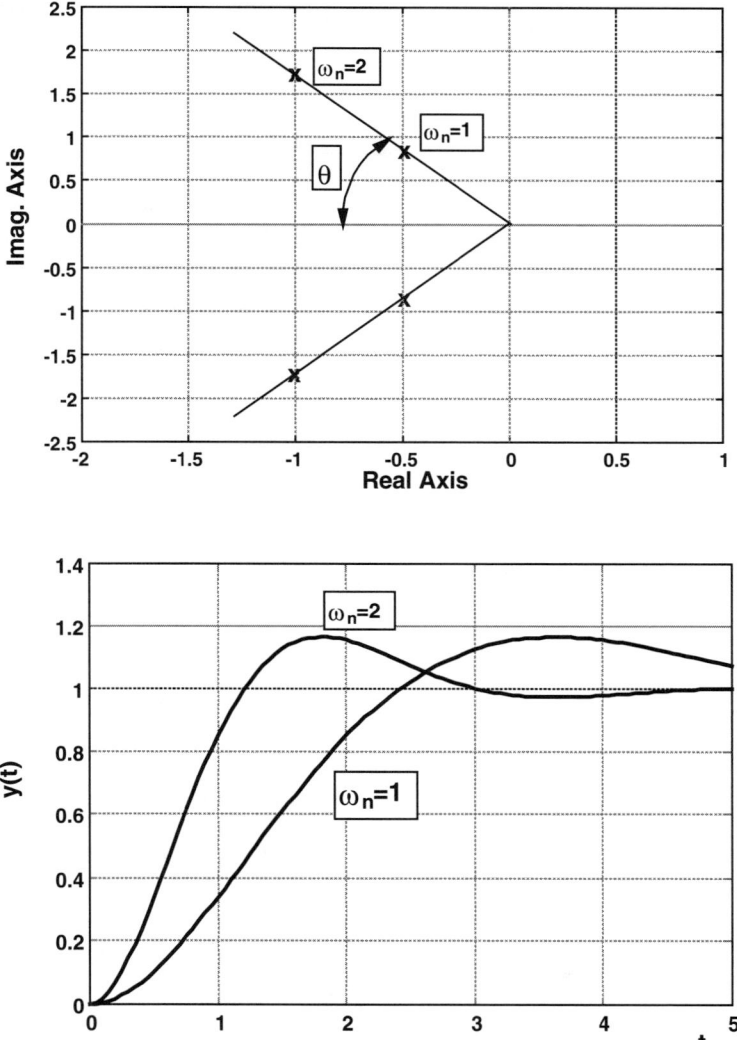

Figure 3–28 Effect of ω_n on the unit-step response of Equation 57

Improving Dynamic Performance by Means of Rate Feedback

Rate feedback is a technique whereby the rate of change of the output is also fed back. Rate feedback is shown in Figure 3–29, where the variable fed back in this configuration is

$$f(t) = y(t) + K_f \frac{dy}{dt}$$

(59)

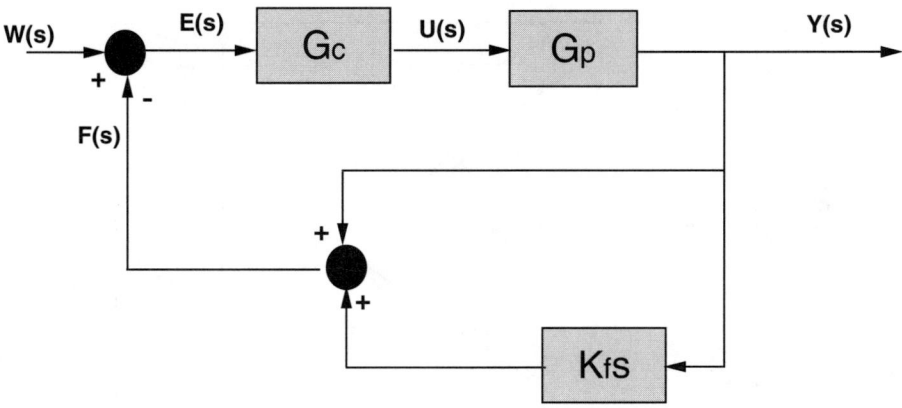

Figure 3–29 Rate feedback

from which

$$F(s) = Y(s) + K_f s Y(s)$$

(60)

The transfer function $H(s)$ is

$$H(s) = \frac{F(s)}{Y(s)} = 1 + K_f s = K_f \left(s + \frac{1}{K_f} \right)$$

(61)

So the effect of the rate feedback is to introduce a zero at $s = 1/-K_f$ in the s-plane. This is a special kind of zero. The closed-loop transfer function corresponding to Figure 3–29 is

$$\frac{Y(s)}{W(s)} = \frac{G_c(s)G_p(s)}{1 + G_c(s)G_p(s)H(s)} = \frac{KG(s)}{1 + KG(s)H(s)}$$

(62)

Suppose that the controlled system is stable for large values of K, so that Equation 62 is simplified to

$$\frac{Y(s)}{W(s)} = \frac{KG(s)}{1 + KG(s)H(s)} \approx \frac{KG(s)}{KG(s)H(s)} = \frac{1}{H(s)}$$

(63)

In this special situation, the zeros of $H(s)$ become the poles of the closed-loop transfer function. The zeros of $H(s)$ therefore do not tend to cancel the poles of $G_c(s)$ or $G_p(s)$.

Control System Design Example

The objective of this example is to design a controller for a (simplified) stirred-tank heater control system. The system is shown in Figure 3–30. The tank contents are well agitated and heated by an electric heater. The control objective is to keep the exit temperature T_{out} at the setpoint by manipulating the heating input, Q. So the simplified transfer function for the system is

$$G_p(s) = \frac{T_{out}(s)}{Q(s)} = \frac{6}{s^2 + 0.9s + 0.08} = \frac{6}{(s+0.1)(s+0.8)} \tag{64}$$

In Figure 3–30, $W(s)$ is the desired outlet temperature, $Y(s)$ the actual system temperature, $G_p(s)$ is the transfer function of the stirred-tank heater system in this case, and $G_c(s)$ and $H(s)$ are the controllers to be designed.

The most important design criteria are zero steady-state error in response to step changes in W and small overshoot in response to this same kind of input. We also seek a system that will respond quickly to a change in desired outlet temperature. In a real-world design exercise, we would also have to look carefully at production cost, system reliability, and trade-off cost and performance.

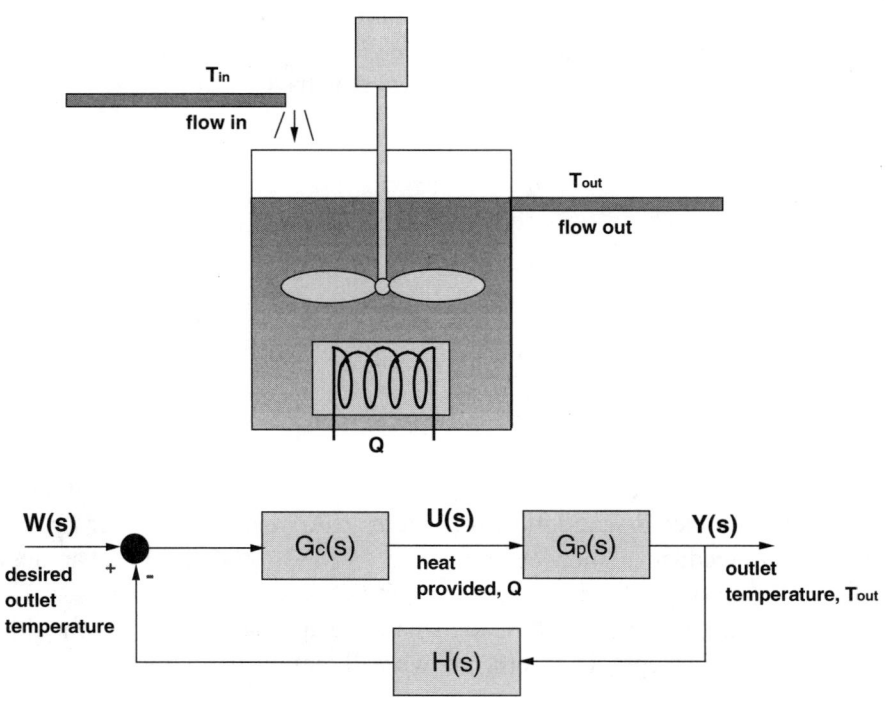

Figure 3–30 Schematic of a continuous stirred-tank heater

We will assume that the design problem is to choose $G_c(s)$ and $H(s)$ such that

1. Steady-state error in response to a step change in desired outlet temperature is zero.
2. Overshoot for a step change in desired speed is less than 10%.
3. For a step input change of 10°C, we require that the time to accomplish 90% of the change be 5 seconds or less (eg, the time to go from 25 to 35°C when a step increase of 10°C is introduced at 25°C).

Application of the final values theorem (FVT) quickly establishes that integral control is required to meet the steady-state error requirement. Tentatively we will specify integral control for steady-state error so that

$$G_c(s) = K_p\left(s + \frac{K_I}{K_p}\right)$$

(65)

and rate feedback for dynamic improvement is

$$H(s) = K_f\left(1 + \frac{1}{K_f}\right)$$

(66)

We need now to determine the values for K_I, K_p, and K_f that will satisfy the design requirements 1 to 3 above. It is reasonable to choose $K_I/K_p = 0.1$ to cancel the undesired slow response due to the pole at $s = -0.1$. Therefore, we now have

$$G_c(s) = \frac{K_p(s + 0.1)}{s}$$

(67)

and

$$G_c(s)G_p(s) = \frac{6K_p}{s(s + 0.8)}, \qquad H(s) = K_f\left(s + \frac{1}{K_f}\right)$$

(68)

To determine the value of K_f, we evaluate the pole-zero plot in the present situation, as shown in Figure 3-31 (see discussion of root locus in Appendix 3–C). For $K_f > 1.25$, the zero due to the rate feedback is between poles at $s = 0$ and $s = -0.8$. This is an undesirable situation, since the zeros tend to become dominant (slow) pole as K_p increases. Therefore, $K_f < 1.25$ gives a faster response. If K_f is too small, we approach the situation without rate feedback, which is not acceptable. Therefore, we will tentatively choose $K_f = 1$, so

$$H(s) = (s + 1)$$

(69)

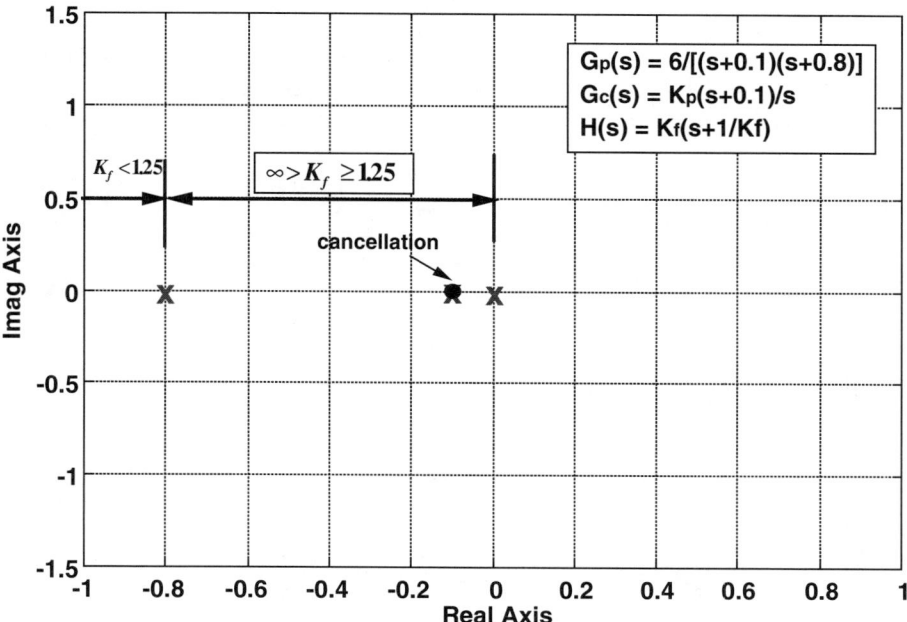

Figure 3–31 Root locus for Equation 68

And the closed-loop transfer function is

$$\frac{Y(s)}{W(s)} = \frac{6K_p}{s^2 + (0.8 + 6K_p)s + 6K_p}$$

(70)

The root locus for Equation 70 is shown in Figure 3–32. The plot shows the root locus for different values of K_p. The requirement for less than 10% overshoot translates to $\zeta > 0.6$ for Figure 3–32 (see Figure 3–27). From Figure 3–27 we also see that for $\zeta = 0.6$, $\omega_n t = 3.5$ for 90% output in response to a step. The value of ω_n to attain 90% in 4 seconds is then

$$\omega_n \approx \frac{3.5}{5} = 0.7$$

(71)

So $\omega_n > 0.7$ is required to satisfy the speed requirement. From this analysis, it seems that $K_p = 0.25$ in Figure 3–32 may satisfy both the speed and overshoot requirements. For $K_p = 0.25$, we have from Equation 70:

$$\frac{Y(s)}{W(s)} = \frac{1.5}{s^2 + 2.3s + 1.5} = \frac{1.5}{s^2 + 2\zeta\omega_n s + \omega_n^2}$$

(72)

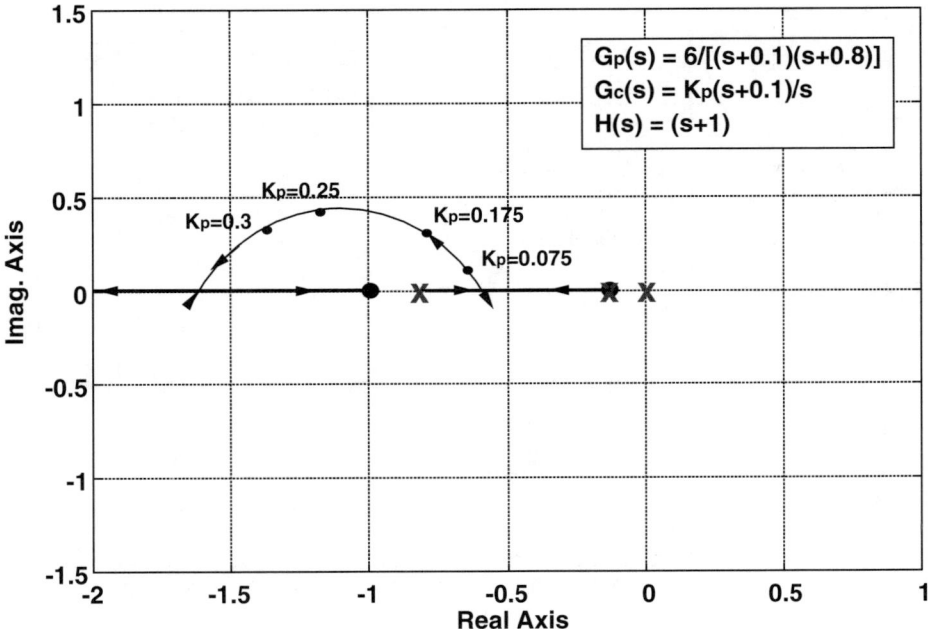

Figure 3–32 Root locus of the system described by Equation 69

$$\frac{Y(s)}{W(s)} = \frac{1.5}{(s+1.15+0.4213\,j)(s+1.15-0.4213\,j)}$$

(73)

and $\omega_n = (1.5)^{0.5} = 1.2247$; $\zeta = 2.3/2\omega_n = 0.939$.

The response of this system to a step change of 10°C is shown in Figure 3–33. This is the shape of the output response in going from 25 to 35°C if the initial value for $y(t)$ is at 25°C at $t = 0$. Clearly, we have met the performance requirements of 1 to 3 above. Note that the solution is not unique. There are many values for K_1, K_p, and K_f that would satisfy the performance requirements. In addition, note that the cancellation of pole with zero at $s = -0.1$ should be checked before final acceptance of this design. In practice we cannot expect exact cancellation due to error in measuring the system transfer function and error in the controller transfer function $G_c(s)$. Sensitivity analysis should be performed to determine the change in $y(t)$ that could be expected due to errors in either pole or zero location. In this case we would expect the design to be relatively insensitive to these errors, since the design of Figure 3–33 well exceeds the performance specifications for dynamic response. Another design check should always be made to ensure that unrealistic values of control $u(t)$ are not called for. In our example this is not a problem.

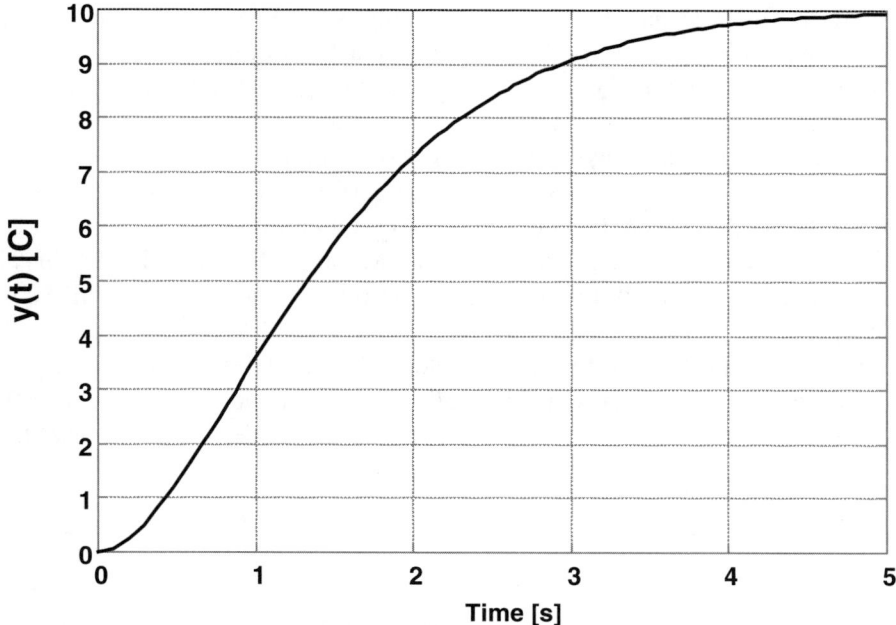

Figure 3–33 Step response of the stirred-tank heater system

ADVANCED CONTROL

The conventional PID controllers have limitations when applied to complex processing plants. Several advanced control strategies exist that can provide improved process control over PIDs. Some of these methods are cascade control, adaptive control, supervisory control, expert control, statistical quality control, time-delay compensation, and selective/override control. With the advances in computer control, many of these techniques have been successfully applied in the industry. In this chapter, only the time-delay compensation, selective/override control, and cascade control techniques will be discussed. The other techniques will be discussed in separate chapters because of their importance for food process control.

Time-Delay Compensation Control

Time delays or dead time are very common in the food-processing industries. Examples of processing systems with time delays include continuous-flow dryers, baking ovens, continuous fryers, and food extruders. In each of these systems, time delay is inherent in the nature of the process. The material within the process must flow through the equipment, and there is an inherent distance/velocity lag or time delay within the processing equipment.

The dead time associated with product (composition) analysis is another example of time delay encountered in the food-processing industry. In this case, the dead time is the time it takes for the analytical equipment (such as viscometers and moisture analyzers) to operate and measure the required property.

The presence of time delays in the process deteriorates the performance of a conventional feedback control system. To improve the performance of these systems, strategies like the Smith predictor have been developed to provide time-delay compensation. Other methods include the Moore analytic predictor (Moore, 1970) and the Dahlin algorithm (Dahlin, 1968), but the Smith predictor is the best known strategy and will be discussed in this chapter.

The Smith predictor algorithm was developed by Smith (1957) and is often referred as a model-based controller because the controller uses the model parameters directly.

Consider a process that can be represented by a first-order lag plus a dead time as

$$\frac{K_p}{1+Ts}e^{-\theta_{ds}}$$

(74)

where K is the process gain, T the process time constant, and θ the dead time. The block diagram shown in Figure 3–34 has the process split into two parts, a first-order lag portion and a dead-time part. A fictitious variable B is then defined as shown in Figure 3–35 that is fed back to the controller. Here, the dead time is outside the loop, so no delay exists in the feedback of B. The variable Y, on the other hand, will do whatever B does, but delayed by θ. However, this schematic arrangement cannot be implemented because the variable B cannot be measured.

By developing a mathematical model of the process and using the variable m as the input to the model and B_m as the feedback signal, as shown in Figure 3–36, the scheme will work as long as the model is perfect and no disturbances enter the loop.

Adding a second feedback loop (Figure 3–37) results in the Smith predictor control strategy. Here, the model of the process without the time delay is used to predict the effect of the control action on the process output. The controller uses the predicted response, B_m, to calculate its output. The predicted process output, B_{m2}, is also delayed by the amount

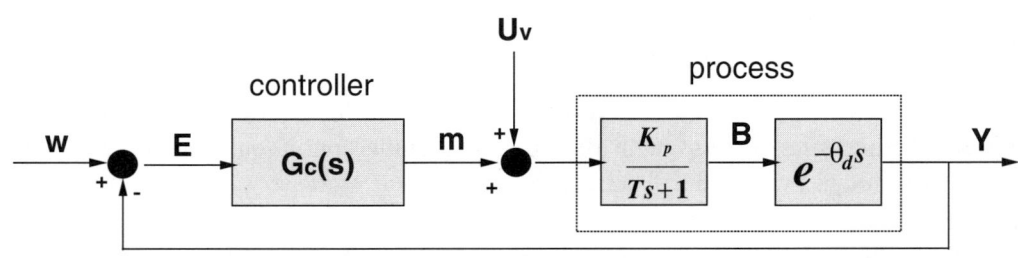

Figure 3–34 Feedback control loop with process dead time

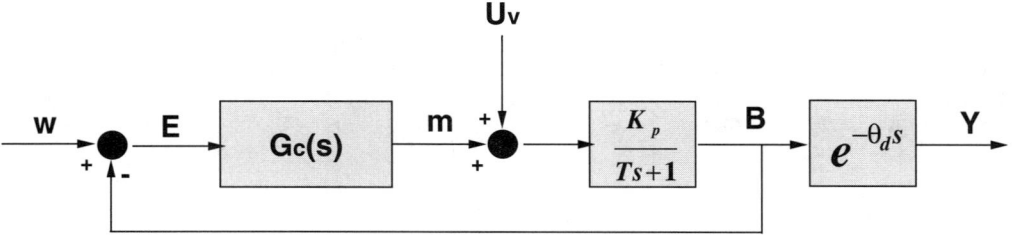

Figure 3–35 Desired feedback loop configuration

of the time delay, θ, for comparison with the actual process output Y. This corrects for modeling errors and load disturbances entering the process. From the block diagram we have

$$E_2 = E_1 - B_{m1} = w - B_{m1} - (Y - B_{m2})$$

(75)

If the process model is perfect and the disturbance is zero, then $B_{m2} = Y$ and the controller will act on the error signal that will occur if no time delay is presented, so

$$E_2 = E_1 - B_{m1}$$

(76)

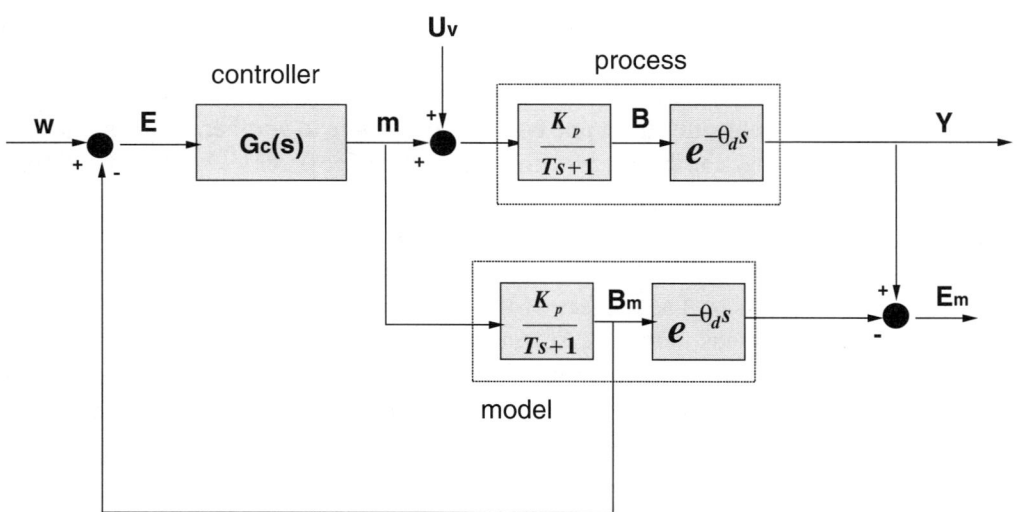

Figure 3–36 Using the manipulated variable as model input and B_m for feedback

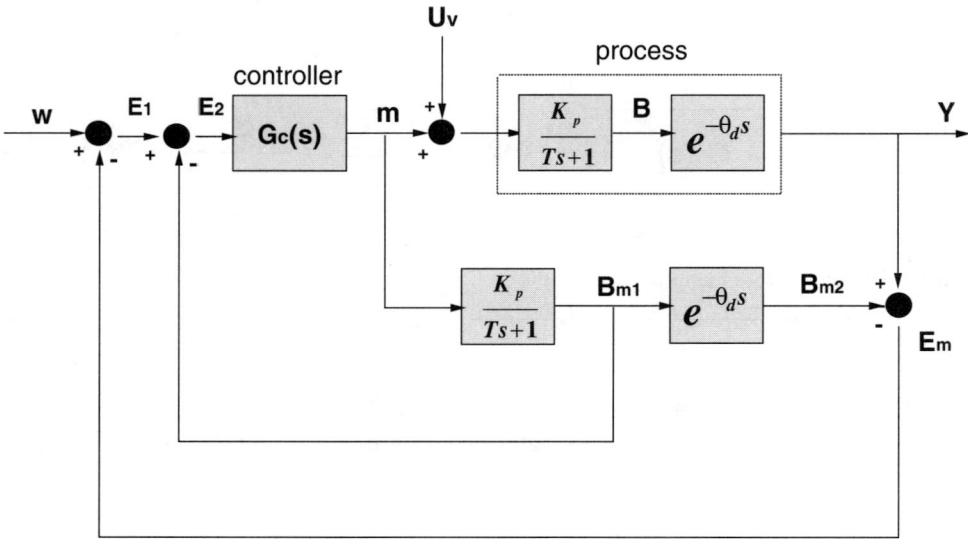

Figure 3–37 Smith predictor control strategy

Selective/Override Control

Each controlled variable in a process must be paired with a specific manipulated variable. Some process control systems have the same number of controlled and manipulated variables. If there are fewer manipulated variables than controlled variables, it is not possible to eliminate offset in all controlled variables for arbitrary load or setpoint changes. For control problems where the number of controlled variables is larger than the number of manipulated variables, a strategy of sharing variables among loops is necessary to improve the control performance. In other situations, there are more manipulated variables than controlled variables.

When the numbers of manipulated and controlled variables is not equal, it may be necessary to switch a controller from one controlled variable to another or one manipulated variable to another. This can be done with selective devices, *selectors*, that have two or more inputs and produce a single output. Depending on the need, this output may be the highest, the lowest, or the median of the inputs. The use of such selective devices is a way to achieve constraint control on flows and operating conditions.

A selector can also be used to select the appropriate process variable from among a number of variable measurements. Selectors can be based on multiple measurement points, multiple final control elements, or multiple controllers. So selectors are used not just to improve the control system performance but also to protect equipment from malfunction.

The *override* control concept is a technique by which process variables are kept within certain limits, usually for protective purposes. The antireset windup feedback controller is a type of override. To illustrate override control, consider the simple process of Figure

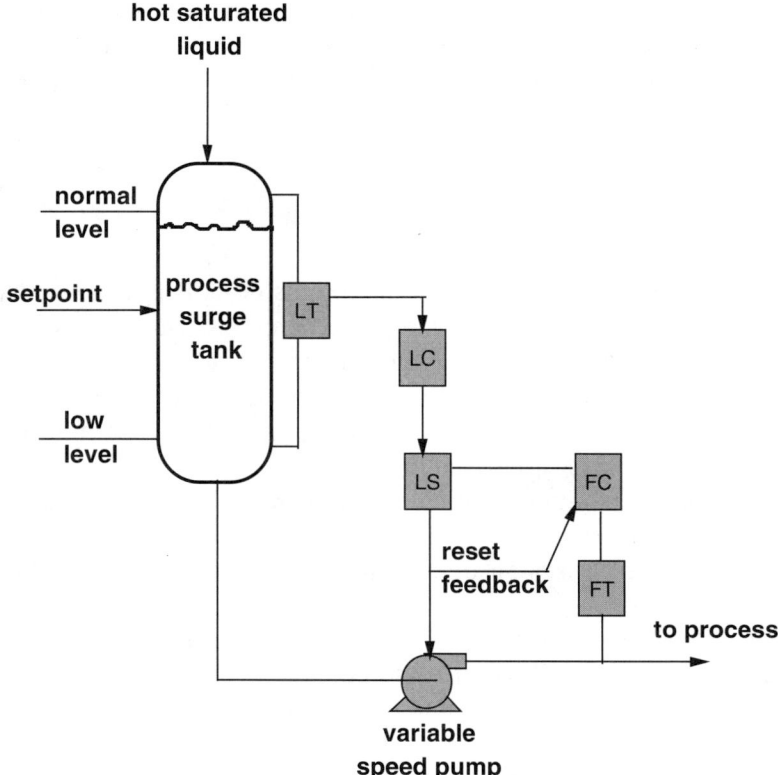

Figure 3–38 Override control system

3–38 (Smith and Corripio, 1985). A hot saturated liquid enters a process surge tank and then is pumped into the process. In general, the tank operates at the normal level; but if the level gets too low, the liquid will not have enough net positive suction heat, and the pump will start to cavitate. Applying an override controls the tank level. The variable-speed pump will pump more liquid as the energy input to it increases. The flow controller must be a reverse-acting controller, and the level controller must be a direct controller. The output of each controller is connected to a low-level selector relay, and its output goes to the pump.

Under normal operating conditions, the actual level is above the setpoint, and the level controller will try to speed up the pump. Normally, the output of the flow controller will be less, and the low-level selector relay will select the flow controller output to manipulate the pump speed.

If the flow of the hot saturated liquid decreases and the level drops, the level controller will attempt to slow down the pump by reducing its output. When the output of the level controller drops below the level of the flow controller, the low-level selector relay will choose the output of the controller to control the pump. Now the level controller "overrides" the flow controller. The override control is also called *constraint control*.

For override control, it is essential that any controller having integral (reset) action also have windup protection, as shown in Figure 3–38 for the flow controller. Without the reset action, the controller action is delayed, and the override protection will probably be too late.

Cascade Control

The design of controllers or control algorithms described up to now assumed that only the control variable y determines the process input u. This results in single control loops. By connecting additional measurable variables to the single loop (ie, disturbances or auxiliary variables), it is possible to obtain improved control behavior. These additions to the single loop lead to interconnected control systems. One of the most important basic interconnected schemes uses cascade control.

In cascade control, additional variables of the process, measurable on the signal path from the manipulated variable to the controlled variable, are fed back to the manipulated variable. Cascade control uses an inner control loop and therefore involves a second controller.

The general concept of cascade control is to nest one feedback loop inside another feedback loop. Figure 3–39 illustrates the cascade control concept. In this case, the process is divided into two parts. An intermediate variable within the process is used as the controlled variable for the inner loop. Cascade control is useful when disturbances are associated with the manipulated variable or when the final control element exhibits nonlinear behavior.

An example of application of cascade control is shown in Figure 3–40 (Mann, 1982). In the sugar industry, beet pulp is a by-product used for cattle feed. The pulp is generally dried using a rotary dryer to a final moisture content of 10% (w.b.) or 90% dry matter. Overdried pulps are brittle and underdrying causes dangerous internal heating to occur during storage and nutrient decomposition. The main objective is therefore to keep the dry matter within a tolerance range of $\pm 1\%$.

A schematic of the rotary dryer is shown in Figure 3–40. The oven is heated by propane. Gases from a steam boiler are mixed with the combustion gases to cool parts of the oven. An exhaust fan blows the gases through the dryer. The wet pulp, with 75% to 85% (w.b.) moisture content, is fed in by an inlet auger. The dryer consists of a long cylindrical shell that rotates slowly (1.5 rpm). The inside of the shell is fitted with a set of flights that repeatedly lift the pulps a certain distance along the periphery of the rotating shell before they fall back to the bottom of the shell. At the end of the drum, a screw conveyor transports the dried pulp to an elevator. The heat transfer is mainly by convection.

The control of the drying process is difficult because of its non–minimum-phase behavior with dead times of several minutes, long settling time (about 1 hour), large variations of the water content of the wet pulp, and unmeasurable changes of the drying properties of the pulp (Mann, 1982).

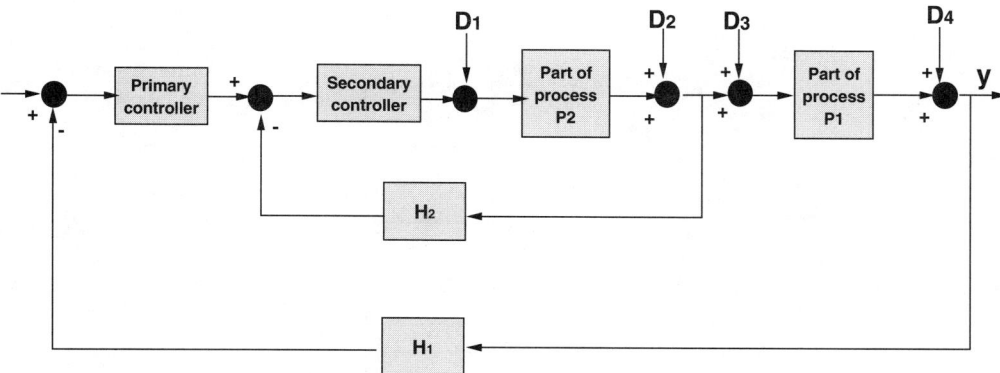

Figure 3–39 Cascade control concept

In this process (Figure 3–41), the main controlled variable is the dry matter of the dried pulp. The gas temperature at the oven outlet, in the middle of the drum, and in the dryer exhaust can be used as auxiliary controlled variables. The main manipulated variable is the fuel flow. The speed of the wet pulp auger can be used as an auxiliary manipulated variable. The water content of the pressed pulp is the main disturbance variable.

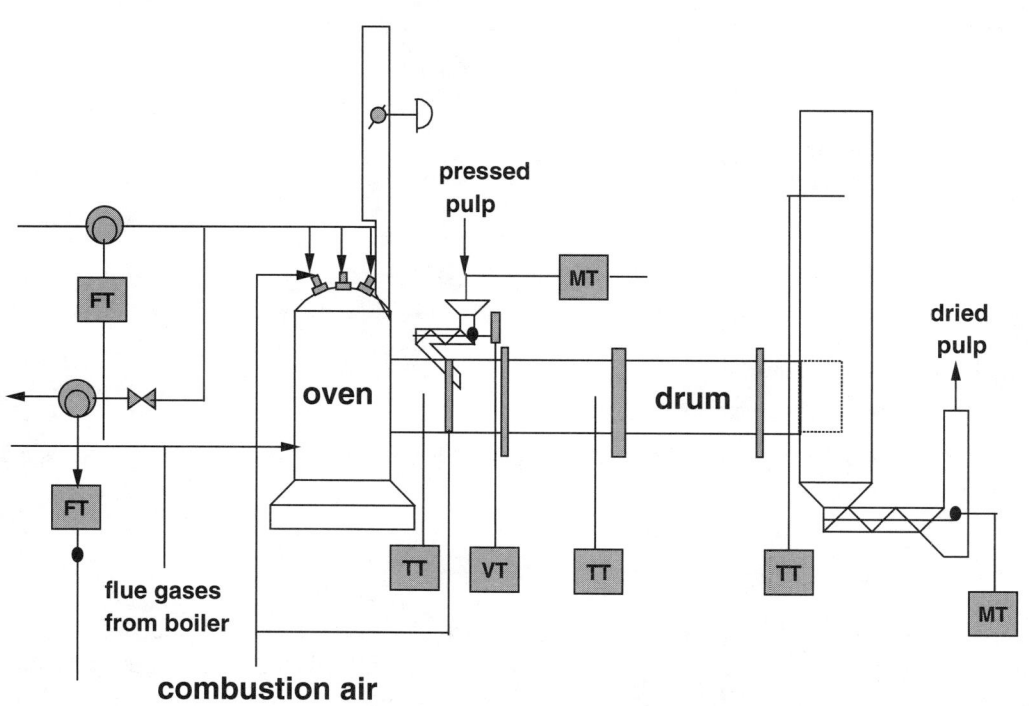

Figure 3–40 Schematic of the rotary dryer.

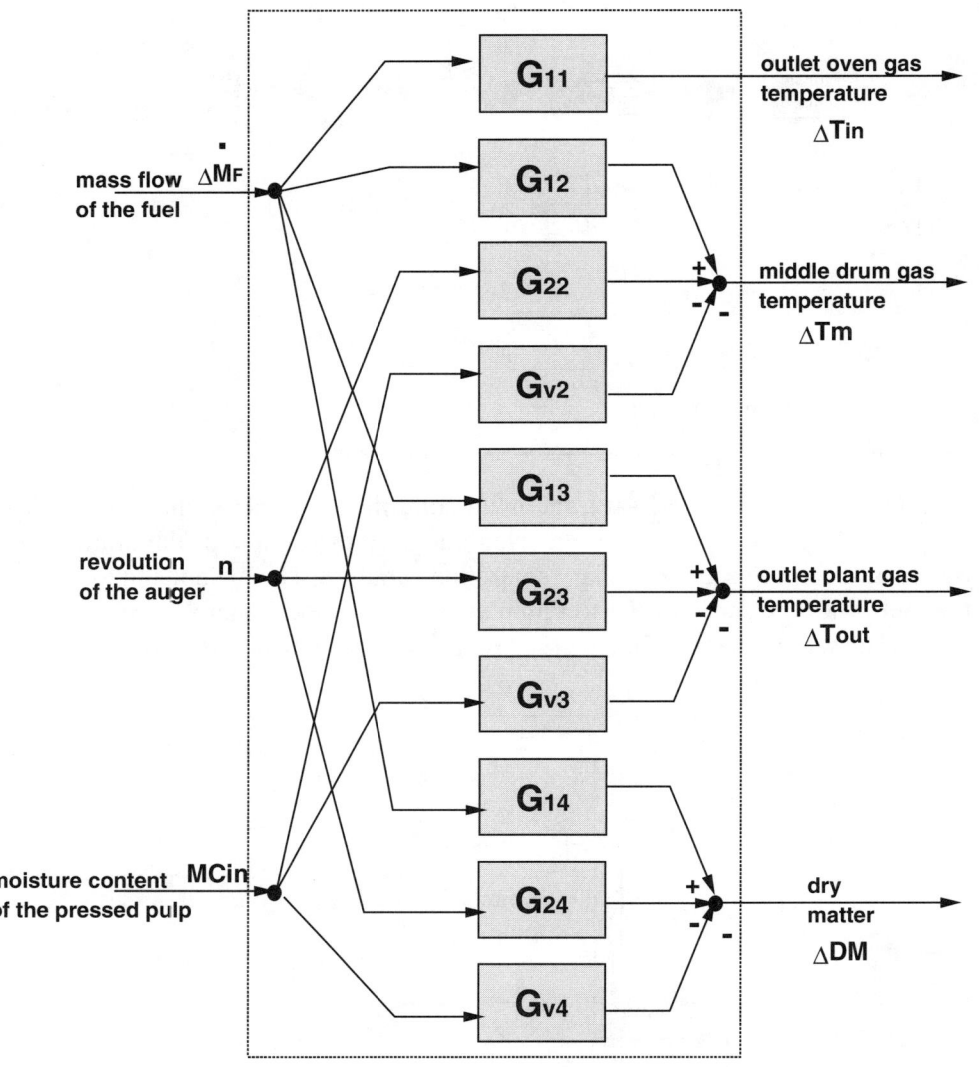

Figure 3–41 Block diagram of the rotary dryer.

A block diagram of the system with additional feedforward, G_{F1}, of the auger speed is shown in Figure 3–42.

CONCLUSION

In this chapter, a short introduction to classical control and a number of advanced control strategies were presented. Several concepts were introduced such as the application of Laplace transform techniques to solve differential equations. Emphasis was given to the important properties of the Laplace transform and its inverse transfer functions.

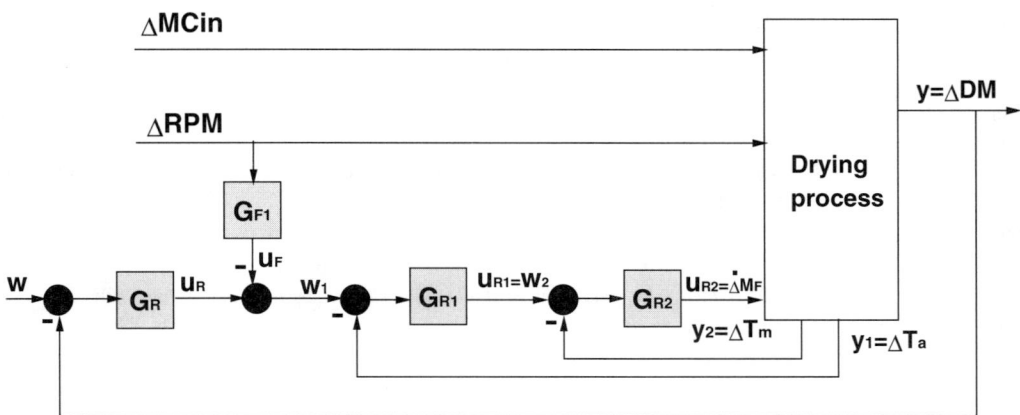

Figure 3–42 Block diagram of the cascade control system.

Other important concepts introduced were the transfer function and block diagrams. The transfer function contains key information about the steady-state and dynamic relations between input and output, such as the process gain and the process time constants, respectively. The transfer function is entered in the corresponding blocks, which are connected by arrows to indicate the direction of the flow signal.

The basic principles of feedback and feedforward control were also introduced in this chapter. The most commonly employed types of feedback controllers were discussed in detail. The food-processing industries have chosen variations of the PID controller and the on-off controller as standards. The Ziegler-Nichols method, known as the ultimate method or loop tuning for feedback control systems, was introduced as a method for field tuning.

In addition, it was shown that feedforward control is a powerful strategy for control problems in which important load variables can be measured on line. Feedforward control is normally implemented in conjunction with feedback control.

Several stability criteria for linear systems that can be described by transfer function models were presented in this chapter. Closed-loop system design for a linear SISO system was also described based on the steady-state error and the transient response.

A number of advanced control strategies, including cascade control, time-delay compensation, and selective/override control, were presented. These techniques are especially useful for difficult control problems that are characterized by long time delays, process constraints, and unmeasured process variables.

REFERENCES

Brogan, W.L. (1982). *Modern Control Theory*. Englewood Cliffs, NJ: Prentice Hall.

Dahlin, E.B. (1968). Designing and tuning digital controllers. *Instrum Control Syst* 41(6), 77-80.

Mann, W. (1982). Digital control of a rotary dryer in the sugar industry. Presented at the 6th International Federation of Automatic Control/International Federation of Information Processing (IFAC/IFIP). Conference on Digital Computer Applications; 1982.

Moore, C.F. (1970). Improved algorithms for direct digital control. *Instrum Control Syst* 43(1), 70–74.

Moreira, R.G. (1989). Adaptive control of continuous flow grain dryers. East Lansing, MI: Michigan State University. PhD dissertation.

Moreira, R.G., Srivastava, A.K. & Gerrish, J.B. (1990). Feedforward control model for a twin-screw food extruder. *Food Control* 1(3), 179–184.

Platt, D., Palazoglu, A. & Rumsey, T.R. (1992). Dynamics and control of cross-flow grain dryers II: a feedforward-feedback control strategy. *Drying Technol* 10(2), 333–363.

Seborg, D.E., Edgar, T.F. & Mellichamp, D.A. (1989). *Process Dynamics and Control*. New York, NY: John Wiley.

Shrinskey, F.G. (1988). *Process Control Systems*. 3rd ed. New York, NY: McGraw-Hill.

Smith, C.A, & Corripio, A.B. (1985). *Principles and Practices of Automatic Control*. New York, NY: John Wiley.

Smith, O.T.M. (1957). Close control of loops with dead time. *Chem Eng Prog* 53(5), 217–219.

APPENDIX 3–A

LAPLACE TRANSFORM

The Laplace transform of a function of time is given by

$$\mathcal{L}[f(t)] = F(s) = \int_o^\infty f(t)e^{-st}dt$$

(a.1)

The symbol $\mathcal{L}$ [ ] denotes the Laplace transformation of the function included in the brackets. The result of the transformation is the function $F(s)$, and a complex variable s is defined as

$$s = \sigma = j\omega$$

(a.2)

where σ is the real and ω the imaginary part of s, and $j = \sqrt{-1}$.

PROPERTIES OF THE LAPLACE TRANSFORM

The following theorems demonstrate the importance of Laplace transform approaches to solve linear time-invariant dynamic systems.

Laplace Transform of Derivatives

$$\mathcal{L}\left[\frac{d^n(f(t))}{dt^n}\right] = s^n F(s) - \sum_{i=1}^{n} f^{(i-1)}(o^-)s^{n-1}$$

(a.3)

Equation a.3 indicates that the derivative of a function is s times the transform of the function minus the initial value of the function at $t = 0$. For example, find the Laplace transform of the system output described in Equation a.4:

$$\frac{d^2y}{dt^2} + 5\frac{dy}{dt} + 3y(t) = u(t)$$

(a.4)

with $y(o) = 0.5$; $y'(o) = 1.5$ and $u(t) = r(t)$ (unit ramp function). Applying to Equation a.3 the Laplace transform of Equation a.4 results in

$$[s^2Y(s) - sy(o) - y'(o)] = 5[sY(s) - y(o)] + Y(s) = \frac{1}{s^2}$$

(a.5)

Inserting the values of the initial and input conditions, we obtain

$$(s^2 + 5s + 1)Y(s) = \frac{1}{s^2} + 0.5s = 4$$

(a.6)

Laplace Transform of Integrals

$$L\left[\int_0^t f(x)dx\right] = \frac{F(s)}{s}$$

(a.7)

where x is the dummy variable of integration and $F(s)$ the Laplace transform of $f(t)$. Equation a.7 states that integration in time corresponds to the division by s. For example, find the Laplace transform of t^2:

$$t^2 = 2\int_0^t r(x)dx$$

(a.8)

The Laplace transform of Equation a.8 is

$$L[t^2] = \frac{2}{s^3}$$

(a.9)

Laplace Transform of Time Shift

This theorem states that the Laplace transform of a time-shifted function is e^{-as} times the transform of the unshifted function:

$$\mathcal{L}[f(t-a)u_1(t-a)] = e^{-as}F(s) \text{ for } a > 0$$

(a.10)

An important application of the time-shift theorem is in finding Laplace transforms of system inputs used in system analysis. For example, find the Laplace transform of the system input shown in Equation a.11:

$$u(t) = 200u_1(t) + 5r(t-30) - 5r(t-70)$$

(a.11)

Taking the transforms term by term results in

$$U(s) = \frac{200}{s} + 5\frac{e^{-30s}}{s^2} - 5\frac{e^{-70s}}{s^2}$$

(a.12)

The Final Value Theorem

The FVT provides a way of determining the long-run behavior of the output of certain linear systems from the transform of the output. It states that if the limit $\lim f(t)$ exists, then

$$\lim_{t\to\infty} f(t) = \lim_{s\to 0} sF(s)$$

(a.13)

This theorem simplifies the task of determining the long-run behavior of a system's output.

Initial Value Theorem

The IVT gives the initial value of a time function directly from the transform of that function:

$$\lim_{t\to 0} f(t) = \lim_{s\to\infty} sF(s)$$

(a.14)

THE INVERSE LAPLACE TRANSFORM

A general way of obtaining the inverse Laplace transform is by means of the *inversion integral*:

$$\mathcal{L}^{-1}[F(s)] = f(t) = \frac{1}{2\pi j} \int_{c-j\infty}^{c+j\infty} F(s)e^{st}ds$$

(a.15)

This integration is performed in the complex plane, since F is a function of the complex variable s. Background in complex variable theory is required by this approach. The approach that will be presented is both adequate for the purpose of this introduction and easier to employ.

Generally, Laplace transforms for linear time-invariant continuous systems involve functions that have numerators and denominators that are polynomials in s. Recall the Laplace transform Equation a.6. To solve for $Y(s)$, Equation a.6 becomes

$$Y(s) = \frac{0.5s^3 + 4s^2 + 1}{s^2(s^2 + 5s + 1)}$$

(a.16)

Equation a.16 can be written in factored form as

$$Y(s) = \frac{(s+8.031)(s-0.0155+j0.4988)(s-0.0155-j0.4988)}{(s+0)(s+0)(s+4.7913)(s+0.2087)}$$

(a.17)

Poles and Zeros

The poles of a Laplace transform $F(s)$ are the values of s for which $F(s) = \infty$. The zeros of Laplace transform $F(s)$ are the values of s for which $F(s) = 0$. In the $Y(s)$ of Equation a.17, the poles of $Y(s)$ are $s = 0$, $s = 0$, $s = -4.7913$, and $s = -0.2087$. The zeros of $Y(s)$ are $s = 8.031$, and $s = -0.0155 \pm j04988$.

Three types of system poles need to be considered when finding inverse transform: (1) simple poles arising from terms of the form $(s + p_i)$; (2) multiple poles arising for terms of the form $(s + p_i)^{k_i}$ where $k_i > 1$ is a positive integer; and (3) complex poles arising from terms of the form $(s + \sigma_i \pm j\omega_i)$. For linear time-invariant systems, these complex poles always occur in complex conjugated pairs.

Inverse Laplace Examples

Applications of inverse Laplace transform for single, multiple, and complex poles are shown below.

Single Poles Case

Find $f(t)$ for

$$F(s) = \frac{2(s+2)}{s(s+1)}$$

(a.18)

Expanding $F(s)$ in partial fractions gives

$$F(s) = \frac{a_1}{s} + \frac{a_2}{s+1}$$

(a.19)

and

$$f(t) = a_1 + a_2 e^{-t} \quad \text{for } t \geq 0$$

(a.20)

$$a_1 = sF(s)\,|_{s=0} = \frac{s2(s+2)}{s(s+1)}\,|_{s=0} = 4$$

$$a_2 = (s+1)F(s)\,|_{s=-1} = \frac{(s+1)2(s+2)}{s(s+1)}\,|_{s=-1} = -4$$

(a.21)

Therefore,

$$f(t) = 4 - 4e^{-t} \quad t \geq 0$$

(a.22)

Multiple Poles Case

Find $f(t)$ for

$$F(s) = \frac{4(s+1)}{(s+2)^2 s(s+4)}$$

(a.23)

Expanding $F(s)$ in partial fractions gives

$$F(s) = \frac{a_1}{s+2} \pm \frac{a_2}{(s+2)^2} + \frac{a_3}{s} + \frac{a_4}{(s+4)}$$

(a.24)

and

$$f(t) = a_1 e^{-2t} + a_2 t e^{-2t} + a_3 + a_4 e^{-4t} \quad \text{for } t \geq 0$$

(a.25)

$$a_4 = (s+4)F(s)|_{s=-4} = \frac{4(s+1)(s+4)}{(s+2)^2 s(s+4)}\Big|_{s=-4} = 0.75$$

$$a_3 = sF(s)|_{s=0} = \frac{4(s+1)s}{(s+2)^2 s(s+4)}\Big|_{s=0} = 0.25$$

$$a_2 = (s+2)^2 F(s)|_{s=-2} = \frac{4(s+1)(s+2)^2}{(s+2)^2 s(s+4)}\Big|_{s=-2} = 1$$

(a.26)

To find a_1 we need to solve the following equation, since the method breaks down when seeking a_1, that is, when $a_1 = \infty$:

$$\frac{4(s+1)}{(s+2)^2 s(s+4)} = \frac{a_1}{s+2} + \frac{1}{(s+2)^2} + \frac{0.25}{s} + \frac{0.75}{(s+4)}$$

(a.27)

So $a_1 = -1$. Therefore,

$$f(t) = -1e^{-2t} + 1e^{-2t} + 0.25 + 0.75e^{-4t} \quad t \geq 0$$

(a.28)

Complex Poles Case

Find $f(t)$ for

$$F(s) = \frac{(s+4)}{s(s^2 + 2s + 5)} = \frac{(s+4)}{s[(s+1)^2 + 4]}$$

(a.29)

The poles in the denominator $(s^2 + 2s + 5)$ are: $s = -1 \pm j2$ (for inverse transforms for complex poles, see Table 3–A–1). Expanding $F(s)$ in partial fractions gives

$$\frac{(s+4)}{s[(s+1)^2 + 4]} = \frac{a_1}{s} + \frac{a_2}{(s+1)^2 + 4} + \frac{a_3(s+1)}{(s+1)^2 + 4}$$

(a.30)

The $f(t)$ corresponding to Equation a.30 is therefore:

$$f(t) = a_1 + a_2 e^{-t} \sin 2t + a_3 e^{-t} \cos 2t \quad t \geq 0$$

(a.31)

Table 3–A–1 Laplace Transform Pairs

$f(t)$	$F(s) = \mathcal{L}[f(t)]$
1 unit impulse $\delta(t)$	1
2 unit step $u_1(t)$	$\dfrac{1}{s}$
3 ramp $r(t)$	$\dfrac{1}{s^2}$
4 $\sin \omega t$	$\dfrac{\omega}{s^2 + \omega^2}$
5 $\cos \omega t$	$\dfrac{s}{s^2 + \omega^2}$
6 $t^n\ (n = 1, 2, 3, \ldots)$	$\dfrac{n!}{s^{n+1}}$
7 $t^n e^{-at}\ (n = 1, 2, 3, \ldots)$	$\dfrac{n!}{(s+a)^{n+1}}$
8 $e^{-at} u_1(t)$	$\dfrac{1}{(s+a)}$
9 $e^{-at} r_1(t)$	$\dfrac{1}{(s+a)^2}$
10 $e^{-at}\sin \omega t$	$\dfrac{\omega}{(s+a)^2 + \omega^2}$
11 $e^{-at}\cos \omega t$	$\dfrac{s+a}{(s+a)^2 + \omega^2}$
12 $\dfrac{1}{b-q}(e^{-at} - e^{-bt})$	$\dfrac{1}{(s+a)(s+b)}$
13 $\dfrac{1}{b-a}(be^{-bt} - ae^{-at})$	$\dfrac{s}{(s+a)(s+b)}$
14 $\dfrac{1}{ab}\left[1 + \dfrac{1}{a-b}(be^{-at} - ae^{-bt})\right]$	$\dfrac{s}{s(s+a)(s+b)}$
15 $\dfrac{1}{a^2}(at - 1 + e^{-bt})$	$\dfrac{1}{s^2(s+a)}$
16 $\dfrac{\omega_n}{\sqrt{1-\zeta^2}}\, e^{-\zeta\omega_n t} \sin \omega_n \sqrt{1-\zeta^2}\, t$	$\dfrac{\omega_n^2}{s^2 + 2\zeta\omega_n s + \omega^2}$
17 $\dfrac{-1}{\sqrt{1-\zeta^2}}\, e^{\zeta\omega_n t} \sin\left(\omega_n \sqrt{1-\zeta^2}\, t\right.$	$\dfrac{s}{s^2 = 2\zeta\omega_n s + \omega^2}$

$$\phi = \tan^{-1} \frac{\sqrt{1-\zeta^2}}{\zeta}$$

$$a_1 = sF(s)\,|_{s=0} = \frac{s(s+4)}{s(s+1)^2 + 4}\,|_{s=0} = \frac{4}{5}$$

(a.32)

For a_2 and a_3 we clear the fraction in Equation a.30 and equate coefficients of like powers of s:

$$s+4 = \frac{4}{5}[(s+1)^2 + 4] + 2a_2 s + a_3 s(s+1)$$

(a.33)

Equating the coefficient of s^2 we get:

$$0 = \frac{4}{5} + 0 + a_3 \Rightarrow a_3 = -\frac{4}{5}$$

(a.34)

Equating the coefficient of s^1 we get:

$$1 = 2 \times \frac{4}{5} + 2a_2 - \frac{4}{5} \Rightarrow a_z = \frac{1}{10}$$

(a.35)

So

$$f(t) = \frac{4}{5} + \frac{1}{10}e^{-t}\sin 2t - \frac{4}{5}e^{-t}\cos 2t \quad \text{for } t \geq 0$$

(a.36)

APPENDIX 3–B

ROUTH'S STABILITY CRITERION

Before discussing the Routh's locus technique, it is important to define explicitly the poles, zeros, and characteristics equation of the system transfer function. Consider the following nth order differential equation:

$$G(s) = \frac{Y(s)}{U(s)} = \frac{b_n s^n + \ldots + b_o}{s^m + a_{m-1} s^{m-1} + \ldots + a_o}$$

(b.1)

Writing Equation b.1 in factored form results:

$$G(s) = \frac{b_n (s + \beta_1)(s + \beta_2)\ldots(s + \beta_n)}{(s + a_1)(s + a_2)\ldots(s + a_m)}$$

(b.2)

Therefore, from Equations b.1 and b.2 we can define the following:

- Poles—are the values of s for which $G(s)$ is infinite. That is, the poles of $G(s)$ are the roots of

$$s^m + a_{m-1} s^{m-1} + \ldots + a_o = 0$$

(b.3)

which are $s = -a_1$, $s = -a_2$, ..., $s = -a_m$.

- Characteristic equation—is the denominator of $G(s)$ set equal to zero. Equation b.3 is the characteristic equation of $G(s)$.
- Zeros—are the values of s for which $G(s)$ is zero: that is, they are the roots of

$$s^n + b_{n-1} s^{n-1} + \ldots + b_o = 0$$

(b.4)

and are equal to $s = -b_1$, $s = -b_2$, ..., $s = -b_n$.

A CRITERION FOR SYSTEM STABILITY

A system is said to be BIBO (bounded input, bounded output) stable if its output is bounded (finite) for all bounded inputs. For linear systems it is possible to establish a criterion for determining if the system is BIBO stable. A linear system described by Equation b.5 is BIBO stable if all roots of the characteristic equation (poles) lie in the left half of the s-plane.

$$\frac{d^m y}{dt^m} + a_{m-1}\frac{d^{m-1} y}{dt^{m-1}} + \ldots + a_o y(t) = b_o u(t) + b_1 \frac{du}{dt} + \ldots + b_m \frac{d^n u}{dt^n}$$

(b.5)

where a's and b's are constant, $n < m$.

The examples in Figure 3–B–1 show that the systems (a) and (c) are BIBO stable and that (b) is unstable. System poles are shown with an "x" in the s-plane and zeros with an "o". Multiple poles and zeros are shown with an associated number giving the multiplicity.

Multiple Poles on the jω Axis

A system with poles in the left half plane and multiple poles on the $j\omega$ axis is clearly unstable, since the multiple $j\omega$ axis poles give rise to time responses of the form t^{p-1}, $t^{p-1}sin(bt)$ or $t^{p-1}cos(bt)$ for $p > 1$.

Simple Poles on the $j\omega$ Axis

In this case the system is stable if the system poles on the $j\omega$ axis are distinct from any input poles that may exist on the $j\omega$ axis: that is, if the system and input do not combine to produce multiple poles in the $j\omega$ axis. A system is marginally BIBO stable if all the roots of the characteristic equation lie in the left half s-plane with the exception of one or more simple poles on the $j\omega$ axis. Figure 3–B–2 shows that the system and input combine to produce a double pole at $s = 0$ and that the system is unstable because $y(t)$ will contain a term proportional to time, t.

To determine whether a given linear system is stable, unstable, or marginally stable is to know where the system poles are located in the s-plane. You can determine the poles by solving the characteristic equation explicitly or using the Routh's stability criterion.

ROUTH'S STABILITY CRITERION

Given the system characteristic equation:

$$a_m s^m + a_{m-1} s^{m-1} + \ldots + a_1 s + a_o = 0$$

(b.6)

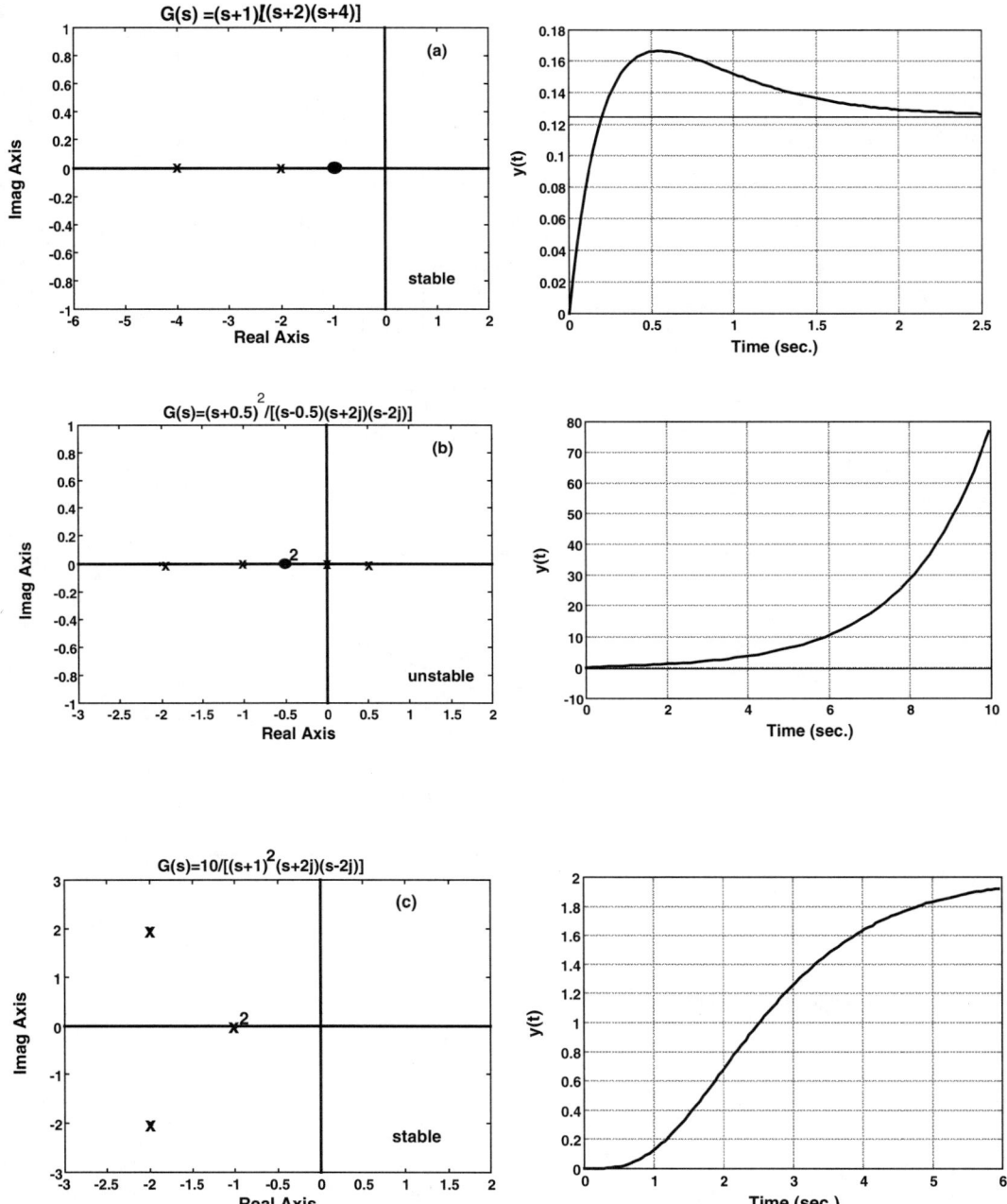

Figure 3–B–1 Examples of stable and unstable systems

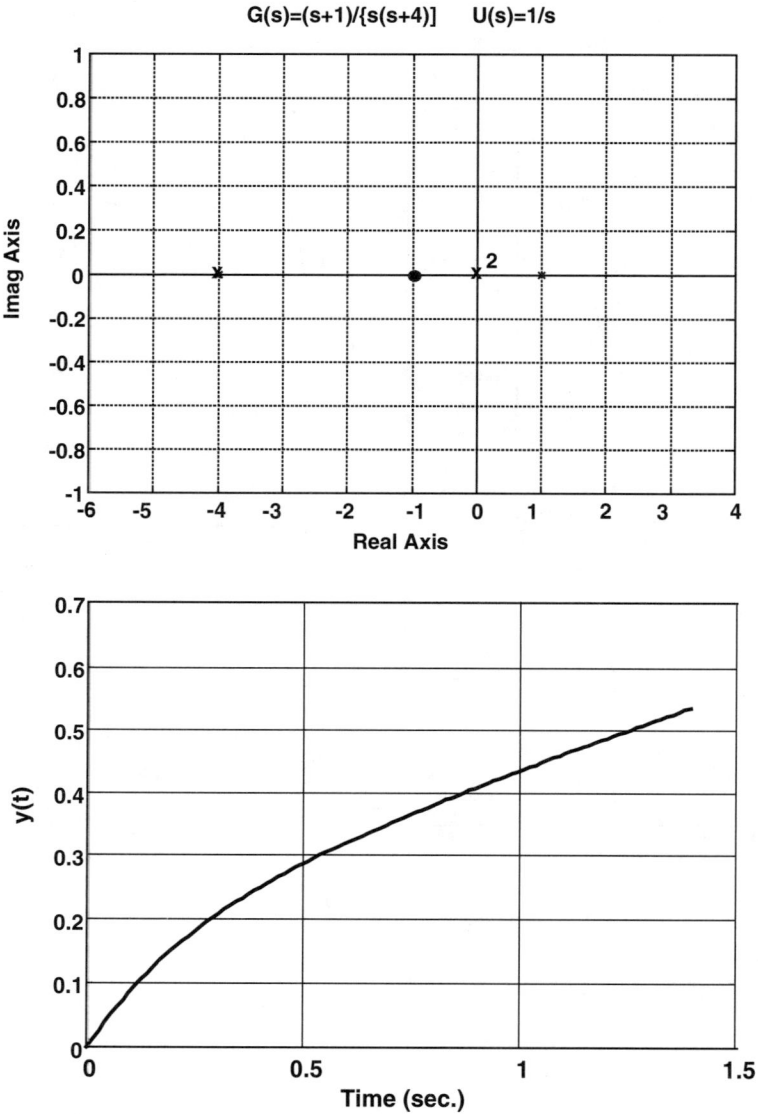

Figure 3–B–2 An example of a marginally stable system

If the a_o term is zero divided by s to arrive at the form of Equation b.6 and if all the coefficients are not of the same sign, the equation has at least one root with a positive real part and the system is known to be unstable without proceeding further with Routh's method.

The Routh array is described as

$$
\begin{array}{cccccc}
s^m & a_m & a_{m-2} & a_{m-4} & \ldots & 0 \\
s^{m-1} & a_{m-1} & a_{m-3} & a_{m-4} & \ldots & 0 \\
s^{m-2} & c_i & c_2 & c_3 & \ldots & 0 \\
s^{m-3} & d_i & d_2 & & \ldots & 0 \\
s^1 & r_i & 0 & & & \\
s^0 & s_i & 0 & & &
\end{array}
$$

$$\textbf{(b.7)}$$

The third row is computed as

$$
c_1 = \frac{a_{m-1} a_{m-2} - a_m a_{m-3}}{a_{m-1}}
$$

$$
c_2 = \frac{a_{m-1} a_{m-4} - a_m a_{m-5}}{a_{m-1}}
$$

$$\textbf{(b.8)}$$

This pattern continues until the last entry in the third row is a zero. The fourth row is computed in similar fashion:

$$
d_1 = \frac{c_1 a_{m-3} - a_{m-1} c_2}{c_1}
$$

$$
d_2 = \frac{c_1 a_{m-5} - a_m c_3}{c_1}
$$

$$\textbf{(b.9)}$$

This process continues until one entry each is computed for the last rows (s^1 and s^0) of the array.

Therefore, Routh's criterion is defined as follows: the number of roots of the characteristic equation (Equation b.6) in the right half of the s-plane is equal to the number of changes in sign in the first column of the Routh array. So if the Routh array for a given linear system has signal changes in the first column, the system is unstable. If no sign changes appear in the first column of the array, the system is at least marginally stable in the BIBO sense.

For an example, determine the values for the parameter K for which the following system characteristic equation is at least marginally stable:

$$
s^4 + 6s^3 + 13s^2 + 4s + K = 0
$$

$$\textbf{(b.10)}$$

$$
\begin{array}{ccccc}
s^4 & 1 & 13 & K \\[1em]
s^3 & 6 & 4 & 0 \\[1em]
s^2 & \dfrac{74}{6} & K & \\[1em]
s & \dfrac{296/6 - 6K}{74/6} & d_2 & \\[1em]
s^0 & K & 0 &
\end{array}
$$

(b.11)

To avoid instability, all entries in the first column of this array must be non-negative. Therefore,

$$
K \geq 0
$$

$$
\frac{296/6 - 6K}{74/6} \geq 0 \Rightarrow K \leq \frac{296}{36} \approx 8.22
$$

(b.12)

This system is at least marginally stable for $0 \leq K \leq 8.22$.

APPENDIX 3–C

ROOT LOCUS

Root locus is the path in the *s*-plane of the closed-loop system poles as some design parameter varies. It is a powerful tool for control system design. For example, consider the control system with the following transfer functions:

$$G_p(s) = \frac{10}{s(s+10)}$$

(c.1)

$$G_c(s) = K_p$$

(c.2)

$$H(s) = 1$$

(c.3)

Then we have

$$\frac{Y(s)}{W(s)} = \frac{\dfrac{10K_p}{s(s+10)}}{1 + \dfrac{10K_p}{s(s+10)}} = \frac{10K_p}{s^2 + 10s + 10K_p}$$

(c.4)

The close-loop system poles are the values of *s* for which

$$1 + \frac{10K_p}{s(s+10)} = 0$$

(c.5)

or

$$s^2 + 10s + 10K_p = 0$$

(c.6)

Equation c.6 is the characteristic equation of the system, and the system pole locations will vary as the parameter K_p changes. Solving Equation c.6, the poles (or roots) are

$$s = \frac{-10 \pm \sqrt{100 - 40K_p}}{2} = -5 \pm \sqrt{100 - 40k_p}$$

(c.7)

From Equation c.7, the system pole locations can be plotted as a function of K_p. This is the root locus for the system and is shown in Figure 3–C–1. The root locus is shown in heavy lines. Note that for $K_p = 0$ the poles are at $s = 0$ and $s = -10$, the open-loop systems values. For $K_p = 2.5$, the poles are together at $s = -5$. For $K_p > 2.5$, the poles become conjugate complex and move up and down that line perpendicular to the real axis at $s = -5$.

Another example is shown in Figure 3–C–2 where the root locus breaks away from the real axis at some point in an interval between a pair of poles and moves into the complex portion of the s-plane as K_p increases. The root locus exists along the real axis for $-4 < s < 0$ and for $s < -10$. As K_p increases, the pole at $s = -10$ moves along the negative

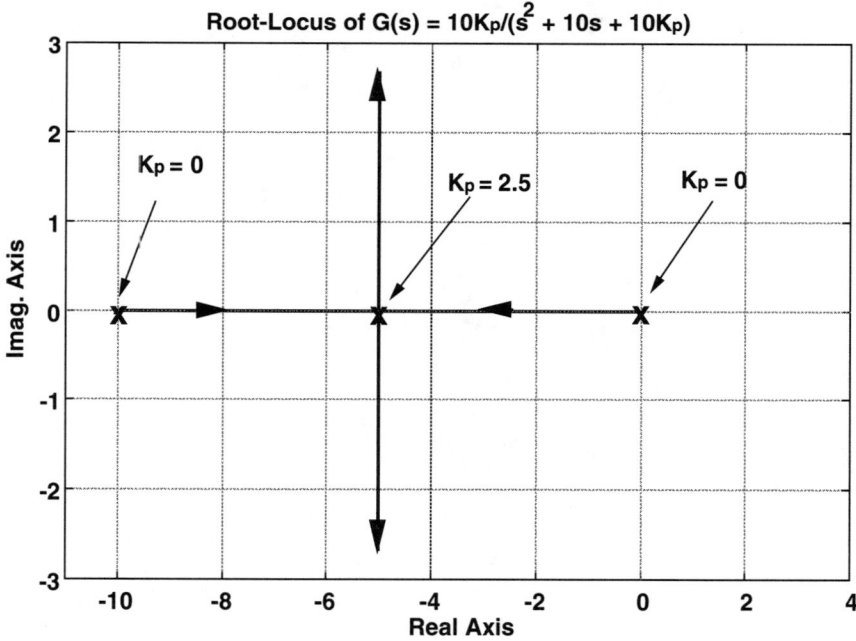

Figure 3–C–1 Root locus example

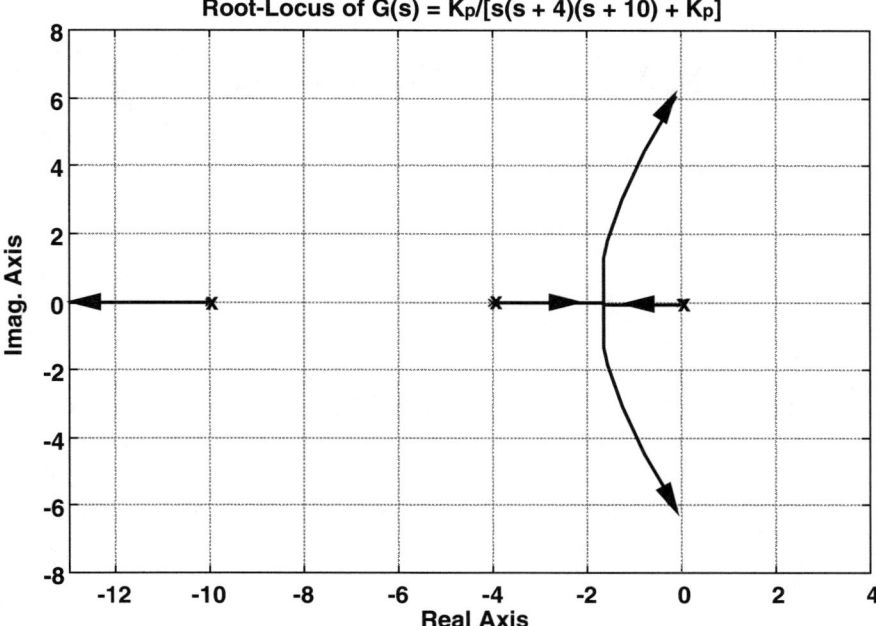

Figure 3–C–2 Root locus example

---

real axis toward $s = -\infty$. The poles at $s = 0$ and $s = -4$ come together, split, and enter the complex portion of the s-plane as K_p increases.

CHAPTER 4

Process Dynamics Modeling

Knowledge of process models is necessary for proper design of complex and well-adjusted control algorithms. Mathematical process models can be obtained by theoretical or experimental process analysis. Theoretical models, which are based on the complete knowledge of the process, are called "white box" models. Experimental models are referred to as "black box" models. Between these two models are the "gray box" (or hybrid) models, which are based on partial knowledge of the process and experimental data.

In this chapter, the concept of transient (dynamic) modeling for process control will be discussed. For detailed discussion on the subject, the reader is referred to Ljung and Glad (1994).

TYPES OF PROCESSES

Food-processing systems are characterized by the transport and transformation of materials and energy. Isermann (1989) classified processing systems by

- Types of signal amplitude-time behaviors
- Types of mass and energy transport
- Classes of mathematical models

Types of Signal Amplitude–Time Behaviors

Process signals can be classified as continuous, discrete, or binary. Figure 4–1 shows the amplitude-time behavior of different signals. In data processing with computers, continuous signals, $y(t)$, are sampled, $y_p(t)$, and digitized (by an analog to digital converter), resulting in discrete (discontinuous) signals, $y_d(t)$, which are measured in *amplitude* and *time*. Figure 4–1 illustrates a combination of a continuous and a discrete-time signal for an amplitude/time combination-signal.

Types of Mass and Energy Transport

Food-processing systems can be classified according to the transport of mass and energy as

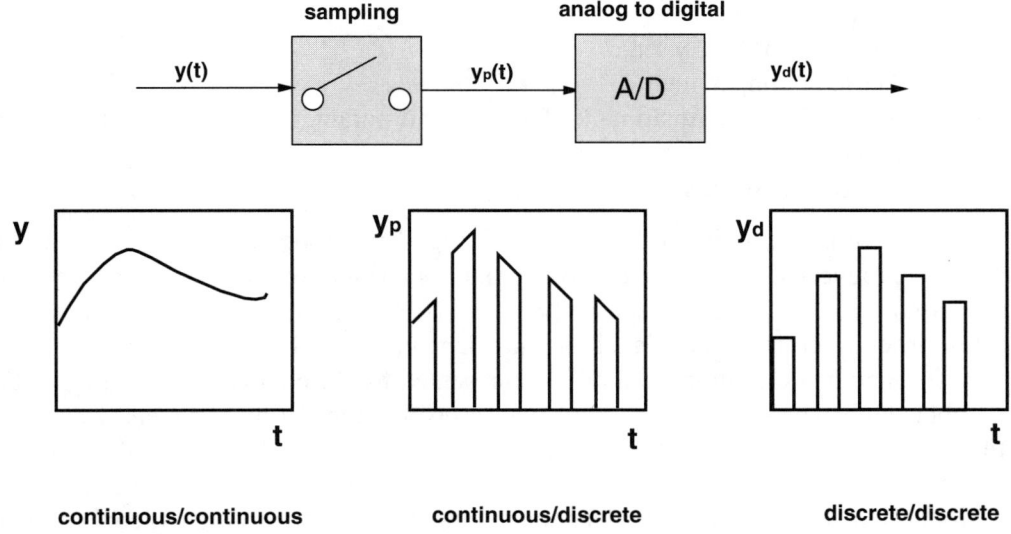

Figure 4–1 Amplitude-time signal generated by sampling and A/D conversion

- Continuous processes
- Batch processes
- Piecewise processes

Continuous processes are processes with continuous feed and product stream flows. Therefore, these processes are once-through operations. They are characterized by (1) signals that can vary in different ways (many combinations, as shown in Figure 4–1) and (2) mathematical models that can be linear or nonlinear ordinary partial differential equations. Examples of continuous processes include the flow of liquid food in pipelines, food extrusion, continuous deep-fat frying and continuous drying.

Batch processes are those processes with discontinuous feed and product stream flows. These processes operate in a closed space. They are characterized by various combinations of signal behavior, and most of the mathematical models used to describe them are nonlinear, ordinary difference or partial differential equations. Examples of batch systems are separation units such as distillation and milling processes.

Piecewise processes are those processes where materials and energy are transported in "pieces" or "discrete samples." These are piecewise operation systems. They are characterized by signals that have mostly discrete (binary) amplitude and discrete or continuous time and are described mathematically by flow schemes and digital simulation programs. Transport of finished/packed products is an example of a piecewise process.

Classes of Mathematical Models

Mathematical models that describe the transient (dynamic) behavior of continuous and batch food-processing systems can be classified as

- Linear or nonlinear
- Time invariant or time variant
- Parametric or nonparametric
- Ordinary difference equations (ODEs lumped parameters) or partial differential equations (PDEs distributed parameters)
- Continuous signals or discrete signals

The system is said to be *linear* if the principle of superposition applies. This means that if $u_1(t)$ produces a response $y_1(t)$, $u_2(t)$ produces a response $y_2(t)$, and so on, the sum $[u_1(t) + u_2(t) + \ldots]$ produces a response $[y_1(t) + y_2(t) + \ldots]$. It is said to be *time invariant* if its response to a certain input signal does not depend on absolute time.

Parametric models are those models represented by differential or difference equations (obtained by step-response or frequency-response fitting or by parameter estimation methods). *Nonparametric* models are, for example, transient functions of frequency responses in tabular form (obtained by using Fourier analysis).

Models may be either static or dynamic in time. Static or steady-state models provide information about the model variables only at a single point in time. A dynamic model is capable of generating the time paths of model variables.

Dynamic models of food processes consist of one or more differential equations, ODE and/or PDE, often combined with one or more algebraic relationships. ODE models are of the *lumped type,* meaning that any dependent variable can be assumed to be a function only of time and not of spatial position. Consider, for example, the stirred steam-jacketed kettle system shown in Figure 4–2. The objective of this system is to heat the liquid food material from an inlet temperature T_{in} to a final temperature T, using steam. The energy supplied by the steam is the input variable. Since the product is well mixed, no temperature gradient exists inside the kettle. Assuming constant product volume and density and instantaneous heat transfer from the steam to the liquid product, the dynamic equation of the system is

$$V\rho C_p \frac{dT}{dt} = \dot{m}C_p(T_{in} - T) + hA(T_s - T)$$

(1)

where $\dot{m}$ is the mass flow rate, T_s is the surface temperature of the kettle, A is the kettle surface area, and V and ρ are the product volume and density, respectively.

Although lumped-parameter models are generally used to describe processes, many food-processing systems are inherently *distributed-parameter* systems: that is, the output variables are functions of both time and position. So these process models contain one or more partial differential equations. Examples include heat exchangers, continuous flow dryers (spray dryers, conveyor dryers, rotary dryers, high-capacity grain dryers), continuous deep-fat frying systems, and continuous cooling and freezing systems.

Consider, for example, the unsteady-state simplified model of a bed of potatoes (Figure 4–3). A simplified two-equation model for cooling of a stationary bed of biological

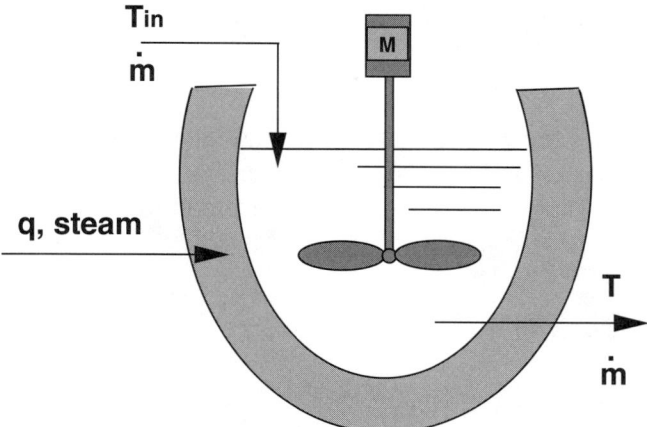

Figure 4–2 Heating of a food product in a stirred steam-jacketed kettle

product illustrates a distributed parameter model. The most important assumptions to develop the equations are (Lerew & Bakker-Arkema, 1978):

- no mass transfer
- no temperature gradients within the individual particles
- no particle-to-particle conduction
- plug-type flow

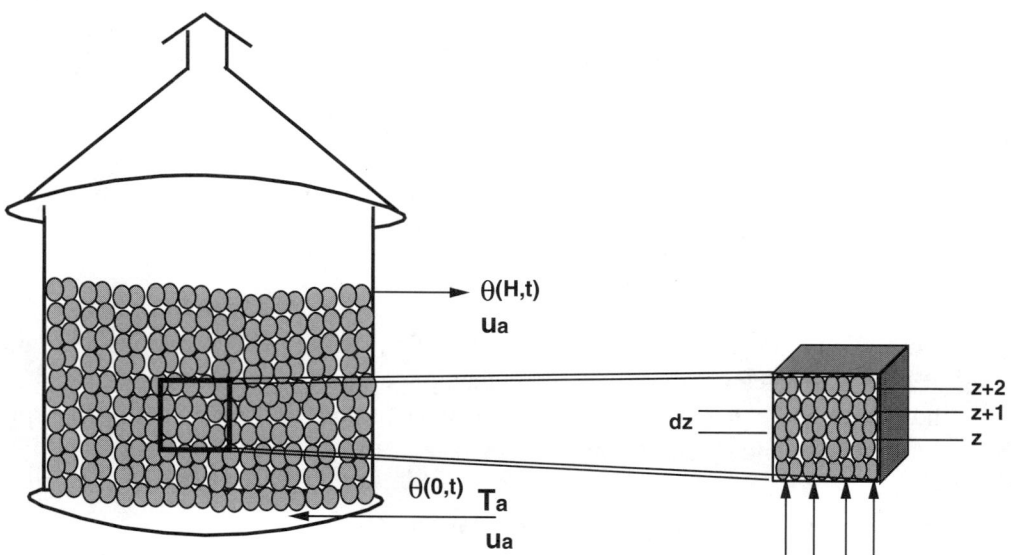

Figure 4–3 Schematic of cooling of a batch of potatoes

- constant thermal properties of the air and the product
- adiabatic bed walls of negligible heat capacity
- one-dimension transfer phenomena

The energy balances are written on a differential volume Adz located at an arbitrary location in the stationary bed. The two unknowns are θ, the product temperature, and T_a, the air temperature. Thus, two energy balances result in the following two differential equations:

- *For the air*

  energy out = energy in + energy transferred by convection + change in energy in the voids

$$u_a \rho_a Cp_a A T dt + u_a \rho_a Cp_a A \frac{\partial T}{\partial x} dx dt = u_a \rho_a Cp_a A T dt + haAdx(\theta - T)dt$$

$$+ haAdx(\theta - T)dt + \varepsilon Adxa(\rho_a Cp_a)\frac{\partial T}{\partial t} dt$$

(2)

Where a in this example is product surface area, (m^2/m^3), and ε is porosity, then

$$\frac{\partial T}{\partial t} = \frac{ha}{\varepsilon \rho_a Cp_a}(\theta - T) - u_a \frac{\partial T}{\partial z}$$

(3)

- *For the product*

  energy transferred by convection = change in internal product energy

$$haAdx(\theta - T)dt = (1 - \varepsilon)Adx(\rho_p Cp_p)\frac{\partial \theta}{\partial t} dt$$

(4)

or

$$\frac{\partial \theta}{\partial t} = \frac{ha}{(1 - \varepsilon)\rho_p Cp_p}(\theta - T)$$

(5)

Thus, Equations 3 and 5 represent a simplified model of a cooling bed of potato. It shows that temperature of the air changes with time and location over the bin height. To solve these equations, numerical techniques are used with the appropriate initial and boundary conditions.

DISCRETE-TIME VARIABLES AND MODELS

Discrete-time system models provide a description of how a system behaves at discrete points in time. Digital computers, for example, are inherently discrete-time systems in that associated system variables change only at discrete points in time. Discrete-time systems are also used to approximate the behavior of systems whose variables vary continuously in time.

A discrete-time variable is one defined only at discrete points in time, as illustrated in Figure 4–4. The discrete-time variable v is written as

$$v(k\Delta t) \text{ for } k = \ldots -2, -1, 0, 1, 2, \ldots$$

(6)

The variable k is the discrete time variable that takes only the values of zero and positive and negative integers. Δt is the time increment between values of v. The time variable, t, is then

$$t = k\Delta t$$

(7)

Sampling Continuous-Time Variables

Discrete-time variables often arise from sampling continuous-time variables. This process is illustrated in Figure 4–5. A continuous-time variable $v(t)$ is shown in (a), a discrete-time variable $v(\Delta t)$ that results when Δt = one time unit in (b), and a discrete-time

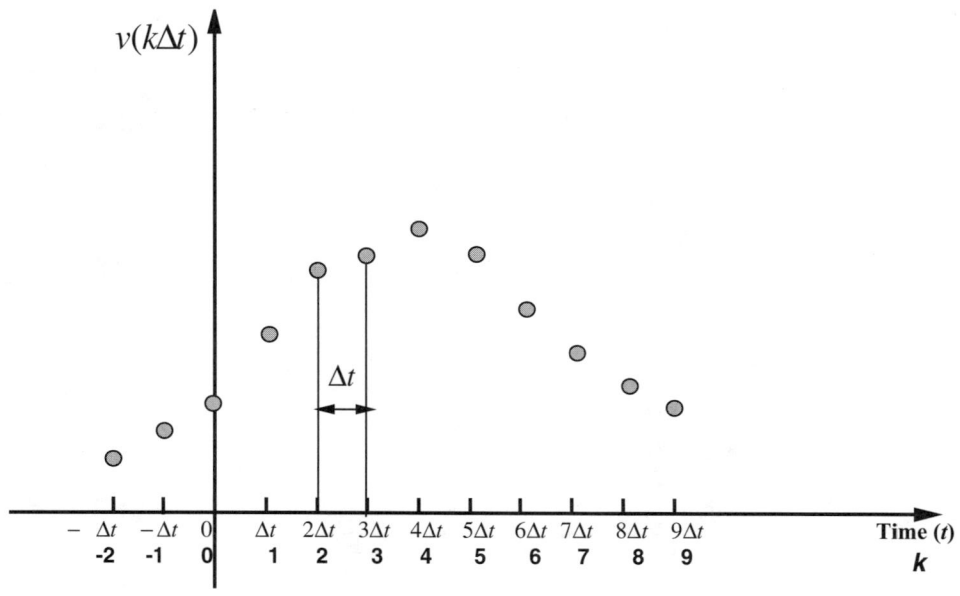

Figure 4–4 Discrete-time variable

variable that results when Δt = two time units in (c). In this case, Δt is called the *sampling interval* or the *sampling period*. Figure 4–5 shows how important the choice of sampling interval is to the quality of discrete representation of a continuous-time variable. See Appendix 4–A for more information on discrete-time variables.

Difference Equation System Models

A discrete-time system can be described by the following equation:

$$y(k) = f[y(k-1), y(k-2), \ldots, y(k-n); u(k), u(k-1), \ldots]$$

(8)

Equation 8 shows that the system output, y, at time $t = k\Delta t$ is a function of past values of the output, and possibly, past values of the input, u. The linear nth-order difference equation can be written as

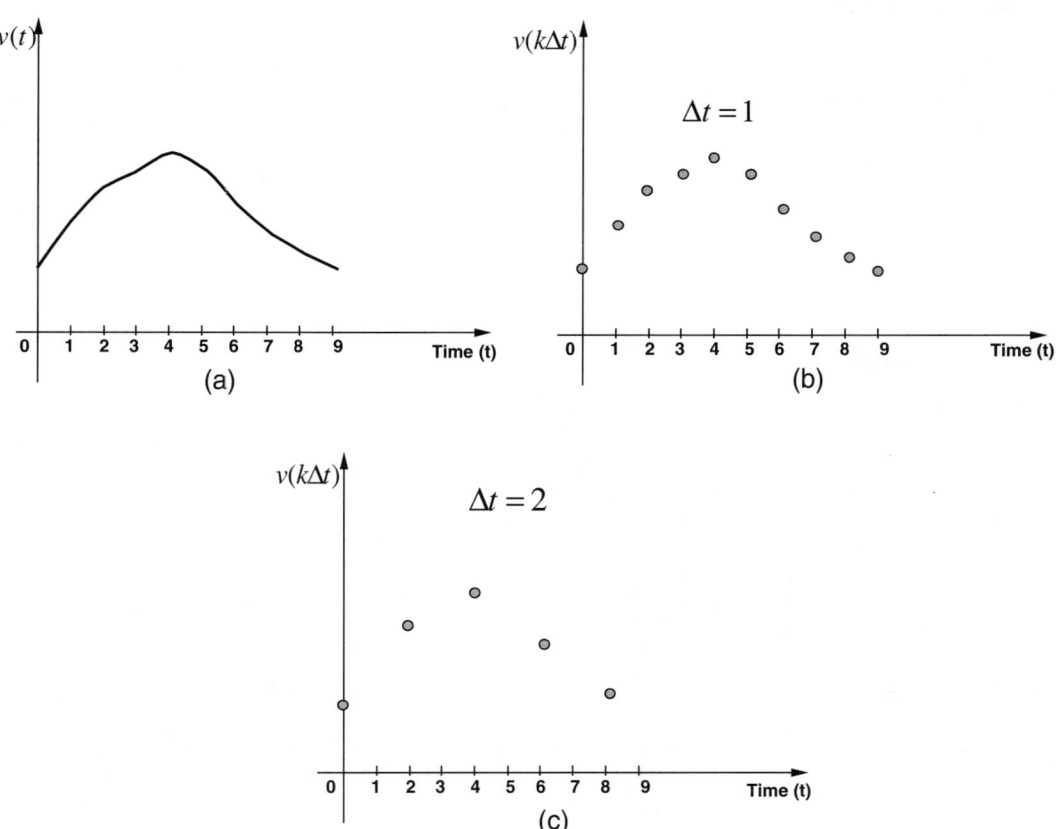

Figure 4–5 Sampling a continuous–time variable

$$y(k) = -a_1 y(k-1) - a_2 y(k-2) - \ldots - a_n(k-n) + b_0 u(k) + b_1 u(k-1) + \ldots + b_m u(k-m)$$

$$(9)$$

Properties of the First-Order Linear Difference Equation

Consider a general solution of the first-order difference equation as

$$y(k) = ay(k-1) + bu(k)$$

$$(10)$$

By giving an initial or starting value for the system output y and the system input u, we can recursively solve for $y(k)$ as k increases. So the general solution for the output of the system described by Equation 10 can be obtained by writing recursively from it:

$$For \ k = 1: y(1) = ay(0) + bu(1)$$
$$For \ k = 2: y(2) = ay(1) + bu(2) = a^2 y(0) + abu(1) + bu(2)$$
$$For \ k = 3: y(3) = ay(2) + bu(3) = a^3 y(0) + a^2 bu(1) + abu(2) + bu(3)$$

$$(11)$$

From this we see the general pattern:

$$y(k) = a^k y(0) + b[a^{k-1}u(1) + a^{k-2}u(2) + \ldots + a^0 u(k)]$$

$$(12)$$

We can write this result in more compact form as

$$y(k) = a^k y(0) + b_i \sum_{i=1}^{k} a^{k-i} u(i), \quad k = 1, \ 2, \ 3, \ \ldots$$

$$(13)$$

Equations 12 and 13 show that the general solution of $y(k)$ is composed of two parts. The first part depends on the initial value $y(0)$ but is independent of the system input. The second part depends on the system input but is independent of the initial value $y(0)$. This is an important general result that applies to all models of dynamic systems. One of the components is called *zero state solution*, described by Equation 14:

$$y_i(k) = b_i \sum_{i=1}^{k} a^{k-i} u(i)$$

$$(14)$$

which is the output due to input only (with $y(0)$ equal to zero). The other component is the *zero input solution*, the output due to the initial state $y(0)$ only (with the system input u equal to zero):

$$y_s(k) = a^k y(0)$$

(15)

So the variables of a dynamic system can change through time due to the autonomous factors within the system, which are forced by the initial state $y(0)$ and/or due to external stimuli that constitute the system input. Appendix 4–B presents the table of z-tranforms and some properties of z-transformations to solve difference equations.

PROCESS DYNAMICS MODELING AND IDENTIFICATION

Mathematical models of food-processing systems can be obtained by theoretical (first-principle) or experimental analysis. *Theoretical modeling* consists of determining the process model by stating balance equations, state equations, and phenomenological laws. Theoretical models are often called *white box models* and are characterized by a set of ordinary and/or partial differential equations (in the case of continuous signals) that represent the fundamental description of the processing system. Derivation of discrete-time models can be made by approximation of continuous-time models, by lumped-parameter models, or by simplification of continuous-time models.

Generally, food-processing systems are very complex and are not well understood from the first-principle point of view, making it difficult to describe them in terms of simple mathematical models. In general, a mathematical model to simulate food-processing systems has to consider physical and chemical data, process variables, and process design and relate them to product quality attributes and process performance. Developing fundamental-based models is a difficult task, so generally experimental-based models are more suitable to model the dynamics and disturbances of food-processing systems.

Another approach used for dynamics modeling is known as *black box modeling*. This method directly connects the observed characteristics of the product with the process variables (set points of the equipment, geometry, ingredients) by equations obtained from statistical analysis. The major limitation of black box modeling is that the resulting equations are usable only in the range covered by the experimental data. In most cases, it does not provide any explanation about the phenomena occurring inside the machine and does not contribute information for scale-up.

Mathematical models for the black box approach are determined using system identification techniques. These models can be obtained by applying step-response, frequency-response, or parameter estimation methods. For identification of discrete-time parametric models, parameter estimation methods are especially suitable.

In the food industry, there often is some knowledge of the structure of processes, but it is incomplete. In this case, the *gray box modeling* approach can be used to describe the process mathematically (Sohlberg, 1998). Gray box modeling requires both theoretical modeling and identification methods: that is, a combination of white and black box modeling approaches.

Gray box models have several advantages compared to black box models. With gray box modeling, it is possible to build nonlinearities into the model and separate different parts of the process into submodels. Gray box modeling is based on available physical knowledge that may be incomplete and uncertain. Experimental data are used to identify the incomplete and uncertain parts that are also influenced by unknown disturbances. As a result, gray box models may achieve more understanding about the functionality and physical characteristics of the process, resulting in a better controlled process with improved quality of the final product.

Theoretical (White Box) Modeling

To develop fundamental models, it is assumed that the a priori information is exact and is based on scientific theory and can be presented by a set of well-defined equations. The main objective of the theoretical modeling approach is to describe the relevant knowledge and form a basic model of the process. The steps required to develop a dynamic white box model are as follows (Sohlberg, 1998):

- *Description of the process*: Detailed description of the process with all operating variables, process variables, and disturbances, including a process schematic
- *Structure of the process*: Definition of manipulated and controlled variables; division of the process into subprocesses with defined functions and connections
- *Assumptions*: Description of the process behavior; list of all assumptions to be used in developing the model
- *Equations*: Formulation of the relations among variables, using balance equations, algebraic equations, or logical expressions. These constitute the basic equations that give a core description of the process.
- *State-space models*: The equations developed above need to be structured so they can be used for identification and process control. The steps required to develop a state-space equation from the basic equations are
  —Selection of the state variables ($x_1, x_2, x_3 \ldots x_n$)
  —The derivatives of the state variables are function of the states and input variables ($u_1, u_2, u_3 \ldots u_n$)
  —Addition of algebraic equations and/or logical expressions
  —Connection of output variables ($y_1, y_2, y_3 \ldots y_n$) to the states and/or to the inputs
- **Model preparation**
  —*Discretization:* use of numerical techniques to solve the equations
  —*Scaling:* use of scale factors when the ranges of input and output data are relatively different (mean value of the data is usually used) to normalize the data

—*Process noise:* fast-varying process disturbance is usually assumed to be white noise and can be modeled with some ad hoc term, but slowly varying process noise requires extra states.

—*Measurement noise:* filtering measurement noise needs to be considered when data are transmitted to the data acquisition system.

• *General form*: continuous-time or discrete-time model forms consisting of differential or difference equations relating process states to outputs. For example, a discrete-time description of a process can be formulated by the state-space form as

$$x(k+1) = F[x(k), \ u(k)] + v_1(k)$$
$$y(k) = G[x(k), \ u(k)] + v_2(k)$$

(16)

where $x(k)$ is the state vector, $u(k)$ is the input vector, $y(k)$ is the output vector, and $v_1(k)$ and $v_2(k)$ are the process and measurement disturbances, respectively.

For an example, consider the transient response of a pilot-scale double-effect vacuum evaporator designed to concentrate sugar solution from 5 wt % to 15 wt % (Ritter & Andre, 1970). Figure 4–6 shows the schematic of the system. It consists of a short-tube first stage and a long-tube forced-circulation second effect. Both feed solution and steam are introduced in the first effect, while the concentrated product is withdrawn from the second effect evaporator. The second effect is equipped with a cyclone separator and a water cooler condenser that is maintained at the desired operating pressure by a steam ejector and pressure controller.

The assumptions made to develop the model are the following (Ritter & Andre, 1970):

• The steam in the calandria is saturated, and the condensate hold-up and heat losses are negligible.
• Longitudinal temperature gradients in the metal tubes of the calandria are negligible.
• The first effect behaves as a perfect mixed cell.
• The resistance to vapor flow between the two effects is low.
• The condensate hold-up in the second effect is negligible as well as the boiling-point rise of the solution.
• The heat transfer surfaces of the heat exchanger tubes are of low thermal capacitance.
• The solution in the second effect experiences a perfect mixing.
• There is a plug flow in the line connecting the two effects.
• The steam enthalpy, the pressure in the second effect, and the heat transfer coefficient remain constant during the process.
• Time delay between the two systems is small and can be considered negligible.

A simplified linear model of the system can be summarized as

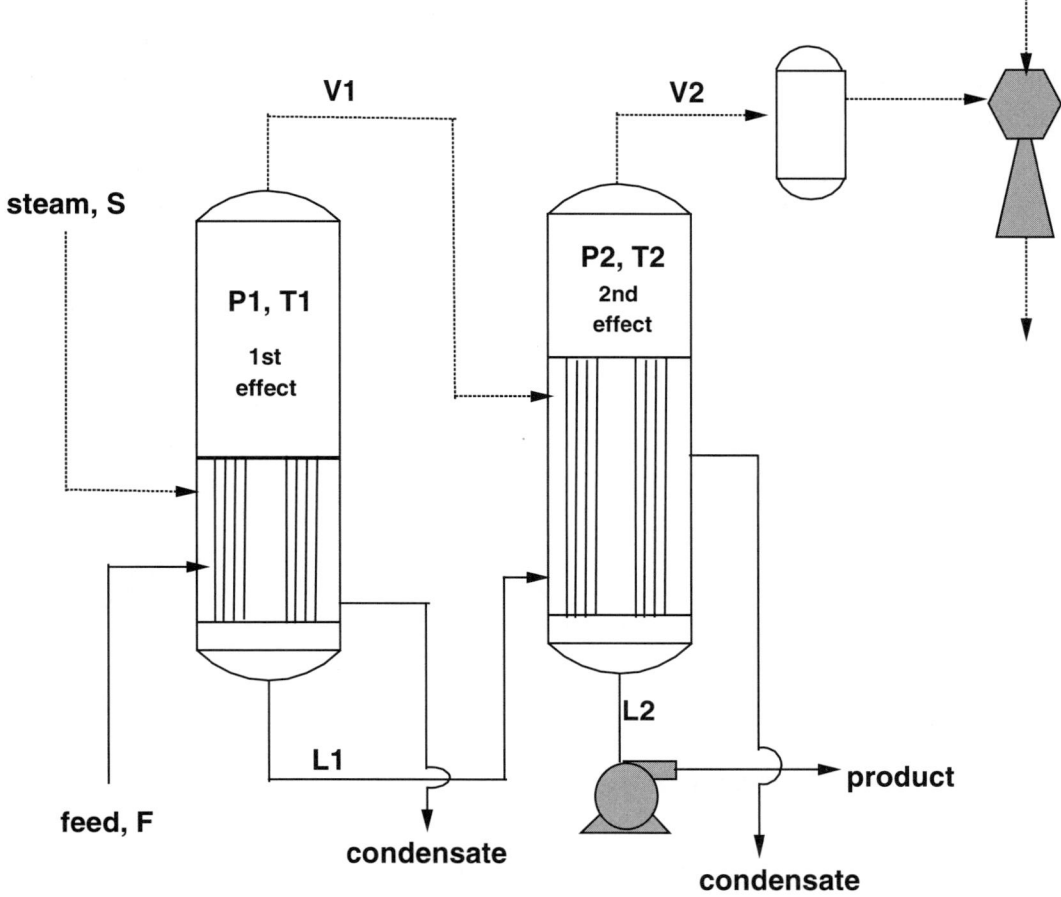

Figure 4–6 Schematic of a double-effect evaporator

$$\frac{dC_1}{dt} = a_1 H_1 + a_2 C_1 + a_3 F + a_4 C_f$$

$$\frac{dH_1}{dt} = b_1 H_1 + b_2 C_1 + b_3 S_i + b_4 F + b_5 H_f$$

$$\frac{dW_1}{dt} = d_1 H_1 + d_2 C_1 + d_3 L_1 + d_4 F$$

$$\frac{dC_2}{dt} = e_1 H_1 + e_2 C_1 + e_3 C_2 + e_4 L_1$$

$$\frac{dW_2}{dt} = g_1 H_1 + g_2 C_1 + g_3 C_2 + g_4 L_1 + g_5 L_2$$

(17)

where C is the dimensionless concentration, H the dimensionless liquid enthalpy, F the dimensionless feed rate, L the dimensionless concentrated liquid flow rate, S_i the dimensionless steam rate, W the dimensionless mass of liquid hold-up, and H_f the dimensionless feed enthalpy. The coefficients a_1 to g_5 refer to the steady-state data for the system. The linearized model described above needs to be transformed so that the model can be handled easily. The five states in this case are

$$x_1 = C_1$$
$$x_2 = H_1$$
$$x_3 = W_1$$
$$x_4 = C_2$$
$$x_5 = W_2$$

(18)

The steam flow rate, S_i, the concentrated liquid flow rates from each effect, L_1 and L_2, and the feed rate, F, are the input variables, so

$$u_1 = F$$
$$u_2 = S_i$$
$$u_3 = L_1$$
$$u_4 = L_2$$

(19)

The outputs of the process are given by

$$y_1 = \alpha x_1$$
$$y_2 = \beta x_3$$
$$y_3 = \alpha x_4$$
$$y_4 = \beta x_5$$

(20)

where α and β are the calibration constants relating the controlled variable to the measured variables (given by a transducer). These equations can be summarized as

$$\frac{dx(t)}{dt} = f[x(t), u(t)]$$
$$y(t) = g[x(t)]$$

(21)

where f is a linear function, $x(t) = [x_1(t)\ x_2(t)\ x_3(t)\ x_4(t)\ x_5(t)]'$ is the state vector, and $u(t) = [F(t)\ S_i(t)\ L_1(t)\ L_2(t)]'$ is the control vector. The output transducers are specified by the matrix $g = \text{diag}[\alpha\ \beta\ \alpha\ \beta]$.

The model can then be discretized and the process outputs scaled (to avoid numerical problems during identification), resulting in the following deterministic equations:

$$x(k+1) = F[x(k), u(k)]$$
$$y(k) = G[x(k)]$$

(22)

It is important to realize that it would be too expensive, resulting in a structure too complex, to make a model a perfect copy of the process. By analyzing the prediction error of the deterministic model given by Equation 22, it is possible to investigate whether there are unmodeled parts of the process. Unmodeled parts can be detected by looking to the properties of the prediction error: that is, the prediction error is not white. Unmodeled parts of the process may originate from wrong assumptions such as incomplete mixing in the system; properties such as enthalpy, pressure, and heat transfer coefficients not being constant during the process; and delay being underestimated. In addition, other disturbances such as feed concentration and steam availability may influence the system.

Unmodeled parts of the process and other disturbances are added as disturbances to the state of the process. During identification, these disturbances are regarded as white noise, that is a simplification. However, coloring of the state noise needs extra states, which is time consuming during identification. Measurement of outputs is also influenced by disturbances that are also regarded as white noise. Generally, transducers' noise specification should be low (around 1%).

Therefore, Equation 22 needs to be extended to include process and measurement noise for all states and outputs, respectively. The final discrete stochastic model is described by Equation 16, where the process and measurement noise, $v_1(t)$ and $v_2(t)$, are independent white noise.

Empirical (Black Box) Modeling

The development of a rigorous theoretical model may be impractical for complex processes if the model has a large number of differential equations and unknown parameters. Black box modeling is an alternative approach to the white box modeling described above.

Black box modeling consists of determining the mathematical model of the process by using measured signals. Inputs and outputs of the process are evaluated using identification methods. This model can be nonparametric.

Parametric models are especially suitable for digital control systems, since modern system theory is based on these models. Parametric models contain the parameters explicitly, and the synthesis of the control algorithms can be performed discretely.

Empirical Dynamic Models from Step Response

To develop empirical dynamic relationships between input and output signals of a process, the first step is to plot the data versus time so that the overall trends of data can be visualized and the model form easily selected. Once the model form has been defined, the unknown model parameters are evaluated by using parameter estimation techniques or *curve fitting*. The problem of estimating the process parameters is simple in the case of a single-input, single-output (SISO) system, but it becomes more complicated when a model involves multiple inputs and outputs.

Consider the case of determining the parameters of a first-order model using step tests. This method consists of applying an instantaneous step change in the process input and measuring the process response over time. If the food-processing system can be approximated by a first- or second-order linear differential equation, the model parameters can be estimated by inspection of the response curve.

For example, a dynamic model for a first-order process is described by

$$T\frac{dy}{dt} + y = Ku$$

(23)

where y (the output) initially is at rest, $y(0) = 0$, when the input $u = 0$. If a step input is imposed on the system at time $t = 0$, a response is then generated. Consider the response of the extrudate's color in a twin-screw food extruder to a step in the water rate from 3.5 to 4.2 kg/h (Figure 4–7). The response of the product's *color* (t) reaches 63.2% of its final value at $t = T$. The steady-state output deviation value is *color* (∞) = KM, where M is the value of the input step change ($4.2 - 3.5 = 0.7$ kg/h) and K is the system gain ($= \Delta y/\Delta u$). The Laplace transform to Equation 22 is

$$\frac{Y(s)}{U(s)} = \frac{K}{Ts+1}$$

(24)

The unit-step response ($M = 1$) of a first-order process can be obtained by substituting $U(s) = 1/s$, the Laplace transform of the unit-step function, so

$$Y(s) = \frac{K}{Ts+1}\frac{1}{s}$$

(25)

Taking the inverse Laplace transform of Equation 25, we obtain

$$y(t) = 1 - Ke^{-t/T}, (t \geq 0)$$

(26)

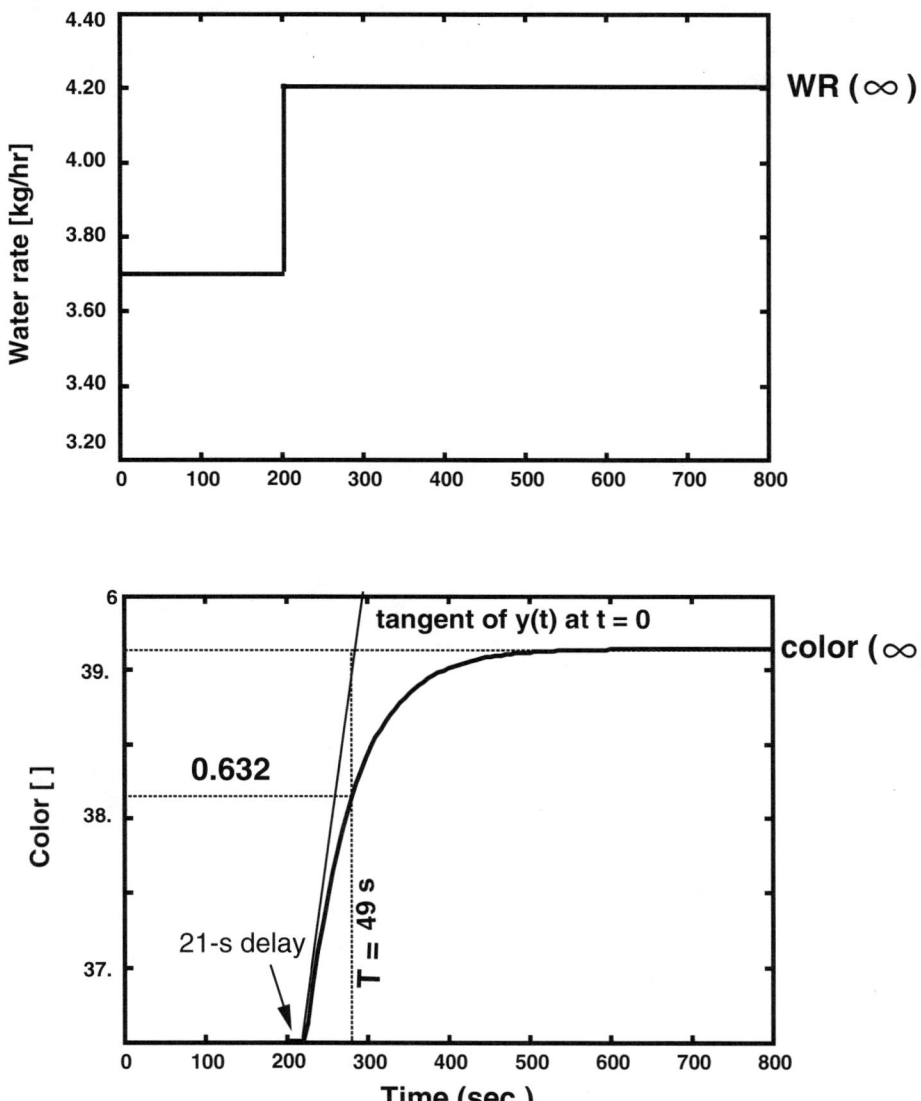

Figure 4–7 Extrudate's color response of twin-screw food extruder for a step change in water flow rate

where T is the time constant of the process. Equation 26 shows that initially the output $y(t)$ is zero and finally it becomes unity. At $t = T$, the value of $y(t)$ is 0.632; that is, the response $y(t)$ has reached 63.2% of its total change. Also, the slope of the tangent line at $t = 0$ is $1/T$, since

$$\frac{d}{dt}\left(\frac{y}{K}\right) = \frac{1}{T}e^{-t/T}\Big|_{t=0} = \frac{1}{T}$$

(27)

From Figure 4–7, the intercept of the tangent at $t = 0$ (at 200 seconds in this case when the step was imposed in the system) with $y/KM = color = 39.11$ occurs at $t = T$. So the time constant T can be calculated by finding the value of t at which the response is 63.2% complete. The plot of Figure 4–7 shows that $T = 49s$; that is,

$$Color = 36.62 + 0.632(39.11 - 36.62) = 38.19$$

(28)

which is the value of the output when 63.2% of the response is complete. The value of the process gain is equal to

$$K = \frac{\Delta color}{\Delta WR} = \frac{2.49}{0.7 \text{kg} / \text{h}} = 3.56 \frac{1}{\text{kg} / \text{s}}$$

(29)

Note that the system shows a time delay, τ, of 21 seconds. The process can then be approximated to the following first-order equation:

$$\frac{Color(s)}{WR(s)} = \frac{3.56 e^{-21s}}{49s + 1}$$

(30)

The time delay in Figure 4–7 can be estimated by drawing a tangent at the point of inflection of the step-response curve. The time delay is the intersection of the tangent line and the time axis.

If the first-order model cannot accurately describe the dynamics of the process, a second-order model, with two time constants, can be estimated using the graphical technique described above. Seborg et al (1989) discussed in detail the different methods used to estimate the parameters of a second-order system like the one presented in Equation 31:

$$G(s) = \frac{k e^{\tau s}}{T^2 s^2 + 2\zeta Ts + 1}$$

(31)

The graphical method described above tends to be subjective. Nonlinear regression methods provide a better fit to the data and are preferred, especially when controlling with process control computers.

The step-response technique described above is very simple and useful for obtaining first-order and second-order dynamic models from step-response data. Although these models are suitable for the design of effective control systems, the resulting models are usually accurate only in a narrow range of operating conditions, close to the nominal steady state.

Some of the limitations of using step tests to determine the parameters of a first-order model for food process systems (Seborg et al, 1989) are as follows:

- Few experimental plots of the step response show purely first-order behavior.
- It is difficult to perform a step input. Control valves or other manipulated variables cannot be changed instantaneously from one condition to another. Generally, the input is ramped over a finite time. A good approximation of a step can be obtained, however, by having the ramp time be smaller than the process time constant. Figure 4–8 shows the actual data of a step input in the water rate of the food extrusion

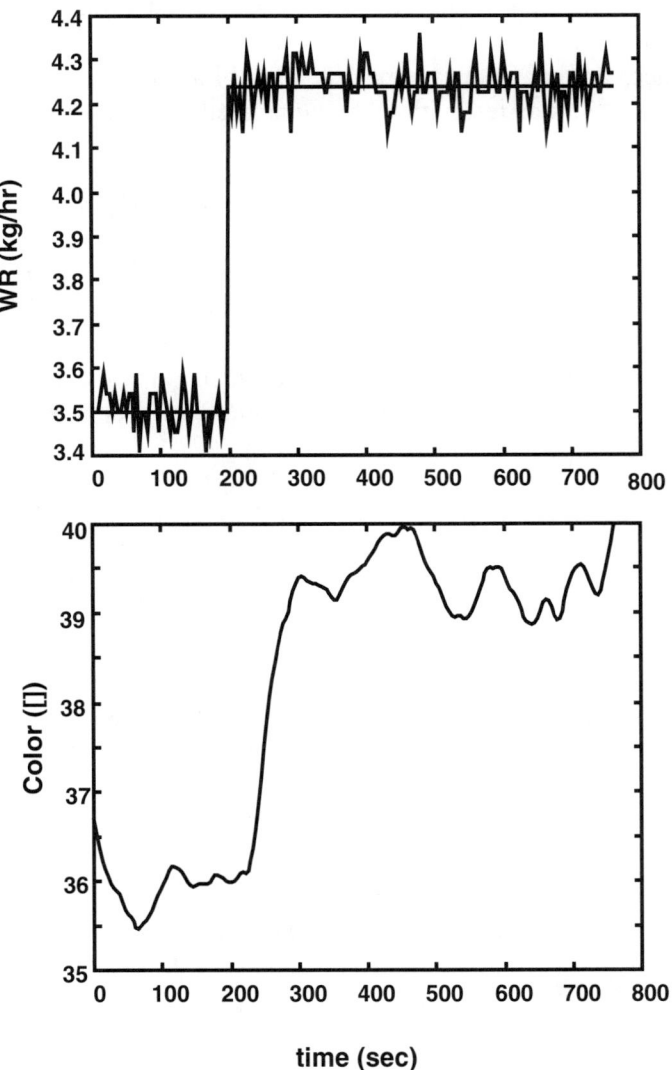

Figure 4–8 Actual data for the extrudate's color response of twin-screw food extruder for a step change in water flow rate

process. The drifted line is the actual measured data, and the smooth line represents the actual data after being filtered using a digital low-pass filter.

- Food-processing systems are very complex; they are not first order or linear in nature.
- The output data generally are corrupted by noise that can arise from instrumentation or normal processing operations. Figure 4–8 shows that the actual data of the extrudate's color response of the food extrusion to a step on the water rate contain noise. A first-order response plot can handle random noise, but nonrandom noise is more difficult to analyze. Nonrandom noise can be caused by linear drifting in the measurement instrumentation: for example, in the case of a food extruder these are feeder variation, water meter variation, screw speed variation, barrel temperature cycling, and so on.
- Disturbances arising from raw material variability or machine wear can affect the results during the step test without being noticed by the operator.

System Identification of Process Dynamics and Stochastic Signals

System identification is the subject of building (or selecting) models of dynamic systems on the basis of observed input/output data. The first step is to determine the class of model that will represent accurately the dynamics of the process. The types of models include transfer-function models, state-space models, and distributed parameter models.

The simplest input-output relationship for a parametric model is described by a linear difference equation as

$$y(k) + a_1 y(k-1) + \ldots + a_n y(k-n) = b_1 y(k-d-1) + \ldots + b_m u(k-d-m)$$
$$+ e(k) + c_1 e(k-1) + \ldots + c_p e(k-p)$$

$$(32)$$

where $e(t)$ is a white noise, a sequence of independent random variables with zero mean values and variances equal to σ^2, and d is the time delay. The parameters of this model are

$$\theta = [a_1 a_2 \ldots a_n, b_1 b_2 \ldots b_m, c_1 c_2 \ldots c_p, d]'$$

$$(33)$$

If we introduce the polynomials $A(z^{-1})$, $B(z^{-1})$, and $C(z^{-1})$ as

$$A(z^{-1}) = 1 + a_1 z^{-1} + \ldots + a_n z^{-n}$$

$$(34)$$

and

$$B(z^{-1}) = b_1 z^{-1} + \ldots + b_m z^{-m}$$

(35)

and

$$C(z^{-1}) = 1 + c_1 z^{-1} + \ldots + c_p z^{-p}$$

(36)

Equation 36 can be written as

$$A(z^{-1})y(k) = B(z)u(k-d) + C(z)e(k)$$

(37)

So

$$y(z) = z^{-d}\frac{B(z^{-1})}{A(z^{-1})}u(z) + \frac{C(z^{-1})}{A(z^{-1})}e(z)$$

(38)

The model of Equation 38 is called the ARMAX (*auto regressive moving average* with the *exogenous input*) model, where *AR* refers to the autoregressive part $A(q)y(k)$, *MA* to the moving average part $C(z^{-1})e(z)$, and *X* to the exogenous input or control part $B(q)u(t)$. Another version uses a forced integration as the ARIMAX (*auto regressive and integrated moving average with exogenous input*) model:

$$A(z^{-1})y(z) = z^{-(d+1)}B(z^{-1})u(z) + C(z^{-1})e(z)/\Delta$$

(39)

where $\Delta(k) = y(k) - y(k-1)$. With $C = 1$ in Equation 39, the model is reduced to the ARX (*auto regressive with the exogenous input*) form. The ARMAX model has become a standard tool in control for both system identification and control design.

There are three basic stages involved in the system identification process: (1) characterization of the nature of the input-output relations (discrete, continuous, causal, etc); (2) determination of the structure of the system; and (3) identification of the parameters within this structure.

Parameter estimation can be carried out in off-line or on-line (recursive) modes. In off-line (batch) parameter estimation, a *least squares* (LS) method is applied to the entire set of input/output data. In on-line identification, which is used in adaptive control, the estimated coefficients of the models are updated at every sampling instance. However, the recursive LS method may fail to track time-varying parameters. Techniques such as covariance resetting, extended LS, and the forgetting factor are often used to overcome these estimation problems (Isermann, 1989). The forgetting factor allows removal of information in all directions but adding of new information in the regressor direction only.

Many of the recursive estimation procedures applied are modifications of the LS procedure. The general recursive LS algorithm is of the form:

$$\hat{\boldsymbol{\theta}}(k) = \hat{\boldsymbol{\theta}}(k-1) + \gamma(k)R^{-1}(k-1)\boldsymbol{\varphi}'(k)[y(k) - \boldsymbol{\varphi}'(k)\hat{\boldsymbol{\theta}}(k-1)]$$

(40)

where $\hat{\boldsymbol{\theta}}$ is the predicted parameter vector, $\boldsymbol{\varphi}$ the measured data vector, R the direction of the parameter update, and the size of the parameter update. The LS algorithm is generally applicable for small noise-to-signal ratios, requires small computational effort, and gives reliable convergence. Another estimation method, the maximum likelihood (ML), is better than the LS in that it usually results in a lower order model even when the signal-to-noise ratio is low. However, the ML is generally more time consuming to apply, which is nonadvantageous for on-line estimation. An example of a recursive prediction error algorithm to estimate the parameters of Equation 38 is presented in Appendix 4–C.

Estimation delay time is very critical to control design. An improper delay time can destabilize the process. One approach for estimating the time delay using the LS algorithm is to evaluate the magnitude between the first and last b_m estimates and either add or eliminate a delay term depending on the magnitude of the difference. Another method is to expand the B polynomial in Equation 38 with d_{max} - d_{min} extra parameters. Although expanding the B polynomial is the most widely used method to identify unknown or time-varying time delay, its drawback is an increased computational effort and its sensitivity to noise.

A number of computer software products are available today that can be used to estimate process parameters based on Equation 38. One of these is the MATLAB System Identification Toolbox (Math Works; Natick, MA). Parametric models can easily be determined to mathematically describe the dynamic behavior of a system. Model structure such as ARX, ARMAX, ARIMAX, Box-Jenkins, and many others can be determined. Additionally, functions are provided that allow identified models to be simulated using test data as inputs.

System Excitation. Obtaining models that adequately describe the process requires excitation of the process across all important frequencies. Pseudorandom binary input signals (PRBSs) are often used to identify stochastic processes. Estimates of the process parameters (time constants, settling time) obtained from initial step tests can be used to determine the frequencies of interest. Davies (1970) and Godfrey (1970) described the proper way to design the signal so that its autocorrelation function resembles that of white noise, which is a requirement for correlation analysis of the signals. Some of the requirements are:

- The period of the PRBS should be greater than the system settling time.
- The bit interval should be less than one half the smallest time constant of interest.

- The signal should contain equal numbers of high and low values, and the lengths of high and low runs should be normally distributed to excite all frequencies.
- The sampling rate should be 5 to 15 times per 95% settling time.
- The number of parameters to be estimated should be considered when determining the period/length of the signal; more parameters require longer data sets.
- For multivariable identification, orthogonal signals are used. This involves shifting each input out of phase so that they are uncorrelated.

PRBSs are preferable for the identification of food-processing systems because they can provide large amounts of information in a short testing period and because time delays can be estimated via correlation techniques.

Practical experience suggests that the PRBS bandwidth should be 10 to 25 times higher than that of the system. The PRBS signal has three fixed parameters: the signal amplitude ($\pm a$), the period ($N \times$ bit interval), and the bit interval. The bit interval is the shortest time between signal state changes (negative to positive). N is the total number of positive and negative states. The signal amplitude should be fixed as the largest amplitude within the linear range that does not unduly disturb normal processing operations. The amplitude needs to be large to yield process sensitivity. The period ($N \times$ bit interval) should be greater than the system settling time. The settling time can be estimated as the dead time plus four or five times the longest time constant. Often the period is fixed as 1.25 to 1.5 times the settling time.

The bit interval should be less than one half the smallest time constant of interest. If reasonable accuracy cannot be obtained at the bit interval selected because of the presence of noise, the bit interval can be increased above the suggested value and the output sampled more often than the bit interval. Also, the number of positive states should be approximately equal to the number of negative states, and short runs of each state should be more frequent than long runs. Appendix 4–D gives a detailed description of the PRBS.

Generally, before starting a dynamic analysis using a PRBS, step function testing should be used to determine the amplitude and the parameters for PRBS. The effective frequency band of the signal is $1/(N \times$ bit interval) to $1/(3 \times$ bit interval) Hz. Sampling should be five to seven times along the rise time of the step response. The number of samples needed depends on the signal-to-noise ratio, with noisy signals requiring more samples.

Figure 4–9 illustrates the PRBS with a bit interval of 25 seconds and N equal to 63. Shift registers were used to design the PRBS signal (see Appendix 4–D).

It is not necessary to use shift registers to define the signal, especially if cross-correlation techniques (impulse fitting) are not being used to define the model. Appendix 4–D gives other types of methods used to design PRBSs. Also of interest and often used is a multilevel random signal (PRMLS). A PRMLS is designed by selecting several levels around the operating point and changing from one level to the other at random. The input signal should be adapted to the specific application. Amplitudes are often constrained by physical limitation of a valve or by the linear operating range.

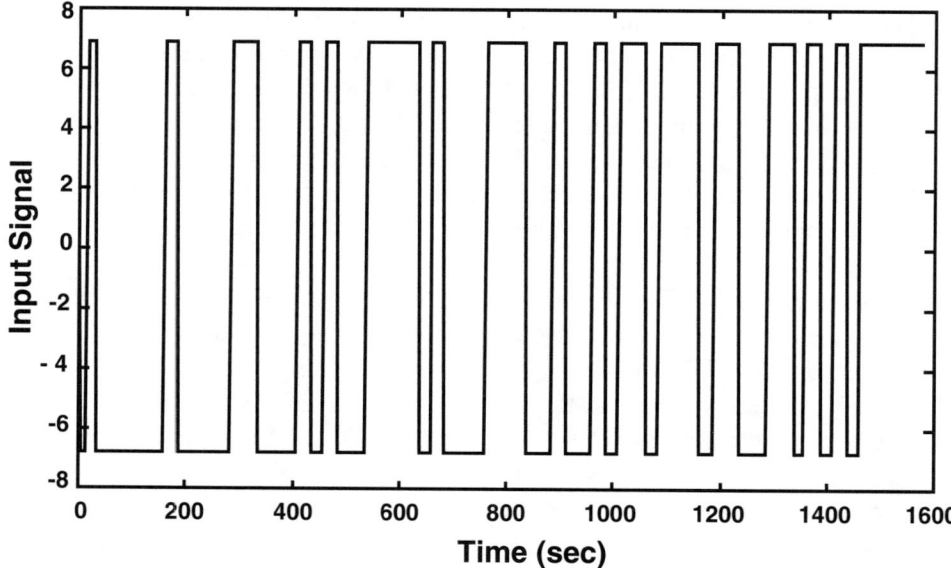

Figure 4–9 A PRBS with $N = 63$ and bit interval = 25 seconds

An alternative approach for identification of systems with multiple inputs and outputs involves a simultaneous multiple-input method. Each system input is excited simultaneously by a different periodic PRBS, each having an impulse-like autocorrelation function so that each signal approximates a periodic white noise. This can be done by using orthogonal signals that are shifted out of phase.

Figure 4–10 shows the multilevel signal. It is 2.5 hours long with a bit interval of 60 seconds. The orthogonal signals shown in Figure 4–10 are formed with the same PRBS signals that are shown in Figure 4–9, with amplitudes adjusted accordingly, and shifted 1/3 out of phase.

Model Analysis. Both objective and subjective tests should be used in the selection of the proper order model (Söderström & Stoica, 1989). Two methods by Akaike, *final prediction-error criterion* (FPE) and *Akaike information criterion* (AIC), are of interest. Akaike's methods address the joint problem of parsimony and good fit when determining model structure by including a term in the minimization criteria that penalizes the number of parameters being estimated. At some point, the decrease in the loss function will no longer be significant compared to the penalty for having higher order models. Parsimonious models are needed for adaptive control to minimize computational time and persistency requirements. The term to minimize is

$$AIC = N \log V_N(\hat{\theta}_N) + 2p$$

(41)

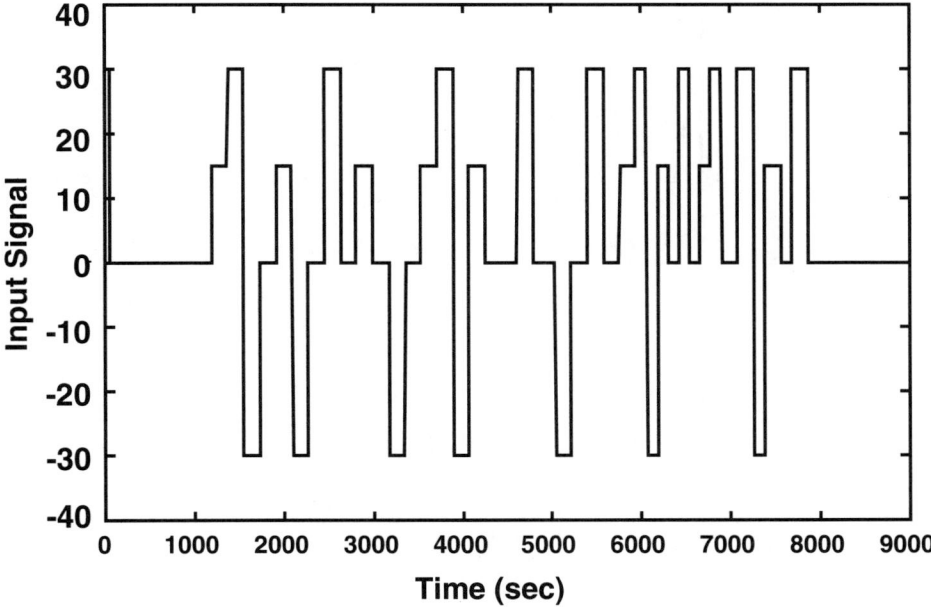

Figure 4–10 A PRMLS with a bit interval of 60 seconds

or

$$FPE = \frac{1 + p/N}{1 - p/N} V_N(\hat{\theta}_N)$$

(42)

where $V_N(\hat{\theta})$ is the loss function (quadratic fit):

$$V_N(\hat{\theta}) = \frac{1}{N} \sum_{k=1}^{N} \varepsilon^2(k, \theta)$$

(43)

and ε is the residual, N the number of data points used in the estimation $\hat{\theta}_N$, and p the number of terms in the model.

Several tools can be used to select the best model in addition to the AIC:

- Determination of whether the model structure is able to describe the dynamic behavior of the process. Prediction values are calculated using the models, and these predictions are compared to the experimental data.
- Plotting of the loss function to address the joint problem of parsimonious models and good fit. Figure 4–11 shows a typical relationship between loss function and model order. It is shown that a second-order ARMAX model is sufficient to describe the process in study, since slight changes in loss function values are observed for higher model orders. In another example, the plot of the loss function (Figure 4–12)

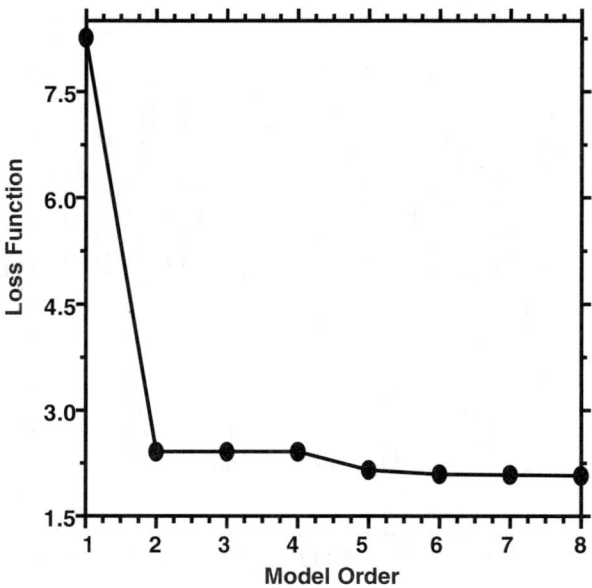

Figure 4–11 The loss function plotted versus model order

indicates that the loss function value continuously decreases as the model order increases, so there is no obvious choice of the model order in this case. The plot of Akaike's test criterion should show a minimum when the correct model order is found. From Figure 4–11, we can say that the AIC does not lead to any obvious choice of the other model either.

- Calculation of the covariance matrix. The covariance matrix gives insight into the quality and reliability of the estimates. It can be used to determine what magnitude of parameter variations one might expect. The diagonal values of the covariance matrix can indicate if the model is overparametrized or the experiment is not

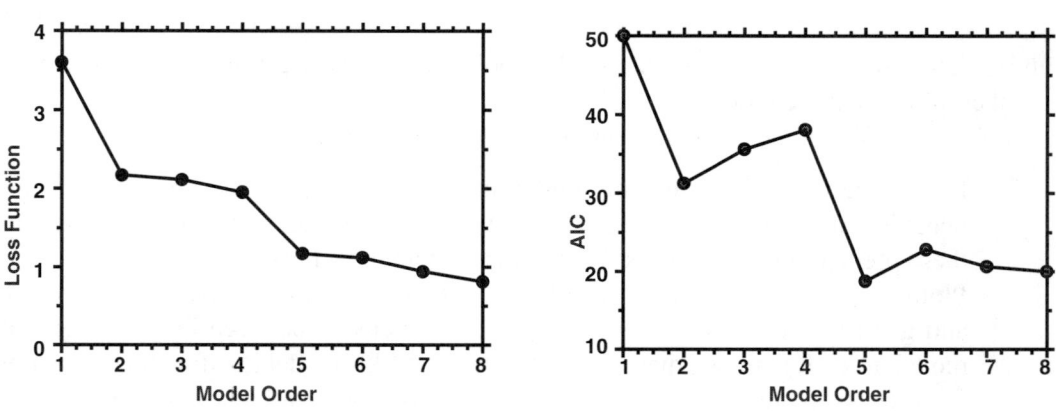

Figure 4–12 The loss function and Akaike's test criterion plotted versus model order

informative enough. The off-diagonal values indicate coupling among parameters. If these values are very large, they are correlated.

- Calculation of the residuals, as well as autocorrelation of the residuals and cross-correlation between the residuals and inputs. The residuals ideally should be white and independent of the input for the model to adequately describe the system. An underestimation of the delays shows up as small values of b parameters in Equation 38 compared to their standard deviation. Overestimated delays have visible significant correlations between residuals and inputs at lags corresponding to the missing b terms. When evaluating the autocorrelation of the residuals and the cross-correlations between residuals and inputs, the general rule is to reject the model if the values are outside the 99% confidence limits. However, if one is more interested in the dynamics of the system, the cross-correlation between residuals and inputs (independence of residuals and inputs) rather than the whiteness of the resisduals should be the focus.

An example of an identification experiment is shown in Figure 4–13. The results were obtained from a pilot-size concurrent-flow grain dryer where the input was the discharge

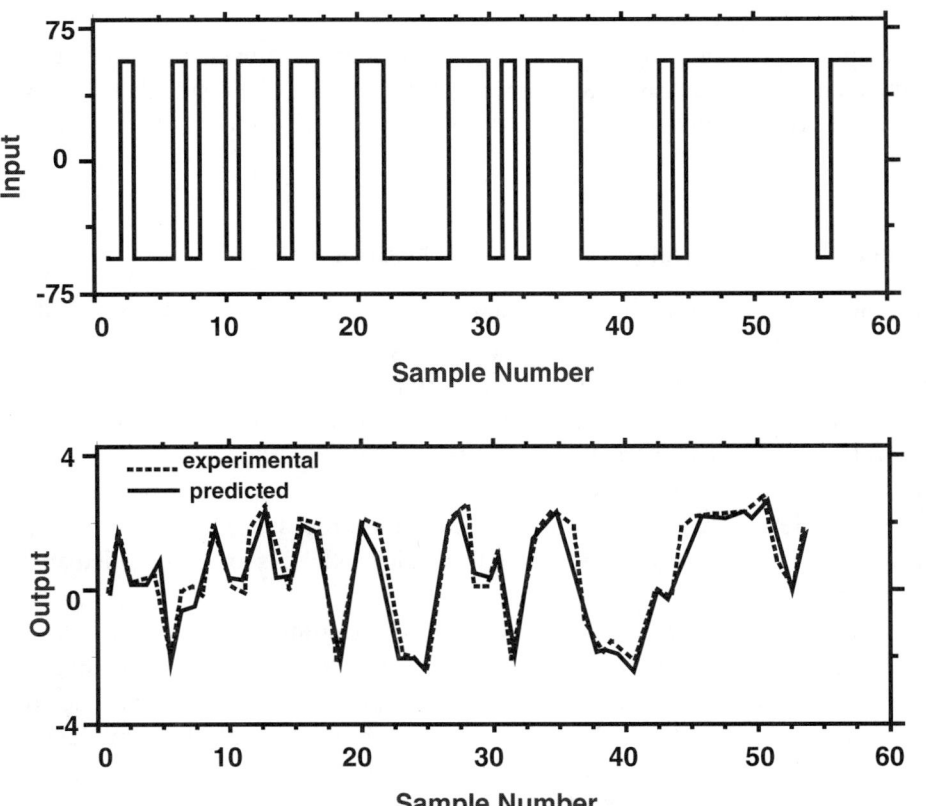

Figure 4–13 Identification experiment with a third-order ARMAX model for a pilot-size concurrent-flow grain dryer

rate (auger speed) and the output was the final grain temperature. The variables had the mean values removed: that is, the variables $u(k)$ and $y(k)$ of the measured signals $U(k)$ and $Y(k)$ were used with the mean values (direct current) $\overline{U}$ and $\overline{y}$ removed as

$$\Delta Y(k = Y(k) - Y(k-1)$$
$$\Delta Y(k) = \left[y(k) + \overline{Y} \right] - \left[y(k-1) + \overline{Y} \right]$$
$$\Delta y = y(k) - y(k-1)$$

(44)

The input $u(k)$ was a PRBS with $N = 43$ and a bit interval of one sample. A third-order ARMAX model with $d = 1$ describes the process well. The simulation data (predicted with the model) show that the model predicted the process well. This work was described in more detail by Nybrant (1986). The process model was described as

$$y(k) - 0.325y(k-1) + 9.51 \times 10^{-2} y(k-2) - 5.14 \times 10^{-3} y(k-3)$$
$$= -4.575 \times 10^{-3} u(k-1) + 1.74 \times 10^{-2} u(k-2) + 1.73 \times 10^{-2} u(k-3) + e(k) - 0.717e(k-1) + 1.27 \times 10^{-2} e(k-2)$$

(45)

The autocorrelation function of the residuals and the cross-correlation function between the residuals and the input are shown in Figure 4–14. The residuals are independent, and the third-order model is relevant to accurately describe the stochastic part (MA) of the model.

Theoretical plus Empirical (Gray Box or Hybrid) Modeling

The main advantage of gray box modeling is that it is possible to build nonlinearities in the model and to separate different parts of the process into submodels. The theoretical model generally presents incomplete and uncertain parts that can be identified by adding experimental data. Therefore, gray models can provide more understanding about the functionality and physical characteristics of the process.

To simplify the complexity that may arise when modeling a process using first principles, the development of a basic model of the process can result in an incomplete model with many uncertainties. Therefore, the basic model has to be expanded and adapted to the measured data.

The expanded modeling phase can be seen as expanding the model to a better description of the real process. The expanded procedure should be carried out in small steps so that one can analyze the influence of a certain expansion. It is also possible to expand the model in several directions, on different levels. Expansion on the same level is made by estimating more parameters or changing the influence of some variables. When it is not possible to find a feasible model by this method, some new components have to be included in the model to make the model feasible. Expansion to a higher level is done by enlarging the model with additional parts.

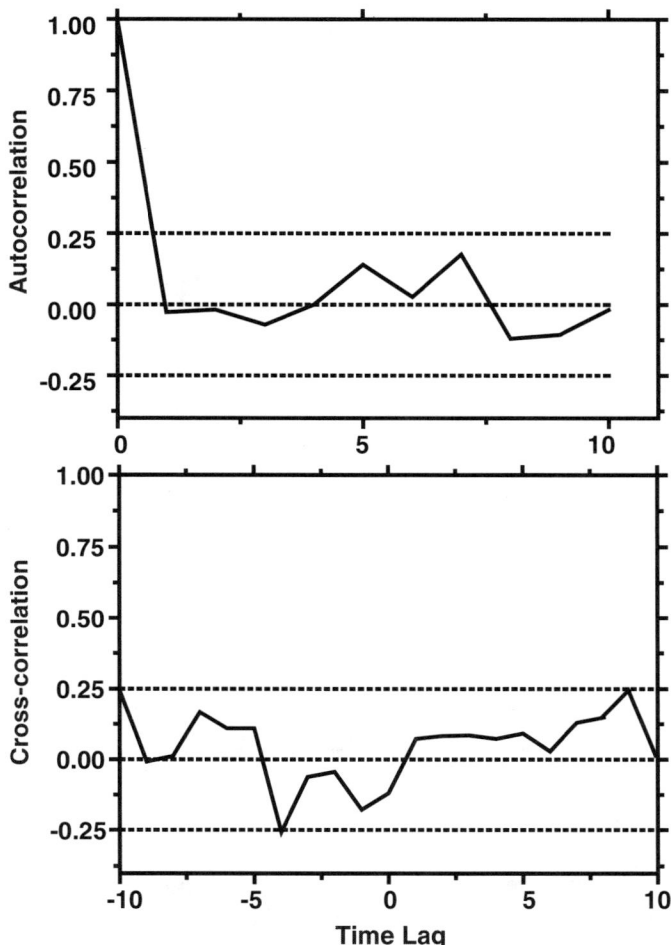

Figure 4–14 Autocorrelation function of residuals and cross-correlation function between residuals and input (with 95% confidence limits) for a third-order ARMAX model for a pilot-size concurrent-flow grain dryer

The decision on which direction to take to expand the model depends on the model analysis (as described in the previous section). This operation is repeated at several levels until the model is acceptable. The expanded modeling procedure is illustrated in Figure 4–15. There is one basic model at level 1 that is expanded further to level 2. Let us say that the model M_{23} is the most promising one and proceed to level 3. Here, the model is expanded further to the models M_{31}, M_{32}, and M_{33}. The model selected is M_{32} and the acceptable model is at level 4, M_{42}.

The expanded modeling procedure gives several models to deal with. Therefore, the results need to be compared from the model analysis. The best model then needs to be selected by taking into consideration the application, or the purpose of the model that decides the necessary characteristics of the model.

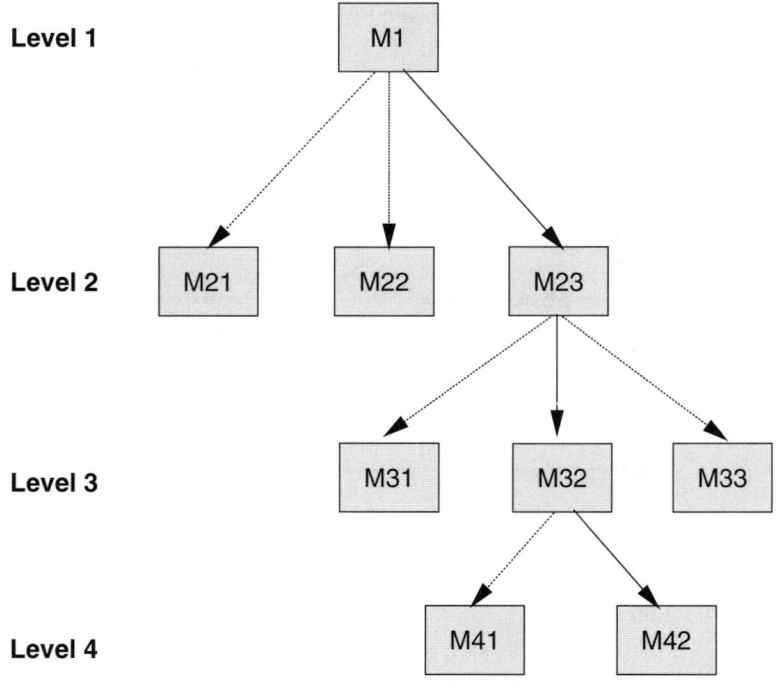

Figure 4–15 Expanded modeling schematic

NONLINEAR MODELING

Many food-processing control systems exhibit nonlinear behavior: that is, the relationship between controlled and manipulated variables depends on the operating conditions. For a process that is only mildly nonlinear (or remains close to the nominal steady state), the effects of the nonlinearities are not severe. In this case, conventional feedback control strategies can provide adequate performance. However, if the dynamic behavior of a nonlinear process system is approximated by a linear model, the model parameters will depend on the nominal operating conditions.

Due to the large variability of raw materials and the complex physicochemical reactions that occur during processing, food-processing systems are characterized by nonlinearities and uncertainties that cannot be handled by classical linear feedback, making the design of a proper controller a difficult task. These processes may be required to operate over a wide range of conditions due to large process upsets or setpoint changes.

The development of nonlinear process models is very important in application of advanced nonlinear control techniques. These models can be based on fundamental laws, an empirical observation, or a combination of both approaches.

It is difficult to derive nonlinear fundamental dynamic models for large-scale processes. An alternative approach for developing rigorous nonlinear models is to use commercial dynamic simulators such as the iGES (ABB Industrial Systems Inc., New York,

NY), SPEEDUP (Aspen Technology Inc., Camarillo, CA), and HYSYS (Hyprotech, Houston, TX) (Henson, 1998).

A fundamental difficulty associated with the empirical modeling approach is the selection of a suitable model structure. Generally, discrete-time models are most appropriate because process data are available at discrete-time instants and most of the advanced control theories are naturally formulated in discrete time. Examples of discrete-time nonlinear models are (Pearson & Ogunnaike, 1997):

- Hammerstein and Weiner models
- Volterra models
- ARMAX models

These models can be represented as nonlinear ARMAX models. For a SISO system, the nonlinear ARMAX form is

$$y(k) = F[y(k-1),\ldots y(k-n), u(k-1),\ldots u(k-m), e(k),\ldots, e(k-p+1)]$$

(46)

where F is a nonlinear mapping. The development of empirical nonlinear models for large-scale processes is very complicated, requiring the availability of suitable software tools. Neural network modeling packages, for example, are offered by several vendors, with Process Insights (Pavilion Technologies, Austin, TX) being one of the most popular (Henson, 1998).

CONCLUSION

Several approaches for modeling system dynamics were presented in this chapter. Fundamental (white box) modeling has several advantages as compared to empirical (black box) modeling. Fundamental models require fewer data for their development and can be extrapolated to operating regions that are not represented in the data set used for model development (important for processes that operate over a wide range of conditions, such as food extrusion). However, fundamental models can be very complex for use in process control design.

Empirical modeling approaches are used when there is lack of knowledge of the process dynamics. Detailed process understanding is not required for empirical modeling development. This is very important for complex process systems. System identification techniques are used to develop empirical models. These consist of several steps: (1) model selection, (2) system excitation, (3) noise modeling, (4) parameter estimation, and (5) model validation.

The combination of fundamental and empirical (hybrid or gray model) approaches allows the advantage of exploiting each modeling approach. This approach allows nonlinearities to be built in the model and makes it possible to separate different parts of the process into submodels.

Nonlinear control systems are used to control highly nonlinear processing systems—that is, those that operate near a fixed operating point—and to control moderately nonlinear process with large operating regimes. For these controllers, the development of nonlinear modeling processes is of paramount importance. Some examples of these models are presented in this chapter.

REFERENCES

Davies, W.D.T. (1970). *System Identification for Self-Adaptive Control*. New York, NY: John Wiley.

Godfrey, K.R. (1970). The application of pseudo-random sequences to industrial processes and nuclear power plants. In *Proceedings of the IFAC Symposium: Identification and Process Parameter Estimation*, pp 1–10. Prague, Czechoslovakia: IFAC.

Henson, M.A. (1998). Nonlinear model predictive control: current status and future directions. *Comput Chem Eng* 23, 187–202.

Isermann, R. (1989). *Digital Control Systems*, Vol 1. New York, NY: Springer-Verlag.

Lerew, L.E. & Bakker-Arkema, F.W. (1978). *Storage of Potatoes: A Simulation Model*. St. Joseph, MI: American Society of Agricultural Engineers. Paper No. 78-4059.

Ljung, L. & Glad, T. (1994). *Modeling of Dynamic Systems*. Englewood Cliffs, NJ: Prentice Hall.

Nybrant, T.G. (1986). *Modelling and Control of Grain Dryers*. Uppsala, Sweden: Uppsala University, Teknikum Institute of Technology. Rep. UPTEC 8625.

Pearson, R.K. & Ogunnaike, B.A. (1997). Nonlinear process identification. In *Nonlinear Process Control*. Edited by Henson & Seborg. pp 10–150. Englewood Cliffs, NJ: Prentice Hall.

Ritter, R.A. & Andre, H. (1970). Evaporator control system design. *Can J Chem Eng* 48, 696–701.

Seborg, D.E., Edgar, T.F. & Mellichamp, D.A. (1989). *Process Dynamics and Control*. New York, NY: John Wiley.

Söderström, T. & Stoica, P. (1989). *System Identification*. Englewood Cliffs, NJ: Prentice Hall.

Sohlberg, B. (1998). *Supervision and Control for Industrial Processes*. New York, NY: Springer-Verlag.

APPENDIX 4–A

DISCRETE TIME VARIABLES

THE ALGEBRA OF DISCRETE-TIME VARIABLES

Simply stated, discrete-time variables, like continuous-time variables, are added, subtracted, multiplied, and divided point by point. See the following examples:

$$\text{Let } z_1(k) = 1 \qquad\qquad \text{for } k = 0, 1, 2, 3, \ldots$$

$$= 0 \qquad\qquad \text{for } k = -1, -2, \ldots$$

(remember that $z_1(k)$ is shorthand notation for $z_1(k\Delta t)$)

$$\text{Let } z_2(k) = (-1)^k \qquad\qquad \text{for } k = 0, 1, 2, 3, \ldots$$

$$= 0 \qquad\qquad \text{for } k = -1, -2, \ldots$$

Then:

$$z_3(k) = z_1(k) + z_2(k)$$

$$= 1 + (-1)^k$$

$$= 2, 0, 2, 0, \ldots \qquad\qquad \text{for } k = 0, 1, 2, 3, \ldots$$

$$= 0 \qquad\qquad \text{for } k = -1, -2, \ldots$$

$$z_4(k) = z_1(k) - z_2(k)$$

$$= 1 - (-1)^k$$

$$= 0, 2, 0, 2, \ldots \qquad\qquad \text{for } k = 0, 1, 2, 3, \ldots$$

$$= 0 \qquad\qquad \text{for } k = -1, -2, \ldots$$

$$z_5(k) = 4z_1(k) + 3z_2(k)$$

$$= 4 + 3(-1)^k$$

$$= 7, 1, 7, 1, \ldots \qquad \text{for } k = 0, 1, 2, 3, \ldots$$

$$= 0 \qquad \text{for } k = -1, -2, \ldots$$

SOME BASIC DISCRETE-TIME VARIABLES

Commonly used discrete-time variables are the unit step, the unit ramp, the Kronecker delta, and the unit alternating sequence. These are used sometimes as test inputs for system models and sometimes as building blocks to synthesize more complex discrete-time functions. They are defined as

1.   Unit step function—$u_1(k)$

$$u_1(k) = 1 \qquad \text{for } k = 0, 1, 2, \ldots$$
$$= 0 \qquad \text{for } k = -1, -2, \ldots$$

$$\textbf{(a.1)}$$

2.   Unit ramp—$r(k)$

$$r(k) = k \qquad \text{for } k = 0, 1, 2, \ldots$$
$$= 0 \qquad \text{for } k = -1, -2, \ldots$$

$$\textbf{(a.2)}$$

3.   Kronecker delta—$\delta(k)$

$$\delta(k) = 1 \qquad \text{for } k = 0$$
$$= 0 \qquad \text{for all other } k$$

$$\textbf{(a.3)}$$

4.   Unit alternating sequence—$u_a(k)$

$$u_a(k) = (-1)^k \qquad \text{for } k = 0, 1, 2, \ldots$$
$$= 0 \qquad \text{for } k = 1, -2, \ldots$$

$$\textbf{(a.4)}$$

TIME DELAY

Simply stated, if $v(k)$ is the value of a discrete-time variable at time $k\Delta t$, then $v(k - 1)$ is the value of the variable at time $(k - i)\,\Delta t$, or i time periods earlier. Therefore, the unit-step function $u_1(k - i)$ becomes

$$
\begin{aligned}
u_1(k-1) &= 1 & &\text{for } k = i \\
&= 0 & &\text{for } k = i-1, i-2, \ldots
\end{aligned}
$$

(a.5)

APPENDIX 4–B

THE z-TRANSFORM

The z-transform approach has the same relationship to linear time-invariant discrete-time systems as the Laplace transform approach does to linear time-invariant continuous-time systems.

Consider the input sampling, where $f^*(t)$ is the sampled signal formed by a sequence of impulses:

$$f^*(t) = \sum_{k=0}^{\infty} f(k\Delta t)\delta(t - k\Delta t)$$

(b.1)

The Laplace transform of Equation b.1 is

$$F^*(s) = \sum_{k=0}^{\infty} f(k\Delta t)e^{-k\Delta t s}$$

(b.2)

Where $e^{\Delta t s} = z$, so Equation b.2 can be written as

$$F(z) = F^*(s) = F^*\left(\frac{1}{\Delta t}\ln z\right) = \sum_{k=0}^{\infty} f(k\Delta t)z^{-k}$$

(b.3)

where $F(z)$ is called the z transform of $f^*(t)$, and the notation for z-transform is $Z[\ ]$. The z-transforms of common time functions are given in Table 4–B–1.

SOLVING DIFFERENCE EQUATIONS

The solution of difference equations by the z-transform method is very useful. By using the z-transform method, we can transform difference equations into algebraic equations in z. For example, we will solve the following difference equation using the z transform:

Table 4–B–1 z Transforms

	$f(t)$ or $f(k)$	$F(s)$	$F(z)$
1	$\delta(t)$	1	1
2	$1(t)$	$\dfrac{1}{s}$	$\dfrac{z}{z-1}$
3	t	$\dfrac{1}{s^2}$	$\dfrac{\Delta t z}{(z-1)^2}$
4	$\sin \omega t$	$\dfrac{\omega}{s^2+\omega^2}$	$\dfrac{z \sin \omega \Delta t}{z^2 - 2z \cos \omega \Delta t + 1}$
5	$\cos \omega t$	$\dfrac{s}{s^2+\omega^2}$	$\dfrac{z(z - \sin \omega \Delta t)}{z^2 - 2z \cos \omega \Delta t + 1}$
6	$e^{-at}u_1(t)$	$\dfrac{1}{(s+a)}$	$\dfrac{z}{(z-e^{-a\Delta t})}$
7	$e^{-at}r_1(t)$	$\dfrac{1}{(s+a)^2}$	$\dfrac{\Delta t z e^{-a\Delta t}}{(z-e^{-a\Delta t})^2}$
8	$e^{-at}\sin \omega t$	$\dfrac{\omega}{(s+a)^2+\omega^2}$	$\dfrac{z e^{-a\Delta t} \sin \omega \Delta t}{z^2 - 2z e^{-a\Delta t} \cos \omega \Delta t + e^{-2a\Delta t}}$
9	$e^{-at}\cos \omega t$	$\dfrac{s+a}{(s+a)^2+\omega^2}$	$\dfrac{z^2 - z e^{-a\Delta t} \sin \omega \Delta t}{z^2 - 2z e^{-a\Delta t} \cos \omega \Delta t + e^{-2a\Delta t}}$
10	$\dfrac{2}{s^3}$	t^2	$\dfrac{(\Delta t)^2 z(z+1)}{(z-1)^3}$
11		ak	$\dfrac{z}{z-a}$
12	$1-e^{-at}$	$\dfrac{a}{s(s+a)}$	$\dfrac{(1-z e^{-a\Delta t})z}{(z-1)(z-e^{-a\Delta t})}$
13		$a^k \cos k\pi$	$\dfrac{z}{z+a}$

$$x(k+2)+4x(k+1)+3x(k)=0, \quad x(0)=0, \quad x(1)=1$$

(b.4)

Taking z-transforms from both sides of Equation b.4 results in

$$z^2 X(z) - z^2 x(0) - zx(1) + 4zX(z) - 4zx(0) + 3X(x) = 0$$

(b.5)

Substituting the initial conditions and simplifying gives

$$X(z) = \frac{z}{z^2 + 4z + 2} = \frac{z^{-1}}{1 + 4z^{-1} + 2z^{-2}}$$

(b.6)

PROPERTIES OF THE z-TRANFORM

The Final Value Theorem

If $f(t)$ has the z-transform $F(z)$ and $F(z)$ has no double or higher order poles on the unit circle centered at the origin of the z plane and no poles outside the unit circle, then the final value of $f(t)$ or $f(k)$ is given by

$$\lim_{k \to \infty} f(k) = \lim_{t \to \infty} f(t) = \lim_{z \to 1}[(z-1)F(z)]$$

(b.7)

Initial Value Theorem

If $f(t)$ has the z-transform $F(z)$ and $\lim_{z \to \infty} F(z)$ exists, then the initial value $f(0)$ of $f(t)$ or $f(k)$ is given by

$$f(0) = \lim_{z \to \infty} F(z)$$

(b.8)

Linearity

$$Z[af_1(k) + bf_2(k)] = aZ[f_1(k)] + bZ[f_2(k)]$$

(b.9)

Shifting to the Right

The discrete-time function $f(k)$ is shifted to the right by d sampling time Δt_o (shift into the past):

$$Z[f(k-d)] = z^{-d} f(z) \quad d \geq 0$$

(b.10)

Shifting to the Left

The discrete-time function $f(k)$ is shifted to the left (shift to the future):

$$Z[f(k+d)] = z^{-d} \left[f(z) - \sum_{q=0}^{d-1} f(q)z^{-q} \right] \quad d \geq 0$$

(b.11)

Damping

The effect of damping (degree of oscillation in the process) is

$$Z[f(k)e^{-ak\Delta t_o}] = f(ze^{a\Delta t_o})$$

(b.12)

THE INVERSE z-TRANSFORMATION

There are three methods of obtaining the inverse z transform. They are infinite power series expansion, partial fraction expansion, and inverse integral.

Infinite Power Series Expansion Method

If $F(z)$ is expanded into a convergent power series in z^{-1}, as

$$F(z) = \sum_{k=0}^{\infty} f(k\Delta t)z^{-k} = f(0) + f(\Delta t)z^{-1} + f(2\Delta t)z^{-2} + \ldots + f(k\Delta t)z^{-k} + \ldots$$

(b.13)

then the values of $f(k\Delta t)$ can be determined by inspection. For example, find the $f(k\Delta t)$ for $k = 1, 2, 3$ when $F(z)$ is

$$F(z) = \frac{10z}{(z-1)(z-2)}$$

(b.14)

$F(z)$ can be written as

$$F(z) = \frac{10z^{-1}}{1 - 3z^{-1} + 2z^{-2}}$$

(b.15)

By long division,

$$F(z) = 10z^{-1} + 30z^{-2} + 70z^{-3} + \ldots$$

(b.16)

This infinite series converges; therefore, by inspection we obtain

$$f(0) = 0$$
$$f(\Delta t) = 10$$
$$f(2\Delta t) = 30$$
$$f(3\Delta t) = 70$$

(b.17)

Partial Fraction Expansion Method

Consider $F(z)$ as

$$F(z) = \frac{\beta_0 z^n + \beta_1 z^{n-1} + \ldots + \beta_{n-1} z + \beta_n}{\alpha_0 z^m + \alpha_1 z^{m-1} + \ldots + \alpha_{m-1} + \alpha_m} \quad (n \le m)$$

(b.18)

First we factor the denominator polynomial of $F(z)$ and find the poles of $F(z)$. Then we expand $F(z)/z$ into small fractions. The inverse z-transform of $F(z)$ is obtained as the sum of the inverse z-transforms of the partial fractions. For example, determine the inverse z-transform of the $F(z)$ function given by Equation b.14 using the partial fraction expansion method:

By first expanding $F(z)/z$ into partial fractions, we get

$$\frac{F(z)}{z} = \frac{10}{(z-1)(z-2)} = \frac{-10}{z-1} + \frac{10}{z-2}$$

(b.19)

Then,

$$F(z) = \frac{10z}{z-1} + \frac{10z}{z-2}$$

(b.20)

From Table 4–B–1, we obtain

$$z^{-1}\left[\frac{z}{z-1}\right] = 1, \quad z^{-1}\left[\frac{z}{z-2}\right] = 2^k$$

(b.21)

So

$$f(k\Delta t) = 10(-1 + 2^k) \qquad (k = 0,1,2,\ldots)$$

(b.22)

$$f(0) = 0$$
$$f(\Delta t) = 10$$
$$f(2\Delta t) = 30$$
$$f(3\Delta t) = 70$$

(b.23)

Inverse Integral Method

By multiplying both sides of Equation b.13 by z^{k-1}, we obtain

$$F(z)z^{k-1} = f(0)z^{k-1} + f(\Delta t)z^{k-2} + f(2\Delta t)z^{k-3} + \ldots + f(k\Delta t)z^{-1} + \ldots$$

(b.24)

Suppose we integrate both sides of Equation b.24 in a counterclockwise direction:

$$\oint F(z)z^{k-1}dz = \oint f(0)z^{k-1}dz + \oint f(\Delta t)z^{k-2}dz + \oint f(2\Delta t)z^{k-3}dz + \ldots + \oint f(k\Delta t)z^{-1}dz + .$$

(b.25)

Applying Cauchy's theorem, we see that all the terms of the right-hand side of this last equation are zero except one term:

$$\oint f(k\Delta t)z^{-1}dz$$

(b.26)

Hence

$$\oint F(z)z^{k-1}dz = \oint f(k\Delta t)z^{-1}dz$$

(b.27)

And we obtain

$$x(k\Delta t) = \frac{1}{2\pi}; \quad \oint F(z)z^{k-1}dz$$

(b.28)

Equation b.28 is the inversion integral of the z-transform and is equivalent to stating that

$$f(k\Delta t) = \sum[\text{residues of } F(z)z^{k-1} \text{ at the poles of } F(z)]$$

(b.29)

For example, obtain $f(k\Delta t)$ by use of the inversion integral when $F(z)$ is

$$F(z) = \frac{10z}{(z-1)(z-2)}$$

(b.30)

From Equations b.28 and b.29 we have

$$f(k\Delta t) = \frac{1}{2\pi j}\oint\left[\frac{10z}{(z-1)(z-2)}z^{k-1}\right]dz = \frac{1}{2\pi j}\oint\left[-\frac{10z^k}{z-1}+\frac{10z^k}{z-2}\right]dz$$

(b.31)

$$(k\Delta t) = \sum \left[\left(\text{residues of } -\frac{10}{z-1} z^k \text{ at the pole } z = 1 \right) + \left(\text{residues of } -\frac{10}{z-3} z^k \text{ at the pole } z = 2 \right) \right]$$

(b.32)

$$(k\Delta t) = 10(-1 + 2^k) \quad (k = 0,1,2,\ldots)$$

(b.33)

APPENDIX 4–C

RECURSIVE PARAMETER ESTIMATION IN AN OPEN LOOP

Consider a linear, stable, time-invariant process that can be described by a linear stochastic z-transfer function with constant parameters:

$$y(k) = \underbrace{\frac{B(z^{-1})}{A(z^{-1})} z^{-d} u(k)}_{G_p(z^{-1})} + \underbrace{\frac{D(z^{-1})}{A(z^{-1})} v(k)}_{G_v(z^{-1})}$$

(c.1)

with the polynomials:

$$A(z^{-1}) = 1 + a_1 z^{-1} + \ldots + a_n z^{-n}$$
$$B(z^{-1}) = b_1 z^{-1} + \ldots + b_n z^{-n}$$
$$D(z^{-1}) = 1 + d_1 z^{-1} + \ldots + d_p z^{-p}$$

(c.2)

where, in Equation c.1, G_p is the process model and G_v is the noise model. For the estimation of the unknown parameters of Equation c.2, the following model is assumed:

$$y(z) = \frac{\hat{B}(z^{-1)}}{\hat{A}(z^{-1})} z^{-d} u(z) + \frac{\hat{D}(z^{-1})}{A(z^{-1})} e(z)$$

(c.3)

where the circumflex accent represents an estimate, and $e(z)$ corresponds to $e(k)$, the equation error, or the one-step-ahead prediction error due to unmeasurable noise $v(k)$.

For recursive parameter estimation in an open loop, several methods can be used, including (1) recursive least squares (RLS), (2) recursive extended least squares (RELS), (3) recursive instrumental variable (RIV), and (4) recursive maximum likelihood (RML) (Isermann, 1989).

For example, the least squares method is based on the minimization of the following loss function:

$$V = \sum_k e^2(k)$$

(c.4)

due to the unknown parameter vector $\hat{\boldsymbol{\theta}}$. The various parameter estimations can be written in a unified form as (Ljung & Glad, 1994):

$$\hat{\boldsymbol{\theta}}(k+1) = \hat{\boldsymbol{\theta}}(k) + \boldsymbol{\gamma}(k)e(k+1)$$

(c.5)

$$\boldsymbol{\gamma}(k) = \mu(k+1)\boldsymbol{P}(k)\boldsymbol{\varphi}(k+1)$$

(c.6)

$$e(k+1) = y(k+1) - \boldsymbol{\phi}'(k+1)\hat{\boldsymbol{\theta}}(k)$$

(c.7)

The definitions of $\hat{\boldsymbol{\theta}}$, $\boldsymbol{\phi}'$, $\boldsymbol{\varphi}$, and $\boldsymbol{P}$ depend on the parameter estimation method. Table 4–C–1 gives the values of those parameters for the RLS, RELS, and RML methods. For RML and RELS, Equation c.3 is assumed. In the case of the RLS method, $\hat{D}(z^{-1}) = 1$.

To obtain unbiased estimates ($E\{\hat{\theta}(N)\} = \boldsymbol{\theta}_0$) and consistent estimates in mean square—that is,

$$\lim_{N\to\infty} E\{\hat{\boldsymbol{\theta}}(N)\} = \hat{\boldsymbol{\theta}}_0 \quad \text{and} \quad \lim_{N\to\infty} E\left\{[\hat{\boldsymbol{\theta}}(N) - \hat{\boldsymbol{\theta}}_0][\hat{\boldsymbol{\theta}}(N) - \boldsymbol{\theta}_0]'\right\} = \boldsymbol{0}$$

(c.8)

the following conditions have to be satisfied (Isermann, 1989):

1. The process order n and dead time d are known.
2. Direct current values U and Y are known.
3. The process input must be persistently exciting of order $m \geq n$.
4. The error signal $e(k)$ is not correlated with $\boldsymbol{\phi}'(k)$ (ie, $e(k)$ must be uncorrelated).
5. $E\{e(k)\} = 0$.

If the process parameters are not constant, a forgetting factor can be used to modify the parameter estimation algorithms (Isermann, 1989).

Table 4–C–1 Parameters for RLS, RELS, and RML Estimation Methods

Method	$\hat{\theta}$	$\phi'(k+1)$	$\phi'(k+1)$	$\mu(k+1)$	$P(k+1)$
RLS	$[a_1, \ldots, a_n, b_1, \ldots, b_n]'$	$[-y(k), \ldots, -y(k-n+1), u(k-d), \ldots, u(k-d-m+1)]$	$\phi'(k+1)$	$\dfrac{1}{\lambda(k+1) + \phi'(k+1)P(k)\phi(k+1)}$	$\dfrac{[I - \gamma(k)\phi'(k+1)]P(k)}{\lambda(k+1)}$
RELS	$[a_1, \ldots, a_n, b_1, \ldots, b_m, d_1, \ldots, d_p]'$	$[-y(k), \ldots, -y(k-n+1), u(k-d), \ldots, u(k-d-m+1), e(k), \ldots, e(k-p+1)]$	$\phi'(k+1)$	as RELS	as RELS
RML	as RELS	as RELS	$[-y^*(k), \ldots, -y^*(k-n+1), u^*(k-d), \ldots, u^*(k-d-p+1), e^*(k), \ldots, e^*(k-p+1)]$	$\dfrac{1}{\lambda(k+1) + \phi^*(k+1)P(k)\phi(k+1)}$	$\dfrac{[I - \gamma^*(k)\phi'(k+1)]P(k)}{\lambda(k+1)}$

Note: $y^*(z) = \dfrac{1}{D(z^{-1})}\, y(z)$; $u^*(z) = \dfrac{1}{D(z^{-1})}\, u(z)$; $e^*(z) = \dfrac{1}{D(z^{-1})}\, e(z)$, and λ is the forgetting factor.

THE RECURSIVE PREDICTION ERROR METHOD

The recursive prediction error (RPE) method as described by Ljung and Söderström (1986) is presented below. The RPE is based on a stochastic Gauss-Newton algorithm and can be expressed as follows. Consider the discrete-time SISO ARMAX model:

$$A(z^{-1})y(z) = B(z^{-1})u(z) + C(z^{-1})e(z)$$

(c.9)

with

$$A(z^{-1}) = 1 + a_1 z^{-1} + \ldots + a_n z^{-n}$$
$$B(z^{-1}) = b_1 z^{-1} + \ldots + b_m z^{-m}$$
$$C(z^{-1}) = 1 + c_1 z^{-1} + \ldots c_p z^{-p}$$

(c.10)

The predicted error (e) is equal to

$$e(k) = Y(k) - \hat{Y}(k)$$

(c.11)

$$S(k) = \varphi'(k)P(k-1)\varphi(k) + \lambda(k)$$

(c.12)

The gain vector (γ) is equal to

$$\gamma(k) = P(k-1)\varphi(k)S^{-1}(k)$$

(c.13)

$$\theta(k) = [a_1 \ldots a_n \,|\, b_1 \ldots b_m \,|\, c_1 \ldots c_p]$$

(c.14)

$$\hat{\theta}(k) = \hat{\theta}(k-1) + \gamma(k)e(k)$$

(c.15)

$$P(k) = \frac{[P(k-1) - \gamma(k)S(k)\hat{\gamma}'(k)]}{\lambda}$$

(c.16)

The residual (ε) is equal to

$$\varepsilon(k) = Y(k) - \hat{c}_1(k)\varepsilon(k-1) - \ldots - \hat{c}_p(k)\varepsilon(k-p)$$

(c.17)

$$\phi'(k+1) = [-Y(k)\ldots - Y(k-n+1)\,|\,U(k)\ldots U(k-m+1)\,|\,\varepsilon(k)\ldots\varepsilon(k-p+1)]$$

(c.18)

$$\hat{Y}(k+1) = \hat{\theta}'(k)\phi(k+1)$$

(c.19)

The filtered signals ($\tilde{Y}$, $\tilde{U}$, $\tilde{\varepsilon}$) are equal to

$$\tilde{Y}(k) = Y(k) - \hat{c}_1\tilde{Y}(k-1) - \ldots - \hat{c}(k)\tilde{Y}(k-p)$$

(c.20)

$$\tilde{U}(k) = U(k) - \hat{c}_1\tilde{U}(k-1) - \ldots - \hat{c}(k)\tilde{U}(k-p)$$

(c.21)

$$\tilde{\varepsilon}(k) = Y(k) - \hat{c}_1\tilde{\varepsilon}(k-1) - \ldots - \hat{c}(k)\tilde{\varepsilon}(k-p)$$

(c.22)

$$\varphi'(k+1) = [-\tilde{Y}(k)\ldots - \tilde{Y}(k-n+1)\,|\,\tilde{U}(k)\ldots\tilde{u}(k-m+1)\,|\,\tilde{\varepsilon}(k-p+1)]$$

(c.23)

The RPE method is based on the minimization of the loss function due to the unknown parameter vector $\hat{\theta}$ (the estimation of the model parameters):

$$LF(k) = \sum_{s=1}^{k} \frac{e^2(s)}{\lambda(k) + \varphi'(s)P(k-1)\varphi(k)}$$

(c.24)

where $LF(k)$ is the loss function and e is the error. The loss function is defined as the sum of the square prediction errors that here is modified to include uncertainties in the transient phase: that is, the use of Equation c.23 allows the estimator to track the time-varying dynamics of the process. In the RPE algorithm used, the covariance matrix P is updated using the U-D factorization algorithm (Thornton & Bierman, 1980) as

$$P = UDU'$$

where D is a diagonal matrix and U an upper diagonal with 1's in the diagonal.

A computer program written to estimate the parameters of Equation c.9 contains the following steps:

- Set the initial conditions at $k = 0$,

$$P(I,I) = 1; \quad P(I,J) = 0$$
$$\hat{\boldsymbol{\theta}}(0) = 0; \quad \lambda(k) = \lambda = 0.98$$

- Compute the prediction error: Equation c.22.
- Update the parameter estimates: Equation c.15.
  To ensure that $C(z)$ contains only zeros inside the unit circle, a stability test needs to be performed in a separate subroutine.
- Compute the residuals: Equation c.17.
- Compute the filtered signals: Equations c.20 through c.22.
- Update the vectors $\boldsymbol{\phi}(k)$ and $\boldsymbol{\varphi}(k)$: Equations c.18 and c.23.
- Compute the gain vector $\gamma(k)$ and update $P(k)$ and $LF(k)$: Equations c.13, c.16, and c.24.

APPENDIX 4–D

PSEUDORANDOM BINARY SIGNALS (PRBSs)

The PRBS is one of the most useful types of periodic signal for process identification. The signal has two levels $\pm V$, with an autocorrelation function, as shown in Figure 4–D–1. A PRBS is based on a pseudorandom binary sequence of length N.

The properties of the PRBS are:

- The signal has two levels and can be switched from one level to another only at certain points $t = 0, \Delta t, 2\Delta t, ...,$
- The PRBS is deterministic, and experiments are repeatable.
- The signal is periodic, with $T = N\Delta t$ where N is an odd integer.
- There are $\frac{1}{2}(N + 1)$ intervals when the signal is at one level and $\frac{1}{2}(N - 1)$ intervals when it is at the other level in any one period.

Before describing the methods used to generate a PRBS, a short introduction to perturbation signals is given below.

THE IMPULSE RESPONSE MODELING

Consider the time-invariant dynamical system with control input $u(t)$ and measured response $y(t)$ contaminated by noise $v(t)$ (Figure 4–D–2). The true process response cannot be measured due to the effect of random disturbances originating in the process, instrumentation, raw material properties, and so on. These effects are lumped together in the noise variable $v(t)$.

For a process unit impulse response $h(t)$, the convolution integral relates the process input and output: that is,

$$x(t) = \int_0^\infty h(\lambda)u(t - \lambda)d\lambda$$

(d.1)

where λ is a time variable and $h(t)$ the weighting function. If the system is time invariant, $h(t)$ is also equal to the unit impulse response of the system. The discrete form of Equation d.1 at time t_r is

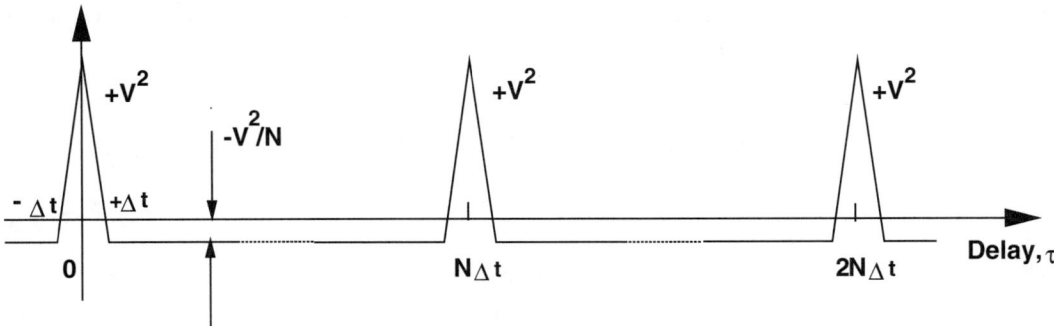

Figure 4–D–1 Autocorrelation signal of a PRBS

$$x(t_r) = \sum_{i=0}^{m} h(i\Delta T)u(t_r - i\Delta T)$$

(d.2)

Adding the noise term in Equation d.2 results in

$$y(j) = \left[\sum_{i=0}^{m} h(i)u(j-i) + v(j) \right], \text{ for } j = 0,1,\ldots,m$$

(d.3)

We define the discrete form of input autocorrelation function as

$$R_{uu}(k) = \frac{1}{N-k} \sum_{j=1}^{N-\tau} u(j)u(j+k)$$

(d.4)

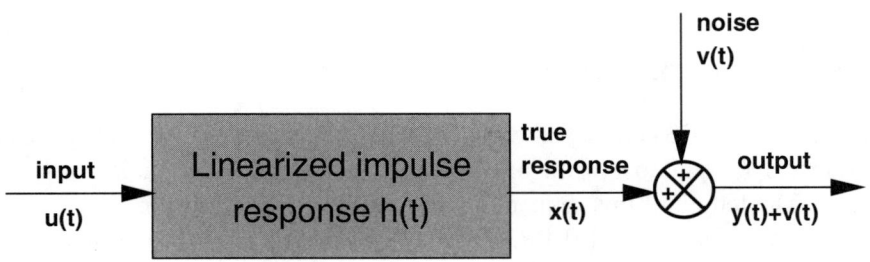

Figure 4–D–2 Block diagram of a system

where N = total number of samples and $\tau = 0, \ldots, m$. Multiply both sides of Equation d.1 by $u(t)$ and correlating gives

$$R_{uy}(i) = \sum_{j=0}^{m} h(j)R_{uu}(i - j) \text{ for } i = 0,1,2,\ldots,m$$

(d.5)

where $R_{uy}(i)$ is the discrete form of the cross-correlation function relating the control input $u(t)$ and output $y(t)$. In Equation d.5, it is assumed that $u(t)$ and $v(t)$ are uncorrelated ($R_{uv}(k) = 0$).

Equation d.5 is the Wiener Hopf equation. If the input autocorrelation function R_{uu} and the cross-correlation function R_{uy} are known, the discrete response vector representing $h(t)$ can be obtained from Equation d.5 that represents a series of $(m + 1)$ simultaneous equations. If $u(t)$ is a pseudorandom binary signal, then R_{uu} approximates a unit impulse at $t = 0$: that is, $R_{uy}(i)=h(j)$ for $j = i \neq 0$, $R_{uy}(i) = \frac{1}{2} h(j)$ for $j = 0$.

PRBSs can be generated using shift registers (Godfrey, 1993). This method is based on maximum-length sequences (m sequences). Binary m sequences exist for $N = 2^n - 1$, where n is an integer (>1). These sequences can be generated using an n-stage shift register with feedback to the first stage consisting of the modulo-2 sum of the logic value of the last stage and one or more of the other stages (Godfrey, 1993). The binary logic values are usually taken as 1 and 0, and modulo-2 addition is given by

$$1 \oplus 1 = 0 \oplus 0 = 0$$
$$1 \oplus 0 = 0 \oplus 1 = 1$$

(d.6)

The sequence logic values can be transformed to signal voltage levels by converting either $1 \rightarrow +V, 0 \rightarrow -V$ or $1 \rightarrow -V, 0 \rightarrow +V$.

In addition to m sequences, other classes of binary sequences on which PRBSs can be based are quadratic residue codes (that exist for $N = 4k - 1$, where k is an integer and N is prime: ie, N = 3, 7, 11, 19, 23, 31, 43 . . .), twin prime sequences (that exist for $N = k(k + 2)$ where N = 15, 25, . . .) , and Hall sequences (that exist for $N = 4k^2 + 27$ where N = 31, 43, 127, . . .) (Everett, 1966).

MULTILEVEL PSEUDORANDOM SIGNALS (MLPRSs)

Most of the MLPRSs are based on m sequences. Multilevel m sequences exist for the number of levels, l, equal to a prime or a power of a prime p (>1): that is, for l = 2, 3, 4, 5, 6, 7, 8, 9, . . . The length of this sequence is l^n-1, where n is an integer.

Consider, for example, an MLPRS based on three-level m sequences with conversions from logic values to signal levels either by $0 \rightarrow 0, 1 \rightarrow +V, 2 \rightarrow -V$ or by $0 \rightarrow 0, 1 \rightarrow -V$, $2 \rightarrow +V$. Properties of these sequences are the following:

- The signal has three levels, 0, $+V$, and $-V$, and can be switched from one level to another only at certain points $t = 0, \Delta t, 2\Delta t, \ldots$.
- The signal is deterministic, and experiments are repeatable.
- The signal is periodic, with $T = N\,\Delta t$ where $N = 3^n - 1$, N is an integer.
- There are 3^{n-1} intervals when the signal has level $+V$, 3^{n-1} intervals when it has level $-V$, $3^{n-1}-1$ intervals when it has level 0.
- The second half is the negative of the first half.
- The autocorrelation function of the signal is shown in Figure 4–D–3.

REFERENCES

Everett, D. (1966). Period digital sequences with pseudonoise properties. *GEC J Sci Technol* 33, 115–126.

Godfrey, K. (1993). *Perturbation Signals for System Identification.* New York, NY: Prentice Hall.

Isermann, R. (1989). *Digital Control Systems.* Vol 1. New York, NY: Springer-Verlag.

Ljung, L. & Glad, T. (1994). *Modeling of Dynamic Systems.* Englewood Cliffs, NJ: Prentice Hall.

Ljung, L. & Söderström, T. (1986). *Theory and Practice of Recursive Identification.* Cambridge, MA: MIT Press.

Thornton, C.L. & Bierman, G.J. (1980). UDU covariance factorization for Kalman filtering. In *Control and Dynamic Systems.* Vol 16. New York, NY: Academic Press.

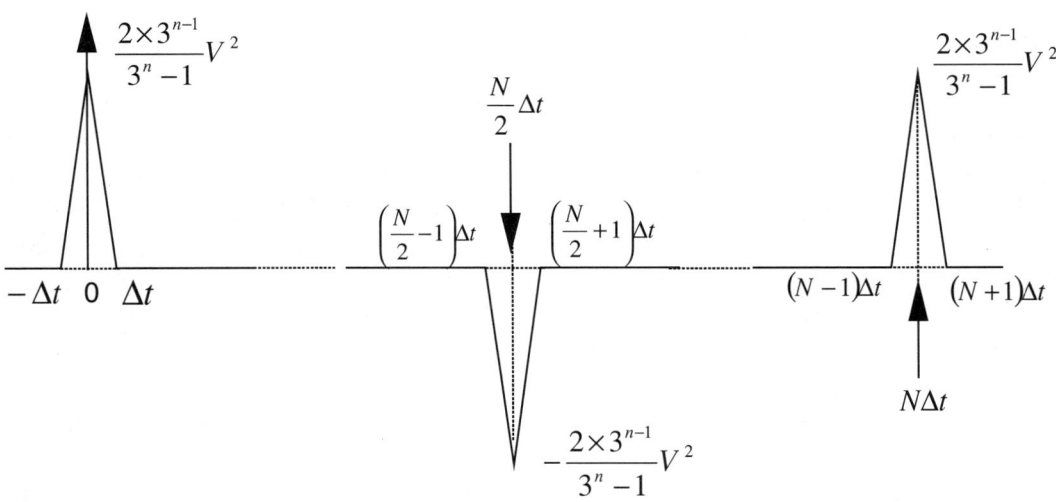

Figure 4–D–3 Autocorrelation function of a signal based on a three-level *m* sequence.

CHAPTER 5

Multivariable Systems

Multivariable control systems are those that have multiple inputs and/or outputs. Most of the control systems discussed up to now dealt with single-input, single-output (SISO) systems. In practice, most food-processing systems are multiple-input, multiple-output (MIMO) systems. MIMO control problems are inherently more complex than SISO control problems because process interactions occur between controlled and manipulated variables. Control loops can interact with one another when there is more than one manipulated and/or controlled variable. The analysis and control of such interaction require mathematical understanding and computational capabilities.

In this chapter, the concepts of control loop interactions, relative gain array, and ways of decoupling loops are discussed.

INTERACTING LOOPS

Control loops interact with each other, and it is not uncommon for a change in a manipulated variable to produce changes in more than one controlled variable. Consider, for example, the blending system shown in Figure 5–1. The system mixes two streams, and both the total flow out of the mixing tank and the composition of the outlet mixture need to be controlled. There are two manipulated variables, the two streams m_1 and m_2. A change in m_1 or/and m_2 will change both the outlet flow and the outlet composition. If you use u_1 to control the outlet flow, the interaction of the two loops is obvious; similarly for u_2.

Several questions arise. Should you use the flow controller to adjust m_1 or m_2? What is the nature of the interaction between the loops once the variables have been paired? Can you decouple the loops? Figure 5–2 illustrates the interaction problem in this 2×2 process.

In general, in the case of MIMO control problems, a change of manipulated variable—for example, m_1—will affect all the controlled variables c_1, c_2, ..., c_n. Due to the process interactions, it is a difficult task to select the best pairing of controlled and manipulated variables for a multiloop control. For instance, if you have a control problem with 5 controlled variables and 5 manipulated variables, there are 120 possible multiloop control configurations.

Consider the 2×2 control system shown in Figure 5–2. The s-domain relations for the linear process can be written as

158

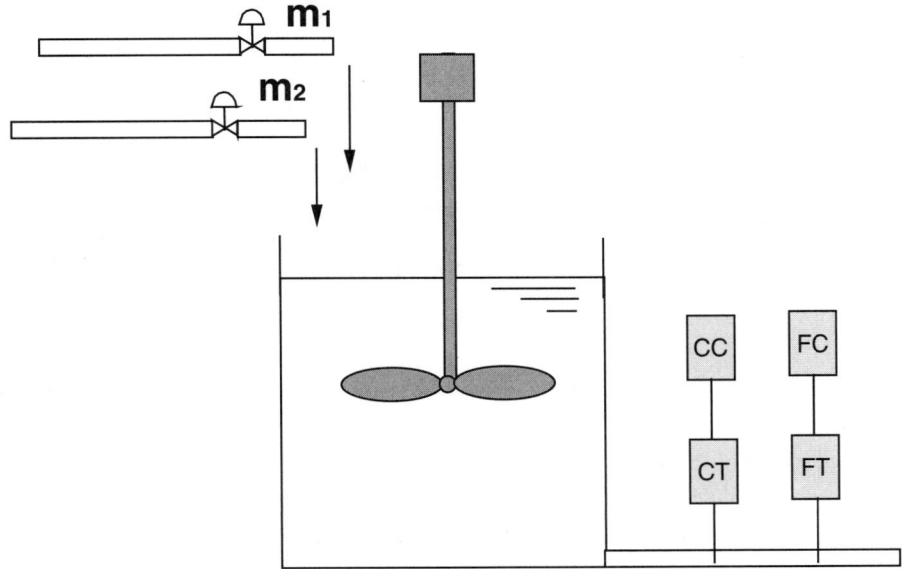

Figure 5–1 A blending system

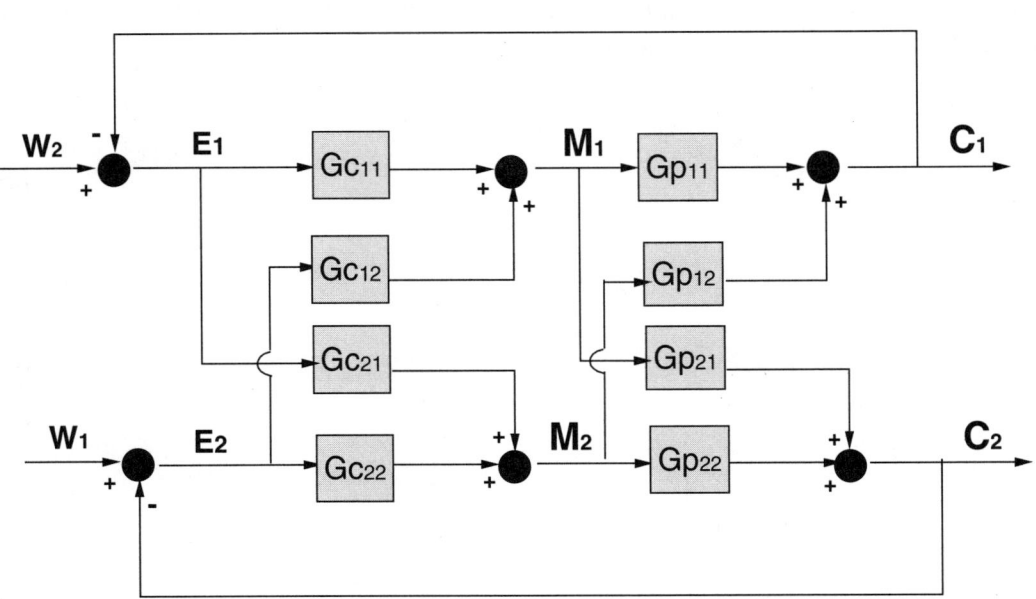

Figure 5–2 Interactions between loops for a two-variable control system

$$C_1(s) = G_{p11}(s)M_1(s) + G_{p12}(s)M_2(s)$$
$$C_2(s) = G_{p21}(s)M_1(s) + G_{p22}(s)M_2(s)$$

(1)

The subscript p in the equations designates process. Equation 1 can be expressed in vector notation as

$$C(s) = G_p(s)M(s)$$

(2)

where $C(s)$ is an m vector, $M(s)$ is a p vector, and the matrix transfer function $G_p(s)$ has m rows and p columns:

$$C(s) = \begin{bmatrix} C_1(s) \\ C_2(s) \end{bmatrix} \quad M(s) = \begin{bmatrix} M_1(s) \\ M_1(s) \end{bmatrix} \quad G_p(s) = \begin{bmatrix} G_{p11}(s) & G_{p12}(s) \\ G_{p21}(s) & G_{p22}(s) \end{bmatrix}$$

(3)

Similarly, we can write an equation for a set of controllers by a matrix transfer function $G_c(s)$ relating the error $E(s)$ of dimension m and variable vector $M(s)$:

$$M(s) = G_c(s)E(s)$$

(4)

where $G_c(s)$ has p rows and m columns. An open-loop matrix transfer function $G(s)$ can be derived by substituting $M(s)$ of Equation 4 into Equation 2:

$$C(s) = G_p(s)G_c(s)E(s) = G(s)E(s)$$

(5)

where

$$G(s) = G_p(s)G_c(s)$$

(6)

where $G(s)$ is the $m \times m$ square matrix (omitting s for simplified notation):

$$\begin{bmatrix} G_{11} & G_{12} \\ G_{21} & G_{22} \end{bmatrix} = \begin{bmatrix} G_{c11}G_{p11} + G_{c21}G_{p12} & G_{c12}G_{p11} + G_{c22}G_{p12} \\ G_{c11}G_{p21} + G_{c21}G_{22} & G_{c12}G_{p21} + G_{c22}G_{p22} \end{bmatrix}$$

(7)

Introducing the *m* vector $W(s)$ as the setpoint vector:

$$E(s) = W(s) - C(s)$$

(8)

For a unit-feedback system [$H(s) = 1$], we can get

$$C(s) = G(s)W(s) - G(s)C(s) \quad \text{or} \quad C(s) = [I + G(s)]^{-1}G(s)W(s)$$

(9)

The characteristic equation of the closed-loop system is

$$\det[I + G(s)] = 0$$

(10)

and the closed-loop matrix transfer function for the setpoint vector is

$$G_{LR}(s) = [I + G(s)]^{-1}G(s)$$

(11)

The dimension of the W and C vectors must be the same since the number of setpoints and measurements are the same. However, the number of manipulated variables does not need to be the same as the number of measurements.

BRISTOL'S RELATIVE GAIN ARRAY

One of the most important problems of multivariable control systems is how controlled and manipulated variables should be paired. An incorrect pairing can result in a controller with poor performance and reduced stability margins.

The most used method for studying interaction is the relative gain array (RGA) proposed by Bristol (1966). The RGA can be used to best pair controlled and manipulated variables and can also help to determine whether decoupling should be undertaken. The main advantages of the RGA technique are that it is easy to calculate and requires only steady-state gain information (Seborg et al, 1989).

The RGA is defined as the ratio of steady-state gain between the *i*th controlled variable and the *j*th manipulated variable when all other manipulated variables are constant, divided by the steady-state gain between the same two variables when all other controlled variables are constant:

$$\gamma_{ij} = \frac{\left[\partial C_i / \partial M_j\right]_M}{\left[\partial C_i / \partial M_j\right]_C} = \frac{\text{open} - \text{loop gain}}{\text{closed} - \text{loop gain}}$$

(12)

for $i = 1, 2, \ldots, n$ and $j = 1, 2, \ldots, n$. In Equation 12, $[\partial C_i/\partial M_j]_M$ is the closed-loop gain between C_i and M_j. The term $[\partial C_i/\partial M_j]_C$ is the closed-loop gain indicating the effect of M_j on C_i when all of the other feedback control loops are closed. The array of the relative gains, the RGA, can be written as

$$\Gamma = \begin{array}{c} \\ C_1 \\ C_2 \\ \cdots \\ C_3 \end{array} \begin{array}{cccc} M_1 & M_2 & \cdots & M_n \\ \left[\begin{array}{cccc} \gamma_{11} & \gamma_{12} & \cdots & \gamma_{1n} \\ \gamma_{21} & \gamma_{22} & \cdots & \gamma_{2n} \\ \cdots & \cdots & \cdots & \cdots \\ \gamma_{n1} & \gamma_{n2} & \cdots & \gamma_{nn} \end{array}\right] \end{array}$$

(13)

The properties of the RGA are: (1) it is normalized—that is, the sum of the elements in each row or column is 1, and (2) the relative gains are dimensionless.

To calculate the relative gains, consider a 2×2 linearized process with steady-state gains K_{ij}:

$$C_1 = K_{11}M_1 + K_{12}M_2$$
$$C_2 = K_{21}M_1 + K_{22}M_2$$

(14)

Or in matrix notation:

$$C = KM$$

(15)

For this system, the gain between C_1 and M_1 when M_2 is constant is

$$\left[\frac{\partial C_1}{\partial M_1}\right]_{M_2} = K_{11}$$

(16)

The gain between C_1 and M_1 when C_2 is constant ($C_2 = 0$) is calculated as

$$C_1 = K_{11}M_1 + K_{12}M_2$$
$$0 = K_{21}M_1 + K_{22}M_2$$

(17)

$$C_1 = K_{11}\left[1 - \frac{K_{12}K_{21}}{K_{11}K_{22}}\right]M_1$$

(18)

It follows that

$$\left[\frac{\partial C_1}{\partial M_1}\right]_{C_2} = K_{11}\left(1 - \frac{K_{12}K_{21}}{K_{11}K_{22}}\right)$$

(19)

Substituting Equation 16 and Equation 19 into Equation 12 results in an expression for the relative gain

$$\gamma_{11} = \frac{1}{1 - \dfrac{K_{12}K_{21}}{K_{11}K_{22}}}$$

(20)

So the RGA for a 2×2 system can be expressed as

$$\boldsymbol{\Gamma} = \begin{bmatrix} \gamma & 1-\gamma \\ 1-\gamma & \gamma \end{bmatrix}$$

(21)

For higher dimension processes, the RGA is calculated as

$$\gamma_{ij} = K_{ij}H_{ij}$$

(22)

where K_{ij} is the (i,j) element of $\boldsymbol{K}$ in Equation 15 and H_{ij} is the (i,j) element of $\boldsymbol{H} = (\boldsymbol{K}^{-1})^1$. Note that the superscript (^{1}) indicates transpose in this book.

For example, consider the mixing system shown in Figure 5–1 (Shrinskey, 1979). The component mass balance can be expressed as

$$x = \frac{m_1}{m_1 + m_2}$$

(23)

and the overall mass balance as

$$w = m_1 + m_2$$

(24)

The RGA for the mixing system can be written as

$$\Gamma = \begin{matrix} & m_1 & m_2 \\ w_1 \\ x_2 \end{matrix} \begin{bmatrix} \gamma & 1-\gamma \\ 1-\gamma & \gamma \end{bmatrix}$$

(25)

The relative gain γ can be calculated from Equation 16 and Equation 20 as

$$K_{11} = \left[\frac{\partial w}{\partial m_1} \right]_{m_2} = 1$$

(26)

$$K_{12} = \left[\frac{\partial w}{\partial m_2} \right]_{m_1} = 1$$

(27)

$$K_{21} = \left[\frac{\partial x}{\partial m_1} \right]_{m_2} = \frac{m_2}{(m_1 + m_2)^2} = \frac{1-x}{w}$$

(28)

$$K_{22} = \left[\frac{\partial x}{\partial m_2} \right]_{m_1} = \frac{-m_1}{(m_1 + m_2)^2} = -\frac{x}{w}$$

(29)

So, from Equation 21, $\gamma = x$, and the RGA is

$$\Gamma = \begin{matrix} w \\ x \end{matrix} \overset{\displaystyle \begin{matrix} m_1 & \quad m_2 \end{matrix}}{\left[\begin{matrix} x & 1-x \\ 1-x & x \end{matrix} \right]}$$

(30)

indicating that the recommended pairing depends on the desired product composition x.

According to Bristol's recommendation, manipulated and controlled variables should be paired so that the corresponding gains are positive and as close to 1 as possible. It is important to note that RGA is based only on steady-state gains; thus, one must consider the dynamic effects of the pairings indicated by the RGA analysis before final selection of loops.

To illustrate the use of RGA to determine controller pairing, consider, for example, the twin-screw food extrusion process described by Figure 2–1. This is a 3×3 system. The three manipulated variables are feed rate (FR), water rate (WR) or moisture in the barrel (MIB), and screw speed (SS). The three controlled variables are the two product quality attributes (PQAs) collet moisture (CM) and color (CB) and one process variable, product temperature at the die (PTDIE). Figure 5–3 shows the block diagram of the process. The results presented here are from the work of Schonauer and Moreira (1997).

The most common way to adjust the extrusion process manually is to make MIB constant when adjusting FR (adjusting water simultaneously) and to adjust MIB independently by changes in WR while maintaining FR constant. However, when one is implementing automatic control schemes, it may be desirable to have FR and WR independent of each other. In fact, MIB as an independent variable may complicate the process, since MIB is dependent on raw material moisture, FR, and WR.

RGAs were constructed using the steady-state gains given in Tables 5–1 and 5–2. Separate RGAs were constructed for each of the step sets conducted. Table 5–3 contains

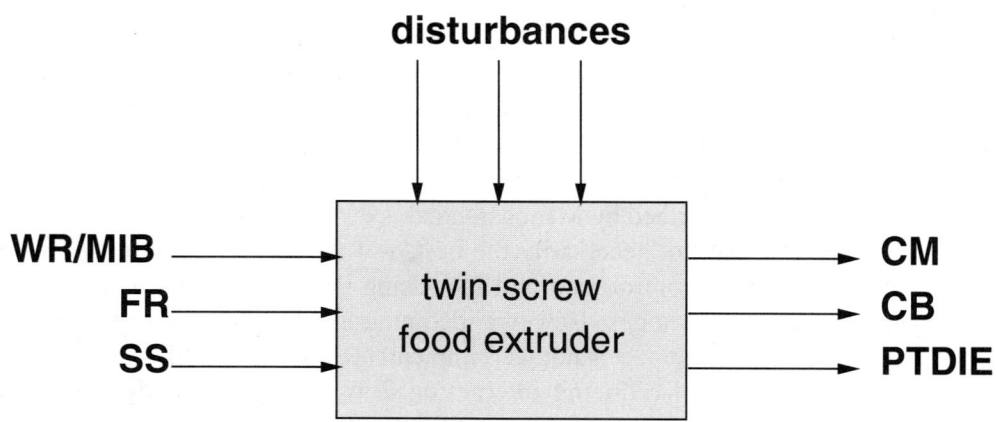

Figure 5–3 Block diagram of 3×3 food extrusion process

data obtained from the smallest positive step tests. The PQAs of interest, CB and CM, along with process variable PTDIE, were included in the analysis.

Table 5–3 (FR and MIB independent) indicates the possible pairings between CM-MIB, CB-FR, and PTDIE-SS. However, there are interactions between the chosen loops (pairs) that would require decouplers (interaction is more severe as values become greater than 1). The RGA for FR and MIB independent for all three sets of step tests indicated the same pairings and need for decouplers.

Table 5–3 (for FR and WR independent) also shows possible pairings between CM-SS, CB-FR, and PTDIE-WR, with severe interaction between the latter two loops. The RGAs from the other sets of step tests (different step sizes) when FR and WR were independent did not replicate these same pairings; in fact, for the negative step tests, there were no possible pairings because of negative elements in the array.

From the results of the RGA analyses, we conclude that FR-MIB had less severe interactions. The difference between these tables was the use of MIB as a manipulated variable formed by combining manipulated variables FR and WR. Combinations of manipulated variables can potentially reduce loop interactions.

If FR is eliminated as a manipulated variable (to maximize throughput), WR and MIB are essentially equivalent. If WR/MIB and SS are used to control the PQAs CM and CB, the resulting 2×2 RGAs from all six sets of step tests indicate that CM-WR and CB-SS could be paired. However, in each case the paired elements had values greater than 1, which would indicate the need for decouplers or use of a multivariable controller.

Overall, the results indicate that a multivariable controller should be used for this process to maximize performance. These results point to multivariable control instead of multiloop control.

REDUCING INTERACTIONS

For some control problems, loop interactions can be significantly reduced by choosing alternative controlled and manipulated variables (Seborg et al, 1989). For example, the new controlled or manipulated variable could be a simple function of the original variables, such as a sum, difference, or ratio. If this does not work, decouplers or multivariable control systems are required. Decouplers that are designed to reduce interaction compromise performance due to their effect of essentially detuning the loop. One could also consider block pairings of inputs and outputs that are not necessarily SISO pairings by using the generalized RGA method described by Manousiouthakis et al (1986).

The most common, but not necessarily the best, way to deal with interaction is by simply detuning (decreasing control gain and increasing the time constant) and, in the extreme, opening one or more loops. However, detuning a controller can result in poor performance of the detuned loop if it is not well implemented (Seborg et al, 1989).

A more powerful way of reducing interaction is by changing manipulated and controlled variables. This approach requires insight, intuition, and analytical study. Consider, for example, the blending system discussed above. The RGA for the system is

Table 5–1 Effect of Step Changes in Moisture in Barrel (MIB) and Screw Speed (SS) on Color-b (CB), Collet Moisture Content (CM), and Product Temperature (PTDIE)

Step in MIB[a] (%)	Gain			Step in SS[b] (rpm)	Gain		
	CM (%)/(%)	*CB* []/(%)	*PTDIE* (°C)/(%)		*CM* (%)/(rpm)	*CB* []/(rpm)	*PTDIE* (°C)/(rpm)
18.5 to 19.2	0.270	3.340	−3.000	470 to 485	−0.004	−0.081	0.068
19.2 to 18.5	0.260	3.520	−2.980	485 to 470	−0.003	−0.064	0.071
18.5 to 19.6	0.210	2.200	−3.200	470 to 500	−0.002	−0.024	0.055
19.6 to 18.5	0.210	2.100	−3.230	500 to 470	−0.001	−0.001	0.055
18.5 to 17.3	0.300	4.300	−4.310	470 to 440	−0.003	−0.060	0.056

a. FR = 47.7 kg/h; BT5 = 135°C; SS = 470 rpm.
b. FR = 47.7 kg/h; BT5 = 135°C; MIB = 18.5%.

Table 5–2 Effect of Step Changes in Feed Rate (FR) (with MIB Constant and WR Constant) on Color-b (CB), Collet Moisture Content (CM), and Product Temperature (PTDIE)

Step in FR[a] (kg/h)	Gain			Step in FR[b] (kg/h)	Gain		
	CM (%)/(kg/h)	*CB* []//(kg/h)	*PTDIE* (°C)//(kg/h)		*CM* (%)/(kg/h)	*CB* []//(kg/h)	*PTDIE* (°C)//(kg/h)
54.5 to 47.7	0.008	0.200	−0.082	54.5 to 47.7	−0.030	−0.272	0.250
47.7 to 61.4	0.006	0.177	−0.122	47.7 to 61.4	−0.023	−0.235	0.240
61.4 to 47.7	0.005	0.128	−0.122	61.4 to 47.7	−0.025	−0.223	0.200
47.7 to 40.9	0.016	0.397	−0.246	47.7 to 40.9	−0.027	−0.243	0.330
40.9 to 47.7	0.016	0.389	−0.239	40.9 to 47.7	−0.036	−0.270	0.330

a. MIB = 18.5%; BT5 = 135°C; SS = 470 rpm.
b. WR = 3.5 kg/h; BT5 = 135°C; MIB = 18.5%.

Table 5–3 RGA Using Gains from Small Positive Steps

Output/Input	FR[a]	MIB	SS
CM	0.18	**2.68**	−1.86
CB	**1.81**	−1.03	0.22
PTDIE	−1.00	−0.65	**2.65**

Output/Input	FR[b]	WR	SS
CM	4.07	−3.74	**0.67**
CB	**12.81**	−32.25	20.44
PTDIE	−15.88	**36.99**	−20.11

a. WR adjusted to maintain MIB constant.
b. WR maintained constant.

presented by Equation 30. If $x = 0.5$, the interaction is terrible. To eliminate this interaction, a new set of manipulated variables is selected:

$$M_1 = m_1 + m_2$$

(31)

$$M_2 = m_1$$

(32)

The model for the process reduces to

$$w = M_1$$

(33)

$$x = \frac{M_2}{M_1}$$

(34)

The open-loop gain matrix is

$$\begin{bmatrix} w \\ x \end{bmatrix} = \begin{bmatrix} 1 & 0 \\ \dfrac{1}{M_1} & \dfrac{M_1}{M_2^2} \end{bmatrix} \begin{bmatrix} M_1 \\ M_2 \end{bmatrix}$$

(35)

and the RGA is

$$\Gamma = \begin{matrix} & M_1\,M_2 \\ \begin{matrix} w \\ x \end{matrix} & \begin{bmatrix} 1 & 0 \\ 0 & 1 \end{bmatrix} \end{matrix}$$

(36)

From Equation 36 it follows that M_1 should be paired with w and that M_2 should be paired with x. The implementation of this control scheme is shown in Figure 5–4 (similar to Figure 5–1 but without the tank).

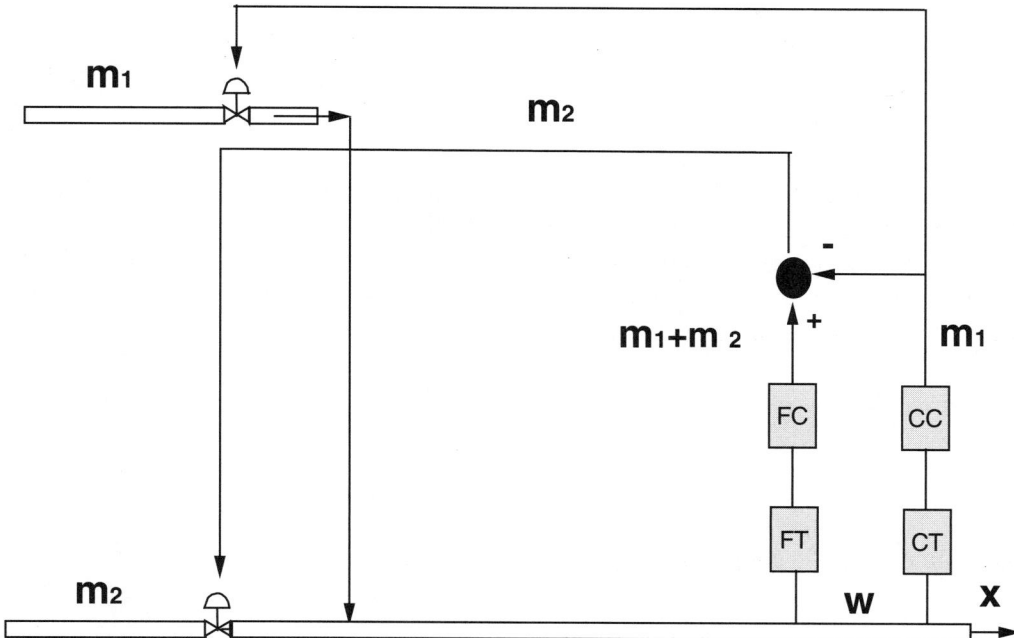

Figure 5–4 A partially decoupled blending system

DECOUPLING

The decoupling approach was developed to achieve simplicity in the design problem. Two important benefits can arise from decoupling control schemes (Seborg et al, 1989): (1) closed-loop stability can be determined by the stability characteristics of individual feedback control loops (no interactions), and (2) a setpoint change for one controlled variable will not affect other controlled variables.

In the case of the system shown in Figure 5–2, decoupling means that each input $M_i(s)$ must affect one, and only one, output $C_i(s)$. For these conditions to be satisfied, the off-diagonal elements of matrix $G(s)$, and $G_{LR}(s)$, must be zero. In Equation 7, this condition will be satisfied if $G_{12}(s) = 0$ and $G_{21}(s) = 0$. Two elements of the controller matrix are therefore fixed by the following conditions:

$$G_{c12}(s) = -\frac{G_{p12}(s)}{G_{p11}(s)}G_{c22}(s), \quad G_{c21}(s) = -\frac{G_{p21}(s)}{G_{p22}(s)}G_{c11}(s)$$

$$(37)$$

When Equation 37 is satisfied in a two-variable system, the decoupled open-loop transfer functions are obtained by substituting Equation 37 in Equation 7. These functions are

$$G_{11}(s) = -\frac{\det G_p(s)}{G_{p22}(s)} G_{c11}(s), \quad G_{22}(s) = -\frac{\det G_p(s)}{G_{p11}(s)} G_{c22}(s)$$

(38)

The system is now reduced to two noninteracting single loops, so $G_{c11}(s)$ and $G_{c22}(s)$ can be easily designed using the frequency method, for instance.

For an example of a decoupling application, consider a 2×2 system. The basic concept is to place a decoupling controller $\boldsymbol{D}$ between the feedback controllers and the process $\boldsymbol{G_p}$ (Figure 5–5). The basic model for this process is

$$\begin{bmatrix} C_1 \\ C_2 \end{bmatrix} = \begin{bmatrix} G_{p11} & G_{p12} \\ G_{p21} & G_{p22} \end{bmatrix} \begin{bmatrix} M_1 \\ M_2 \end{bmatrix}$$

(39)

It is assumed that the gain matrix $\boldsymbol{G_p}$ vary between 0 and 1. The decoupler model is

$$\begin{bmatrix} M_1 \\ M_2 \end{bmatrix} = \begin{bmatrix} d_{11} & d_{12} \\ d_{21} & d_{22} \end{bmatrix} \begin{bmatrix} m_1 \\ m_2 \end{bmatrix}$$

(40)

Substituting Equation 40 into Equation 39 results in

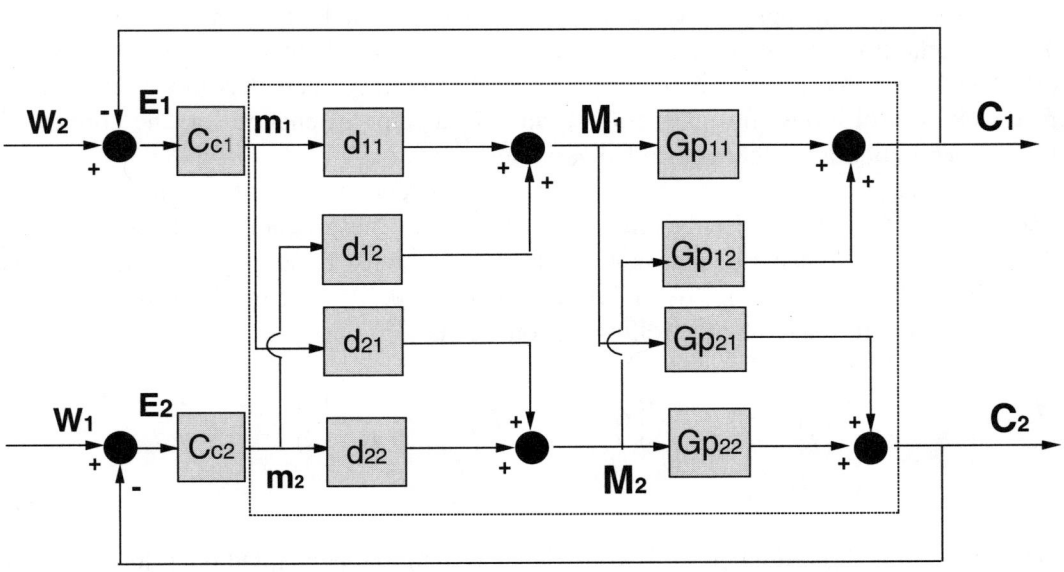

Figure 5–5 A decoupling control system

$$\begin{bmatrix} C_1 \\ C_2 \end{bmatrix} = \begin{bmatrix} b_{11} & b_{12} \\ b_{21} & b_{22} \end{bmatrix} \begin{bmatrix} m_1 \\ m_2 \end{bmatrix}$$

(41)

where

$$\boldsymbol{B} = \begin{bmatrix} d_{11}G_{p11} + d_{21}G_{p12} & d_{12}G_{p11} + d_{22}G_{p12} \\ d_{11}G_{p21} + d_{21}G_{p22} & d_{12}G_{p21} + d_{22}G_{p22} \end{bmatrix}$$

(42)

In Figure 5–5, the block within the dotted line can be reviewed as a new process with a gain matrix $\boldsymbol{B}$. If either b_{12} or b_{21} is zero, the system is partially coupled. If both b_{12} and b_{21} are zero, complete decoupling is achieved. For simplified decoupling, let d_{11} and d_{22} be 1. Then, Equation 41 can be solved with $b_{12} = b_{21} = 0$. The resulting decoupler is (Figure 5–6):

$$\boldsymbol{D} = \begin{bmatrix} 1 & \dfrac{-G_{p12}}{G_{p11}} \\ \dfrac{-G_{p21}}{G_{p22}} & 1 \end{bmatrix}$$

(43)

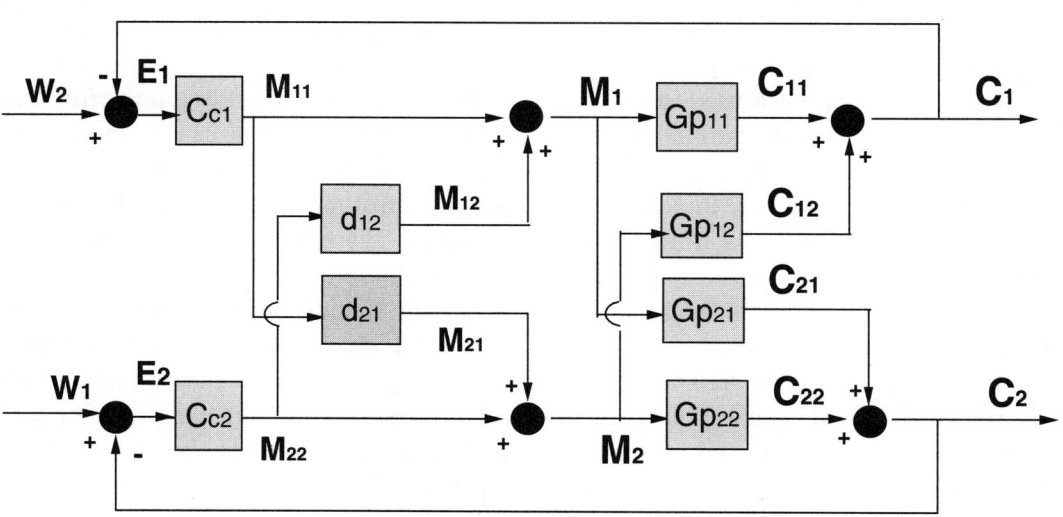

Figure 5–6 Simplified decoupling system

This decoupler can be interpreted as a type of feedforward controller with an input signal that is a manipulated variable rather than a disturbance variable.

MULTIVARIABLE CONTROL

Multivariable control is defined as a class of control strategy in which each manipulated variable is adjusted on the basis of errors in all the controlled variables. Consider, for example, a 2×2 process with a multivariable proportional control strategy:

$$M_1(t) = K_{c11}E_1(t) + K_{c12}E_2(t)$$

(44)

$$M_2(t) = K_{c21}E_1(t) + K_{c22}E_2(t)$$

(45)

The multivariable control system shown by Equations 44 and 45 can be reduced to a multiloop control system if $K_{c12} = K_{c21} = 0$ or $K_{c11} = K_{c22} = 0$. In these cases, each manipulated variable is adjusted on the basis of a single error signal, unlike multivariable control systems.

For additional information on multivariable control systems, the reader is referred to Takahashi et al (1972), Friedland (1986), and Isermann (1989).

STATE-SPACE REPRESENTATION OF SYSTEMS

Unlike classical control theory, modern control theory is applicable to MIMO systems. Modern control theory is essentially a time-domain approach, while classical control theory is a complex frequency-domain approach.

The design approach used in classical control theory is based on trial and error that generally will not yield optimal control systems. However, system design in modern control theory enables the engineer to design optimal control systems with respect to given performance indexes. In addition, in modern control theory, initial conditions can be included in the design, and it can be carried out for any class of inputs (instead of a specific input such as impulse, step, or sinusoidal functions).

The state-variable approach is a direct approach to multivariable control system design and analysis. A dynamic system consisting of a finite number of lumped elements may be represented by ordinary differential equations. By use of vector-matrix notation, an nth-order differential equation may be expressed as a first-order vector-matrix differential equation. If n elements of the vector are a set of state variables, then the vector-matrix differential equation is called a *state equation*. Consider the following second-order system:

$$\ddot{y} + a_1\dot{y} + a_2 y = u$$

(46)

Let us define:

$$x_1 = y$$
$$x_2 = \dot{y}$$
$$x_3 = \ddot{y}$$

(47)

So Equation 47 can be written as

$$\dot{x}_1 = x_2$$
$$\dot{x}_2 = x_3 = -a_2 x_1 - a_1 x_2 + u$$

(48)

Therefore, the state equation is

$$\dot{x} = Ax + Bu$$

(49)

where

$$x = \begin{bmatrix} x_1 \\ x_2 \end{bmatrix}, \quad A = \begin{bmatrix} 0 & 1 \\ -a_2 & -a_1 \end{bmatrix}, \quad B = \begin{bmatrix} 0 \\ 1 \end{bmatrix}$$

(50)

The output equation is

$$y = \begin{bmatrix} 1 & 0 \end{bmatrix} \quad \text{or} \quad y = Cx$$

(51)

where

$$C = \begin{bmatrix} 1 & 0 \end{bmatrix}$$

(52)

Transfer Matrix

The concept of a transfer matrix is an extension of the transfer function (defined in Chapter 3). The transfer matrix $G(s)$ relates to the output $Y(s)$ to the input $U(s)$ as

$$Y(s) = G(s)U(s)$$

$$(53)$$

If the input vector u is n dimensional and the output vector y is m dimensional, the transfer matrix is an $m \times n$ matrix. The state-space representation for this system is:

$$\dot{x} = Ax + Bu$$
$$y = Cx + Du$$

$$(54)$$

The Laplace transform of Equation 54 is

$$sX(s) - x(0) = AX(s) + BU(s)$$
$$Y(s) = CX(s) + DU(s)$$

$$(55)$$

Assuming that $x(0)$ is zero in Equation 55 results in

$$X(s) = (sI - A)^{-1}BU(s)$$

$$(56)$$

Substituting Equation 56 into Equation 55 we obtain

$$Y(s) = [C(sI - A)^{-1}B + D]U(s)$$

$$(57)$$

By comparing Equation 57 with Equation 53, we obtain the transfer matrix for MIMO systems as

$$G(s) = C(sI - A)^{-1}B + D$$

$$(58)$$

Transfer Matrix of Closed-Loop Systems

Consider the MIMO system shown in Figure 5–7. The transfer matrix of the feedforward path is $G_o(s)$, and that of the feedback path is $H(s)$. The transfer matrix between the feedback signal vector $B(s)$ and the error vector $E(s)$ is obtained as follows:

$$B(s) = H(s)Y(s) = H(s)G_0(s)E(s)$$

(59)

Then the transfer matrix between $B(s)$ and $E(s)$ will be $H(s)G_o(s)$. Thus, the transfer matrix of the cascade elements is the product of the transfer matrices of the individual elements.

The transfer matrix of the closed-loop system is obtained as

$$Y(s) = G_0(s)[U(s) - B(s)] = G_0(s)[U(s) - H(s)Y(s)]$$

(60)

So we obtain

$$Y(s) = [I + G_o(s)H(s)]^{-1}G_o(s)U(s)$$

(61)

The closed-loop transfer matrix $G(s)$ is then given by

$$G(s) = [I + G_0(s)H(s)]^{-1}G_0(s)$$

(62)

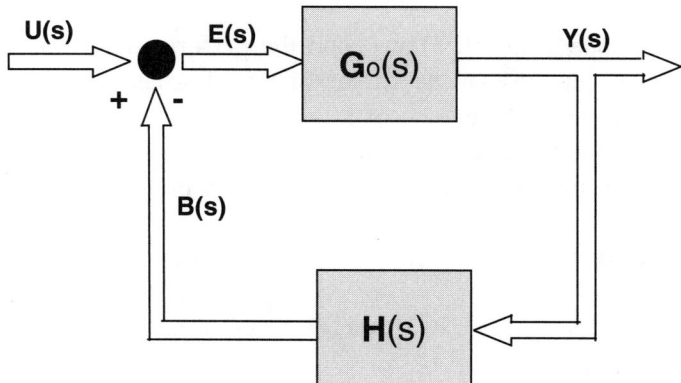

Figure 5–7 Block diagram of a multiple-input, multiple-output system

State-Space Representation of Discrete-Time Systems

The most general state-space representation of linear time-invariant, deterministic, discrete-time systems is

$$x(k+1) = Ax(k) + Bu(k)$$

(63)

$$y(k) = Cx(k) + D(k)$$

(64)

where $x(k)$ is an $(m \times 1)$ state vector, $u(k)$ a $(p \times 1)$ control vector, $y(k)$ an $(r \times 1)$ output vector, A an $(m \times m)$ systems matrix, B an $(m \times p)$ control matrix, C an $(r \times m)$ output matrix, and D an $(r \times p)$ input-output matrix.

The state representation of multivariable systems has several advantages over the transfer matrix notation. For instance, the analysis and design of controllers for SISO processes can easily be extended to MIMO processes, and arbitrary internal structures with a minimal number of parameters can also be described.

OPTIMAL CONTROL

Problems of optimal control received a great deal of attention during the 1960s due to increasing demand for systems of high performance and to the availability of digital control (Ogata, 1970).

Control system optimization involves selection of a performance index and a design that yields the optimal control system within limits imposed by physical constraints.

Performance Index (Cost Function)

A performance index indicates how good the performance of a system is. Design of an optimal control depends on the choice of a performance index to be minimized (or maximized). In general, the choice of a performance index involves a compromise between analytical feasibility and practical application. Therefore, it is very difficult to select the most appropriate performance index for a given problem, especially for complex systems.

Consider, for example, the quadratic performance index:

$$J = \int_0^\infty [\tilde{x}(t) - x(t)]Q[\tilde{x}(t) - x(t)]dt$$

(65)

where $\tilde{x}(t)$ is the desired state, x(t) the actual state, and Q a positive-definite matrix. Note that $[\tilde{x}(t) - x(t)]$ is the error vector.

The Optimal Control Problem

Given a system state equation and output equation, a control vector, constraints of the problem, a performance index, and system parameters, the optimal control problem is to determine the optimal vector $u(t)$ over the interval $0 < t < \infty$ that will transfer the system from an initial state $x(0)$ to some final state $x(\infty)$ in such a manner as to minimize a performance index. The vector $u(t)$ generally depends on (1) initial state or initial output, (2) desired state or desired output, (3) nature of the constraints, and (4) nature of the performance index.

A comprehensive review of optimal control is beyond the scope of this work. The reader is referred to Ogata (1970), Takahashi et al (1972), and Chen (1970). The objective here is to introduce this control theory and show with an example how it can be applied to food-processing systems.

Optimal State-Feedback Controller for a Rotary Dryer

Consider the rotary dryer system shown in Figure 3–39 (Mann, 1980). The drying process can be described by the following state-space model (in discrete time):

$$x(k+1) = Ax(k) + bu(k) + gv(k)$$

(66)

$$y(k) = Mx(k)$$

(67)

where $x(k)$ is the n-dimensional state vector, $y(k)$ the s-dimensional output vector, $u(k)$ the control variable, $v(k)$ the disturbance variable, b the n-dimensional control distribution vector, g the n-dimensional disturbance distribution vector, A the $n \times n$—dimensional system matrix, and M the $s \times n$—dimensional output matrix.

Based on Equations 66 and 67, the model for the rotary dryer can then be formulated as

$$\begin{bmatrix} x_m(k+1) \\ x_n(k+1) \end{bmatrix} = \begin{bmatrix} A_M & 0 \\ 0 & A_n \end{bmatrix} \begin{bmatrix} x_M(k) \\ x_n(k) \end{bmatrix} + \begin{bmatrix} b_M \\ 0 \end{bmatrix} u(k) + \begin{bmatrix} 0 \\ b_n \end{bmatrix} v(k)$$

(68)

$$y(k) = \begin{bmatrix} M_M & M_n \end{bmatrix} \begin{bmatrix} x_M(k) \\ x_n(k) \end{bmatrix}$$

$$(69)$$

where x_M, the mass flow rate of the fuel, is the manipulated variable and x_n, revolutions of the screw conveyor, is the disturbance variable. The controlled variable (y_R, or dry matter) is calculated as

$$y_R(k) = \begin{bmatrix} c'_M & c'_n \end{bmatrix} \begin{bmatrix} x_M(k) \\ x_n(k) \end{bmatrix} = c'x(k)$$

$$(70)$$

where y_R is contained in the vector y.

We shall now consider the optimal control problem of, given the system Equations 68 through 70, determining the vector K of the optimal control vector:

$$\Delta u(k) = K' \Delta x(k)$$

$$(71)$$

so as to minimize the performance index:

$$J = \Delta x'(N) Q \Delta x(N) + \sum_{k=0}^{N-1} [\Delta x'(k) Q \Delta x(k) + \Delta u'(k) R \Delta u(k)]$$

$$(72)$$

where Q is the weighting matrix of the state variable x, R is a weighting factor of the control variable u, and $\Delta u(k) = u(k) - \bar{u}$ and $\Delta x(k) = x(k) - \bar{x}$ where $\bar{u}$ and $\bar{x}$ are steady-state values to eliminate remaining control errors after a step change of the disturbance variable v.

The steady-state values can be calculated by the following equations:

$$\bar{x}_n = \begin{bmatrix} I & A_n \end{bmatrix}^{-1} b_n \bar{v}$$

$$(73)$$

$$\bar{u} = \frac{-c'_n \bar{x}_n}{c'_M (I - A_M)^{-1} b_M}$$

$$(74)$$

$$\overline{x}_M = \begin{bmatrix} I & A_M \end{bmatrix}^{-1} b_M \overline{u}$$

(75)

In this example, the state variables Equations 68 through 70 cannot be measured (they have no physical meaning). However, the matrices A and M and vectors b and g (from Equation 66) are completely known. The problem, then, is that of estimating $x(t)$ from the available input u and output y with the knowledge of the matrices A and M and vectors b and g.

The unknown state variables can be estimated recursively by a state-variable filter (Kalman filter) that measures u and y and applies the algorithm (Isermann, 1989):

$$\hat{x}(k+1) = A\hat{x}(k) + bu(k) + gv(k) + H(k+1)[y(k+1) - MA\hat{x}(k)]$$

(76)

where $H(k+1)$ is the correction matrix. Assuming that the disturbance is unmeasurable, $v(k)$ also must be reconstructed by the estimator. Equation 76 can be extended to:

$$\begin{bmatrix} \hat{x}_M(k+1) \\ \hat{x}_n(k+1) \\ \hat{v}(k+1) \end{bmatrix} = \begin{bmatrix} A_M & 0 & 0 \\ 0 & A_n & b_n \\ 0 & 0 & 1 \end{bmatrix} \begin{bmatrix} \hat{x}_M(k) \\ \hat{x}_n(k) \\ \hat{v}(k) \end{bmatrix} + \begin{bmatrix} b_M \\ 0 \\ 0 \end{bmatrix} u(k) +$$

$$H' \left[y(k) - [M_M, M_N, 0] \begin{bmatrix} \hat{x}_M(k) \\ \hat{x}_n(k) \\ \hat{v}(k) \end{bmatrix} \right]$$

(77)

The values of H and H' are determined using the transpose equation of the state-space models (Isermann, 1989). The final structure of the state-feedback control system is shown in Figure 5–8.

CONCLUSION

In this chapter we discussed multivariable systems and techniques used to decouple multiloops. One way of reducing interaction among loops is to select different manipulated or controlled variables.

The relative gain array approach can be used to determine the degree of process interaction and the best pairing of controlled and manipulated variables for a multiloop control scheme.

A convenient way of formulating multivariable systems is by using state-variable approach. System design using state-space representation enables the engineer to design

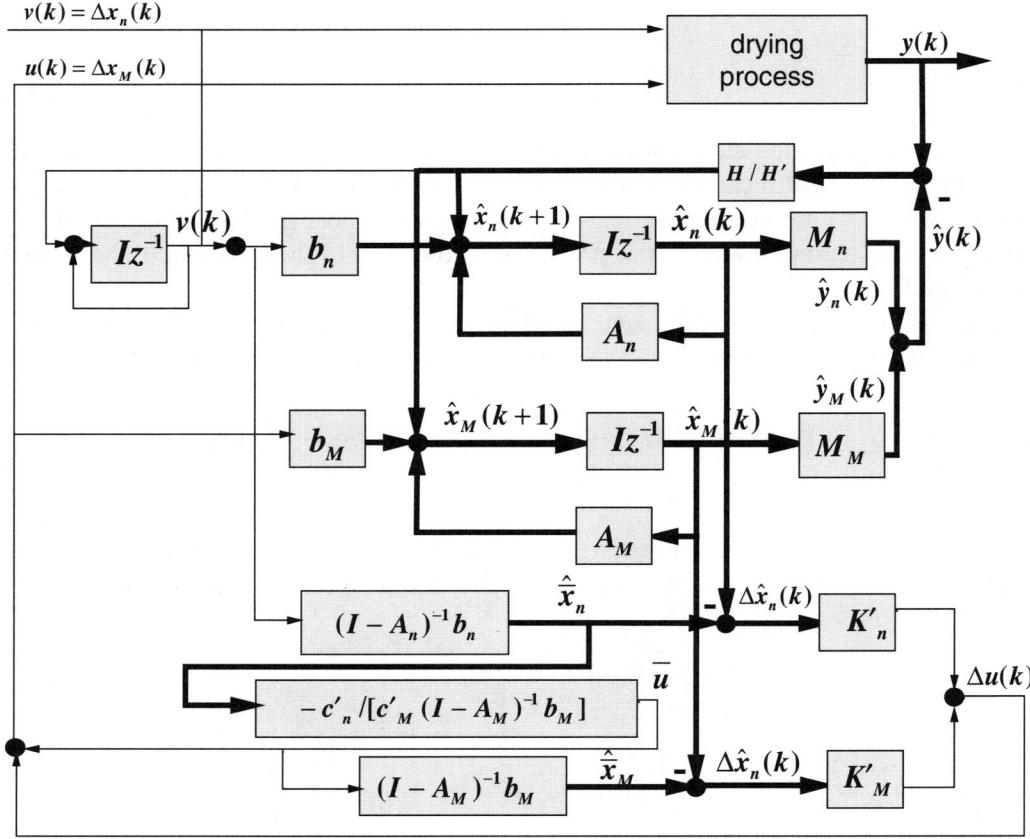

Figure 5–8 Block diagram of the state feedback control system of a rotary dryer.

optimal control systems with respect to given performance indexes. An example of optimal control technique is also demonstrated as applied to a rotary dryer.

REFERENCES

Bristol, E.H. (1966). On a new measure of interactions for multivariable process control. Institute of Electrical and Electronics Engineers (IEEE) *Trans Automatic Control* AC-1, 133–154.

Chen, C. (1970). *Linear System Theory and Design.* New York, NY: Holt, Rinehart & Winston.

Friedland, B. (1986). Control System Design. New York, NY. McGraw-Hill

Isermann, R. 1989. *Digital Control Systems. Vol 2. Stochastic Control, Multivariable Control, Adaptive Control, Applications.* New York, NY: Springer-Verlag.

Mann, W. (1980). Digital control of a rotary dryer in the sugar industry. Presented at the 6th IFAC/IFIP Conference on Digital Computer Applications, October, 1980, in Düsseldorf, Germany.

Manousiouthakis, V., Savage, R. & Arkun, Y. (1986). Synthesis of decentralized process control structures using the concept of block relative gain. American Institute of Chemical Engineers (JAICHE) 32, 991–1003.

Ogata, K. (1970). *Modern Control Engineering.* Englewood Cliffs, NJ: Prentice Hall.

Schonauer, S. & Moreira, R.G. (1997). Dynamics analysis of on-line product quality attributes for automation of food extruders. *Food Sci Technol Int* 3, 413–421.

Seborg, D.E., Edgar, T.F. & Mellichamp, D.A. (1989). *Process Dynamics and Control*. New York, NY: John Wiley.

Shrinskey, F.G. (1979). *Process Control Systems*. 2nd ed. New York, NY: McGraw-Hill.

Takahashi, Y., Rabins, M.J. & Auslaner, D.M. (1972). *Control and Dynamic Systems*. Menlo Park, CA: Addison-Wesley.

CHAPTER 6

Model Predictive Control

Control design using long-range prediction based on a dynamic model of the plant has become an important contender for high-performance applications. These controllers are referred to as model predictive controllers (MPCs). With MPCs, a dynamic process model is used to predict and optimize process performance. The objective function is defined in terms of both present and predicted system variables and is evaluated using an explicit model to predict future process outputs. Note that not all process controllers use a model explicitly in implementation. However, *all* controllers are "model based," since some characteristics of the process must be known in order to design an efficient controller. For example, to design a PID controller, at least the sign of the open-loop process gain must be known.

MPC is appropriate for high-performance control of constrained multivariable processes because explicit pairing of input and output variables is not required and constraints can be incorporated directly into the open-loop optimal control problem (Henson, 1998).

The objectives of this chapter are to introduce the basic concepts of MPC and to present design methods based on discrete-time models such as the dynamic matrix control (DMC) and the generalized predictive control (GPC) techniques. Examples of applications to food-processing systems will also be presented. For detailed information on MPC, the reader is referred to Prett and Garcia (1988) and Meadows and Rawlings (1997).

INTRODUCTION

Several techniques/algorithms have emerged in the literature with distinguishing differences in the form of underlying plant models, choice of cost functions, assumptions about future control actions, assumptions about command signals, and prediction methods. Byun and Kwon (1988) have reviewed various aspects of these methods and have focused on a unifying approach to MPC.

One form of MPC, dynamic matrix control (DMC), has been used extensively in the petrochemical industry for controlling complex processes that are multivariable and have operating constraints (Morari et al, 1988; Garcia & Morshedi, 1986). The method calculates moves on manipulated variables that minimize future projections of controlled variable errors and constraint violations in the least-squares sense.

A comparison of the primary forms of MPC as originally developed and reviewed by Byun and Kwon (1988) is given in Table 6–1. The features of the cost (objective) function being compared are described in Equation 1:

$$\frac{\min}{u(k), u(k+1), \ldots, u(t+NU-1)} \sum_{j=N_1}^{N_2} \|y(k+j) - w(k+j)\|_{Q_{(j)}}^2 \sum_{j=1}^{NU} \|\xi(z^{-1})u(k+j-1)$$

(1)

where $|x|_q^2$ means $x'qx$, N_1 is the minimum prediction horizon, N_2 the maximum prediction horizon, $Q(j)$ and $R(j)$ are the weighting functions, $\xi(z^{-1})$ is the costing function for $u(k)$, NU is the control horizon ($NU < N_2$), and w is the reference trajectory.

A primary difference between DMC and generalized predictive control (GPC) is the underlying process model (Table 6–1). Nearly every type of model (state space, X, continuous, ARX, etc) can be converted into the appropriate model form for the controller. DMC is based on X models or coefficients of the step response; GPC is based on ARX or ARMAX models, but the algorithm also simplifies if an X model is available. The use of ARX and ARMAX models (see Chapter 4) in GPC means fewer coefficients to adapt than the X model-based algorithms. When using ARIMAX models (representing the disturbance as random steplike or Brownian motion), the controller will automatically have an integrator in it that will prevent offset. Although step/pulse models are easily obtained and make few assumptions about the system, the purpose of a model is to emulate the dynamic plant behavior so that accurate predictions can be made. The ratio of polynomials (ARX) will give a better fit than a simple polynomial (X) when approximating functions (Clarke & Mohtadi, 1988).

Table 6–1 Various Model-Based Predictive Controllers as Originally Derived

Type	Model	N_2	N_1	NU	ξ	Q	R
Model algorithmic control (MAC)	Impulse response, state space	N_2	1	N_2	[a]	1	0
Dynamic matrix control (DMC)	Step response	N_2	1	NU	[a]	1	0
Generalized predictive control (GPC)	ARIMAX	N_2	N_1	NU	$1 - z^{-1}$	Q	R
Extended horizon adaptive control (EHAC)	ARMAX	N_2	1	N_2	1	0 $Q(N_2) = \infty$	
Predictive control algorithm (PCA)	Impulse response	N_2	1	N_2	$z^{j-1} - z^{-1}$	Q	R
Receding horizon tracking control (RHTC)	State space, ARMAX	N_2	1	N_2	1 $1 - z^{-1}$	Q	R

a. Not included in the cost function.

MPC controllers are especially well suited for food processes because of their ability to deal with delays. Other controllers, such as the Smith predictor, can easily become unstable if the delay is incorrect (Seborg et al, 1989). The main difficulty with internal model control (IMC), which is based on a transfer function model and is a combination of MAC (model algorithm control) and DMC forms, is the need for symbolic inversion, which becomes very difficult for multiple-input, multiple-output (MIMO) systems, especially if on-line estimation is being performed. IMC can be designed for robust disturbance rejection and tracking performance (Prett & Garcia, 1988).

MPC designs can yield high-performance control systems. MPCs are not inherently more or less robust than classical feedback, but can be adjusted more easily for robustness (Morari et al, 1988). MPC techniques provide the only methodology to handle constraints in a systematic way during the design and implementation of the controller. Computationally, this is more involved than standard linear time-invariant algorithms. Without constraints, the problem becomes one of a standard linear least-squares problem that can be solved explicitly. With the moving-horizon assumption, a linear-time invariant controller is found. Model predictive control schemes can be represented through Q parameterization as a classical feedback controller (Prett & Garcia, 1988; Morari & Zafiriou, 1988). With constraints, quadratic programming techniques are required to solve the optimization (Garcia & Morshedi, 1986). Incorporating constraints becomes more feasible as computers become faster and allows optimization of entire systems.

THE MPC IDEA

Because MPC is implemented with digital computers, in this chapter the solution of MPC problems proposed will be considered only in discrete time. The reader is referred to Meadows and Rawling (1997) for discussion of continuous-time systems.

The main purpose of predictive control is to predict the effects of potential control actions on the future values of the process output and to find the best control actions that minimize the squared sum of deviations of predicted outputs from command signals.

The idea of the MPC is described below. Assume that the following are known (Clarke & Mohtadi, 1988):

1. An appropriate model, $M(\theta)$
2. A vector of past process outputs, $y = [y(k), y(k-1), \ldots]$
3. A vector of future setpoints, $w = [w(k+1), w(k+2), \ldots]$
4. A vector of previous controls, $u = [u(k-1), u(k-2), \ldots]$
5. A vector of potential future controls, $u = [u(k), u(k-1), \ldots]$

Then the predicted output $\hat{y}(k+j)$ over a prediction horizon can be computed as shown in Figure 6–1. At the present time k, the behavior of the process over a horizon N is considered. Using a model, the process response to changes in the manipulated variable is predicted. The moves of the manipulated variables are found that minimize a cost criterion. Only the first computed action of the manipulated variables is implemented. At time $k + 1$, the computation is repeated with the horizon moved by one time interval. Corresponding to these predictions are the predicted system errors, $e(k + j) = w(k + j) - y(k + j)$, and the vector of predicted errors is $e = [e(k + 1), e(k + 2) \ldots]$.

One key question in MPC design is what assumptions to make about future control actions. All future controls (except for the last k controls) must be taken into account since they affect the vector e. By considering *increments* of control (moves) instead of *full-valued* controls, it is assumed that any increments after a control horizon NU are taken to be zero, so that

$$\Delta u(k + j) = u(k + j) - u(k + j - 1) = 0, \quad j > NU$$

(2)

Equation 2 implies a constant control after the horizon and a future control increment vector $\bar{u}$ as

$$\tilde{u} = [\Delta u(k), \Delta u(k - 1), \ldots, \Delta(k + NU - 1), 0, 0 \ldots]$$

(3)

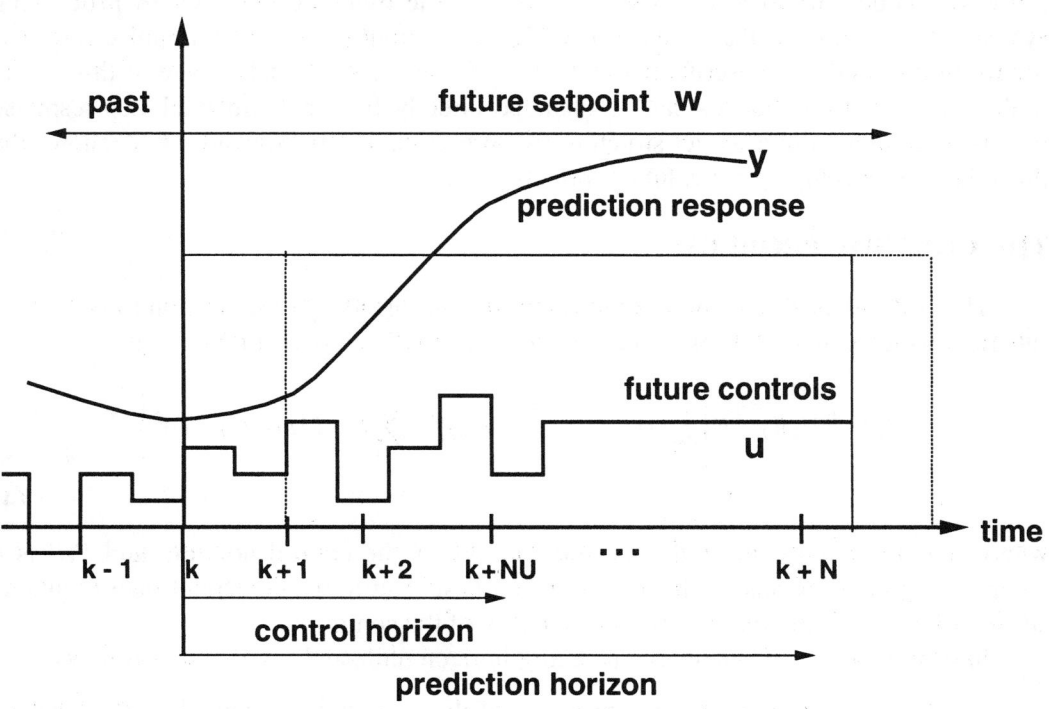

Figure 6–1 The moving horizon approach of MPC

Note that in Equation 2, if $NU = 1$, only one control increment is to be calculated: $\Delta(k) = u(k-1)$.

THE MODEL

The heart of MPC is the process model itself. Models can be classified by several features (see Chapter 4). Since MPC requires the solution of a model to predict future process outputs, the form of the model selected has a large impact on one's ability to implement MPC. Another aspect of modeling for MPC is the fact that since MPC is an on-line control, the speed of the computer algorithm used is important. Linear models are very well suited to MPC because they are fast to calculate and because the optimization problem can be solved easily (by linear or quadratic programming) using commercially available software that is robust and reliable. Continuous models for MPC also affect the speed of solution (they usually require fewer parameters than discrete models). Distributive parameter models are more complicated to solve, requiring more time for solving the partial differential equation models (Meadows & Rawling, 1997).

Among the input-output models, step- or impulse-response models have been popular in industrial applications. In conventional industrial MPC, installation of an MPC starts with identification of a linear input-output model in discrete time (Meadows & Rawling, 1997).

The concept of discrete-time modeling was introduced in Chapter 4. The parametric ARMAX model structure was used to describe the dynamic behavior of processing systems. One approach that is used in DMC is to employ a discrete–impulse-response (convolution) model to describe the dynamics of a process. The advantage of this model is that the model coefficients can be obtained directly from experimental step-response data without assuming a model structure (Seborg et al, 1989). Appendix 6–A shows the procedure for developing convolution models.

THE CONTROL PROBLEM

The MPC algorithm computes the vector of controls using optimization of some cost function, which generally is quadratic. In the case of GPC, this cost function is

$$J(N, NU, R) = E\left\{\sum_{j=1}^{N}[y(k+j) - w(k+j)]^2 + \sum_{j=1}^{NU} R(t)[\Delta u(k+j-1)]^2\right\}$$

(4)

where N is the maximum prediction horizon, NU is the control horizon, and $R(j)$ is a control-weighting sequence. This cost function is subject to the constraint that u remains constant for $j > NU$ in order to maintain stability of the controller.

In general, the MPC involves a receding-horizon philosophy, defined as follows:

1. At sample time k, the free response of the process is computed on the basis of known data $[y(k), y(k - 1), \ldots, u(k - 1), \ldots]$.

2. The control increment vector is computed using the optimization routine with given future setpoints.
3. The first control input $u(k)$ is extracted and applied to the process.
4. All sequences are shifted so that the next sample and procedure can be repeated.

The k-step-ahead prediction is

$$\hat{y} = G\tilde{u} + p + e$$

(5)

where $\hat{y}$ is the vector of future outputs, $\tilde{u}$ the vector of future controls, p the vector of output predictions, e the error vector, and G the coefficient matrix. The first j terms in $G_j(z^{-1})$ are the parameters of the step response, so $g_{ij} = g_j$ for $j = 0, 1, 2, \ldots < i$ independent of the particular G polynomial.

The minimizing solution is

$$\tilde{u} = [G'G + RI]^{-1} G'[w - p]$$

(6)

but only the one-step-ahead control move is made.

The general optimization scheme, including the forced and free responses, is depicted in Figure 6–2. Feedforward signals can easily be incorporated into the algorithm.

PROCESS CONSTRAINTS

Constraints are functions of variables to be kept within bounds. There are two kinds of constraints: (1) *hard constraints* —those constraints for which no violations of the bounds are allowed at any time, and (2) *soft constraints*—those for which violation of bounds can be allowed temporarily for satisfaction of other criteria (Prett & Garcia, 1988).

An important characteristic of process control problems is the presence of constraints on input and output variables. Input constraints arise due to actuator limitations such as saturation and rate-of-change restrictions. Input constraints are hard constraints: that is, they must be satisfied. Input constraints are of these forms:

$$u_{\min} \leq u \leq u_{\max}$$

(7)

$$\Delta u_{\min} \leq \Delta u \leq \Delta u_{\max}$$

(8)

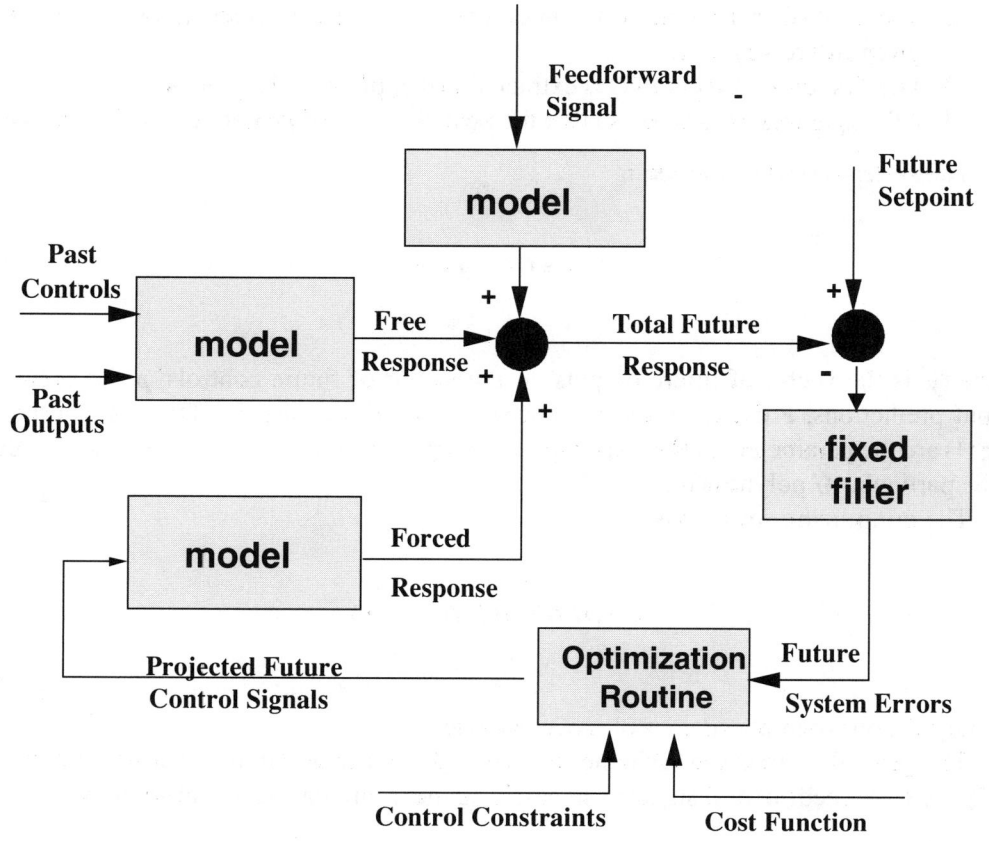

Figure 6–2 Block diagram of optimization in GPC

where u_{min} and u_{max} are the minimum and maximum values of the inputs, respectively, and Δu_{min} and Δu_{max} are the minimum and maximum values of the rate of change of the inputs, respectively.

Output constraints are related to operational limitations such as equipment specifications and safety considerations. Output constraints are considered soft constraints because their violation may be necessary to obtain a feasible optimization problem (Henson, 1998). They take this form:

$$y_{min} \leq y \leq y_{max}$$

(9)

where y_{min} and y_{max} are the minimum and the maximum values of the outputs.

One of the advantages of MPC is that it can handle constraints explicitly (Qin & Badgwell, 1997). By including constraints in the optimization problem, the controller can predict future constraint violations and respond accordingly.

DYNAMIC MATRIX CONTROL (DMC)

The heart of DMC is the discrete-time linear state-space model (Prett & Garcia, 1988):

$$y(k+1) = \sum_{i=1}^{\infty} S_i \Delta u(k+1-i) + v(k)$$

(10)

where the term v collects all the noise effects, and the coefficient matrix S_i is equal to

$$S_i = C \left\{ \sum_{j=1}^{i} A^{j-1} \right\} B$$

(11)

and can be interpreted as the step-response coefficients of the following system in Equations 12 and 13:

$$x(k+1) = Ax(k) + Bu(k) + v_1(k)$$

(12)

$$y(k) = Cx(k) + v_2(k)$$

(13)

where $x(k)$ is the $(n \times 1)$ state vector, $u(k)$ is the $(s \times 1)$ input vector, A is the $(n \times n)$, B $(n \times s)$, C $(r \times n)$ are matrices, $y(k)$ is the $(r \times 1)$ output vector, and $v_1(k)$, $v_2(k)$ are noise inputs.

Since the manipulated variable $m(k)$ and the disturbance variable $d(k)$ are included in the input vector u, the step-response model Equation 10 can then be formulated as

$$y(k+1) = \sum_{i=1}^{\infty} S_{mi} \Delta m(k+1-i) + \sum_{i=1}^{\infty} S_{di} \Delta d(k+1-i) + w(k)$$

(14)

Equation 14 is used to build the prediction equation, assuming that the disturbance variables may be measured at the current time k but do not change over the future time intervals $k + \lambda$. The predicted values of the output can be obtained by writing Equation 14 for all future times from $k + 1$ to $k + N$, as

$$\begin{bmatrix} \hat{y}(k+1) \\ \hat{y}(k+2) \\ \cdots \\ \hat{y}(k+N) \end{bmatrix} = \begin{bmatrix} \hat{y}^*(k+1) \\ \hat{y}^*(k+2) \\ \cdots \\ \hat{y}^*(k+N) \end{bmatrix} + \begin{bmatrix} \hat{v}(k) \\ \hat{v}(k+1) \\ \cdots \\ \hat{v}(k+N-1) \end{bmatrix} + \begin{bmatrix} S_{m1} & 0 & \cdots & 0 \\ S_{m2} & S_{m1} & \cdots & 0 \\ \cdots & \cdots & \cdots & \cdots \\ S_{mN} & S_{mN-1} & \cdots & S_{m1} \end{bmatrix} \begin{bmatrix} \Delta m(k) \\ \Delta m(k+1) \\ \cdots \\ \Delta m(k+N-1) \end{bmatrix}$$

$$\underbrace{\qquad\qquad}$$
"Dynamic Matrix" (15)

So the *dynamic matrix* of the process is the matrix of step-response coefficients. The term $\hat{y}^*$ refers to the contribution to future values ($\lambda = 1,..., N$) of the output due to past input moves up to time k-1 and the most recent measured disturbance change:

$$\hat{y}^*(k+\lambda) = \sum_{i=\lambda+1}^{\infty} Sj\Delta u(k+\lambda-j) + \sum_{j=1}^{\infty} S_{d_j}\Delta d(k)$$

(16)

To predict the output values over future time intervals using Equation 14, the future values of the noise effects $\hat{v}(k)$ must be estimated. In DMC they are estimated as

$$\hat{v}(k+\lambda) = \hat{w}(k) = y(k) - \hat{y}^*(k)$$
$$\lambda = 1,..., N-1$$

(17)

In Equation 17, the estimate of the "unmodeled" effects ($\hat{v}$) is the difference between the feedback measurement $y(k)$ and the model output predicted for the current time k:

$$\hat{y}^*(k) = \sum_{i=1}^{\infty} S_j\Delta u(k-j)$$

(18)

Having an expression for the prediction, the manipulated variable moves are calculated by solving the optimization problem Equation 1 subjected to Equation 15. This problem is a standard linear square problem that can be explicitly solved easily. With the moving horizon assumption, a linear time–invariant controller is obtained. In Prett and Garcia (1988), it is shown how to obtain the controller transfer function and the control law that is

$$\Delta m(k) + \sum_{\lambda=1}^{p} L_\lambda \Delta m(k-\lambda) = \sum_{\lambda=1}^{N} R_\lambda [w(k+\lambda) - \hat{v}(k)] + \sum_{\lambda=1}^{p} R_{d\lambda}\Delta d(k-\lambda)$$

(19)

Note that the controller Equation 19, or the manipulated-variable–move equation, is a standard recursive expression dependent on past input moves, the setpoint, and the estimate of the unmodeled effects.

Equation 19 is solved by DMC for the unconstrained case, where the objective function is a quadratic function and the prediction equation is based on a step-response model with no uncertainties. Generally, constraints can be added ad hoc. In the case of constrained DMC, constraints are included as inequalities in the optimization problem. This problem then becomes a standard quadratic program (Prett & Garcia, 1988).

DMC Applied to a Continuous Frying Process

In this example, the unconstrained multivariable control of a continuous fryer is presented. The results presented here are from the work of Haarsma (1994). The frying process is described in Chapter 2 and illustrated in Figure 2–3.

The reasons for using DMC for the continuous frying process were the following:

- The process is open-loop stable—the future values of the noise effects are assumed to be constant over the entire horizon.
- The process has a large and variable dead time.
- The system is a MIMO process.
- The unconstrained DMC can be implemented easily with little on-line computation effort.
- Feedforward compensation can be obtained naturally because the effect of measurable disturbances can be explicitly included in the design.
- The process is easy to implement using available software (eg, MATLAB, Mathworks, Natick, MA).

The frying process was modeled using a black box approach with an auto-regressive exogenous input model (ARX). A pseudorandom binary signal (PRBS) was designed for identification of the process (Chapter 4).

The DMC algorithm was set up by considering the moisture content (MC) and oil content (OC) as controlled variables. The oil temperature (OT), the submerger speed (SS), and the takeout conveyor speed (TC) were the manipulated variables. The predicted model had a total of 192 parameters. This model was then converted to a state-space model using some of the MATLAB functions.

The control inputs were then calculated using the procedure described above. The complete simulation of the fryer and the controller was built in MATLAB. A graphical user interface was used to display the simulation results. Figures 6–3 and 6–4 show the control and display panels of the virtual fryer. In the control window (Figure 6–3a), we have four buttons located at the top of the panel: (1) tuning, (2) disturbances, (3) uncertainties, and (4) stop. Once one of these buttons is pressed, a new window will be displayed. The simulation results are shown in the display panel, where changes in the setpoint as well as switching from automatic to manual operation are possible.

(a)   **(b)**   **(c)**

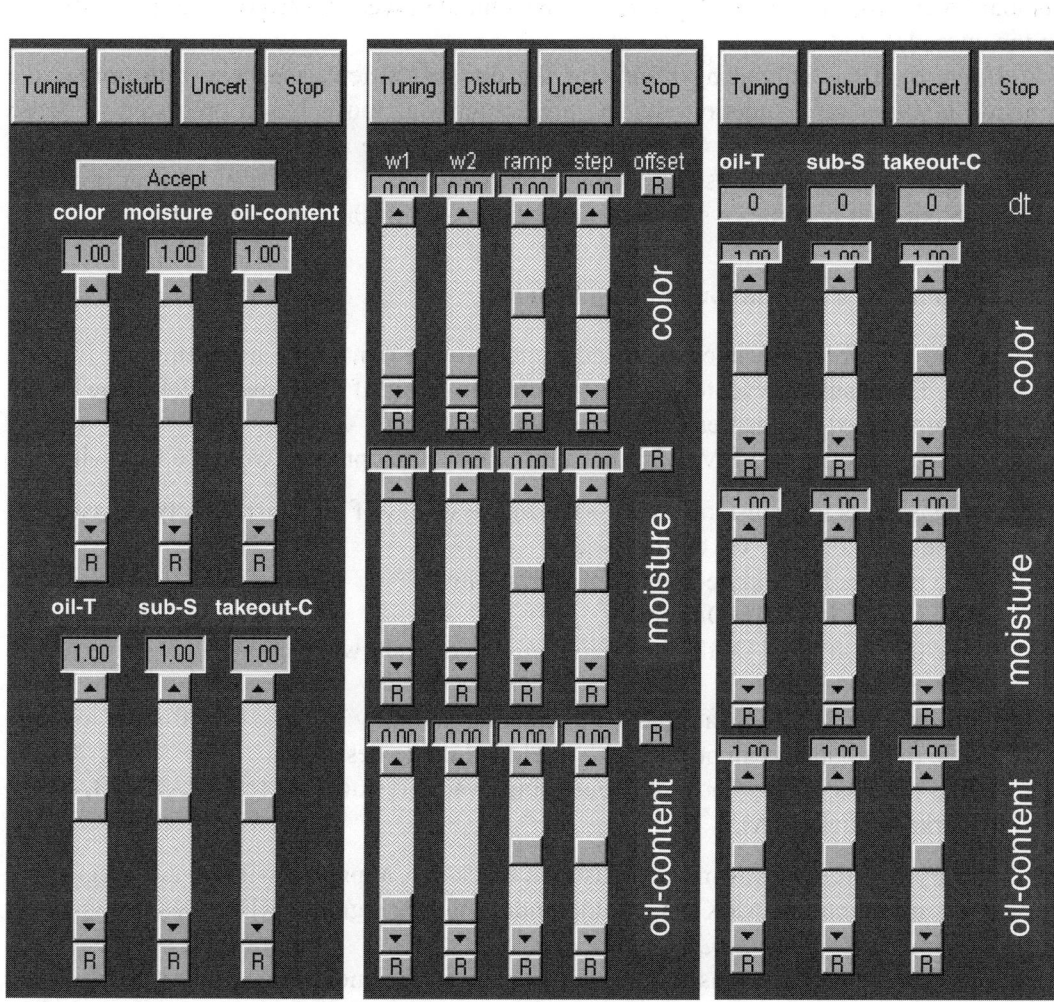

Figure 6–3 The control panel. (a) The tuning window; (b) the disturbance window; (c) the uncertainties window.

The tuning window (Figure 6–3a) allows the operator to select the best parameters, the weighting functions Q and R (Equation 1), that will result in the best control performance. The sliding up-down buttons are used to select the changes in these parameters. Under these buttons we will find a series of push-buttons used to reset the changed values to their original values. The "accept" button must be used to activate the changes made.

The disturbance window (Figure 6–3b) is used for simulating extra disturbances in the frying system. For each output there are four kinds of disturbances. The first two, w_1 and w_2 are random noise from a normal distribution. The first noise, w_1, acts in the process, and the second noise, w_2, acts on the measurement device. The two other disturbances in

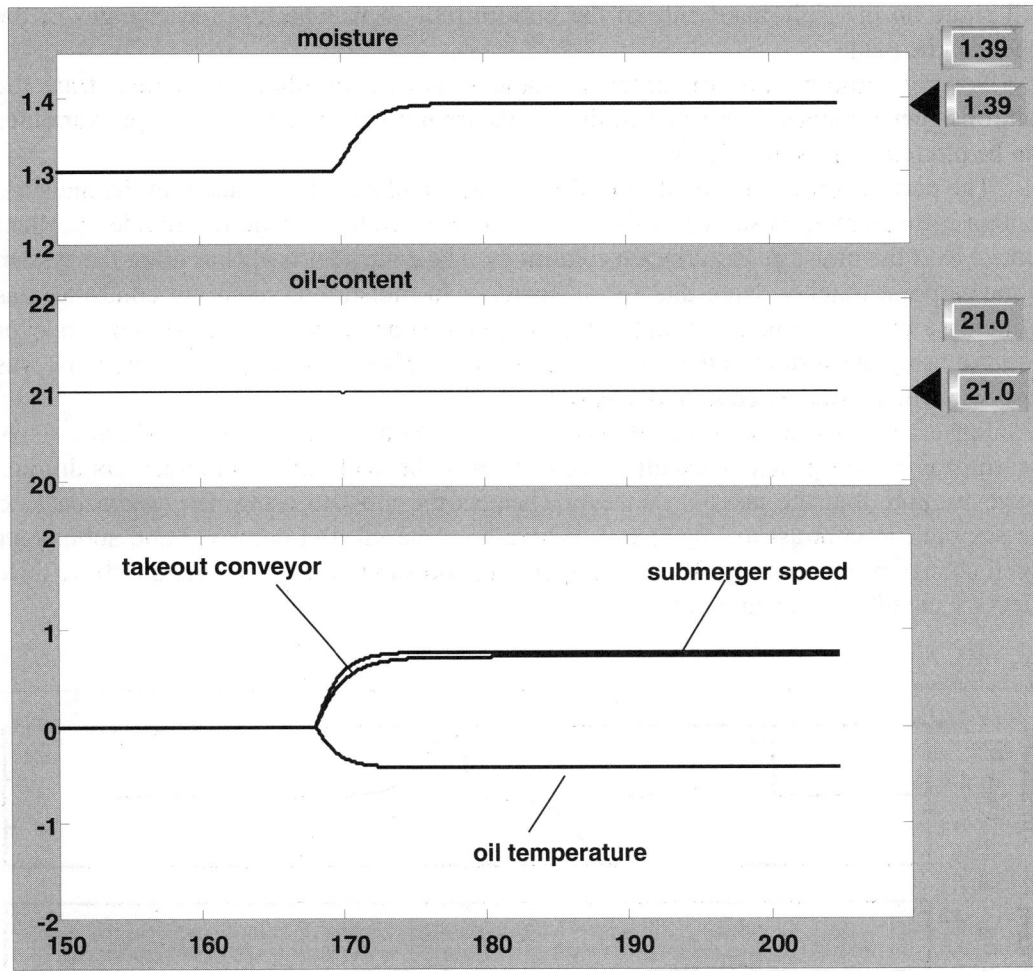

Figure 6–4 The display window

this window are the ramp and step disturbances. The ramp disturbance simulates the slow change in outputs with time, such as the temperature effects on the process. The offset buttons are used to remove offsets caused by the ramp and step disturbances.

The uncertainty window (Figure 6–3c) allows the operator to simulate the case when an underpredicted or overpredicted gain or delay occurs in the process model. The top boxes are for changing the values of the delays, and the sliding up-down buttons are for the gains.

The display window (or the results panel) is illustrated in Figure 6–4. In this window, the input and output results of the process are displayed. There are two arrows at the right-hand side of the output plots that are used to simulate setpoint changes by sliding up and down. It is also possible to change the control system from automatic to manual by clicking twice on the words *oil temperature*, *submerger*, and *takeout*. Three arrows will show on the right-hand side of the bottom plot so that the desired changes in the input can be made.

Several setpoints and disturbance changes will be simulated to demonstrate the controller performance. The value of the inputs are normalized so that all input variables can be plotted in the same figure.

The performance of the controller for a sequence of step disturbances in the moisture content measurement is shown in Figure 6–5. The controller is able to provide excellent control over the moisture disturbance conditions. The controller is able to bring the system to the setpoint quickly. Note that for an increase in moisture content the controller has to increase the oil temperature and reduce both the submerger and the takeout conveyor speed to bring the system to the setpoint again. Similar behavior (opposite directions) was observed when moisture content decreased.

For a sequence of step disturbances in the oil content measurement (Figure 6–6), the controller also provides excellent control over the wide oil disturbance conditions. However, note that the takeout conveyor changes the most to bring the system back to the setpoint, in contrast to Figure 6–5, where both the takeout conveyor and submerger speed change proportionally. The oil temperature and submerger speed seem to have little effect on the oil content changes.

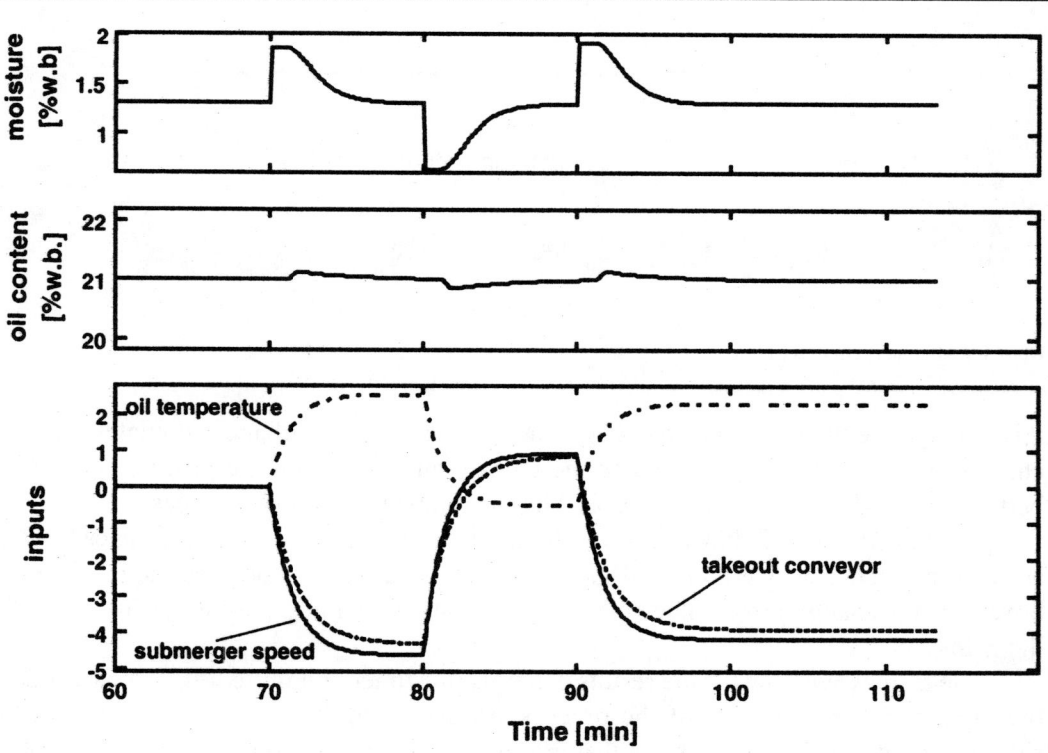

Figure 6–5 Closed-loop response for a sequence of step disturbances in the moisture content measurement

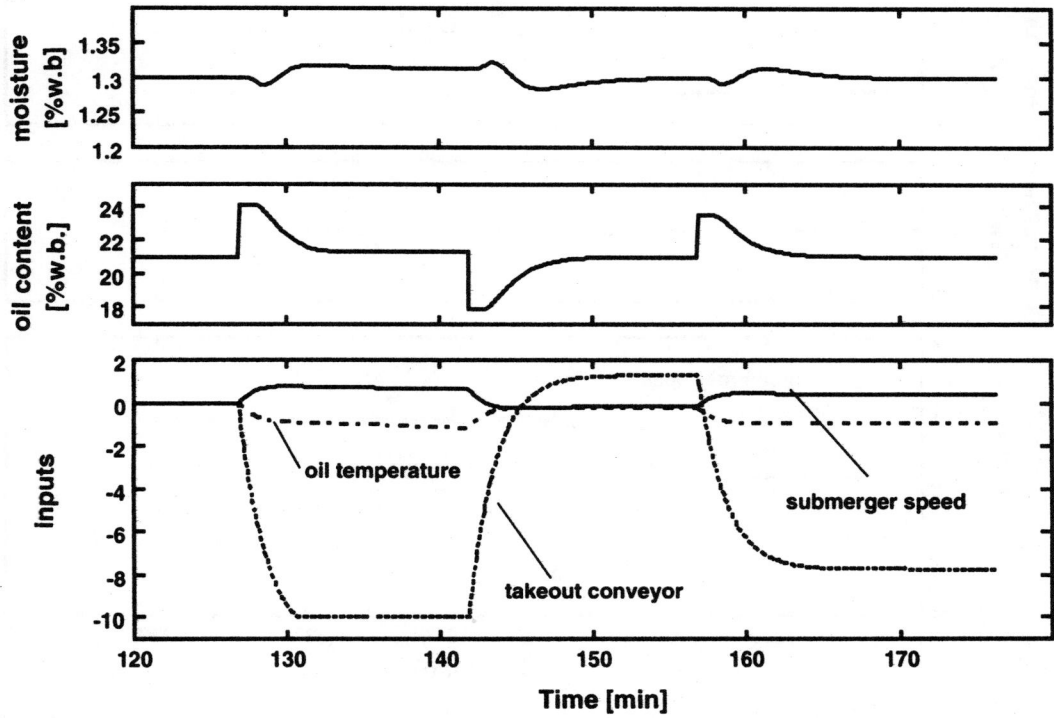

Figure 6–6 Closed-loop response for a sequence of step disturbances in the oil content measurement

Figures 6–7 and 6–8 show the DMC performance for process model uncertainties in the gains and time delay, respectively. Figure 6–9 shows the case in which no uncertainties are present in the process model. The closed-loop response for a step change in the moisture setpoint for overpredicted gains (Figure 6–7) shows that the DMC is faster than in Figure 6–9, but the response shows overshoot and oscillates around the setpoint. For the overpredicted time delay in submerger speed (Figure 6–8), the closed-loop response for a step change in the moisture setpoint shows little overshoot compared to Figure 6–9. Despite using poor estimates of the gains and time delay, the DMC controller provides excellent moisture content responses.

THE GENERALIZED PREDICTIVE CONTROL (GPC)

The model adopted by GPC is of the auto-regressive and integrated moving average with exogenous input (ARIMAX) form:

$$A(z^{-1})y(k) = B(z^{-1})u(k-1) + C(z^{-1})e(k) / \Delta$$

(20)

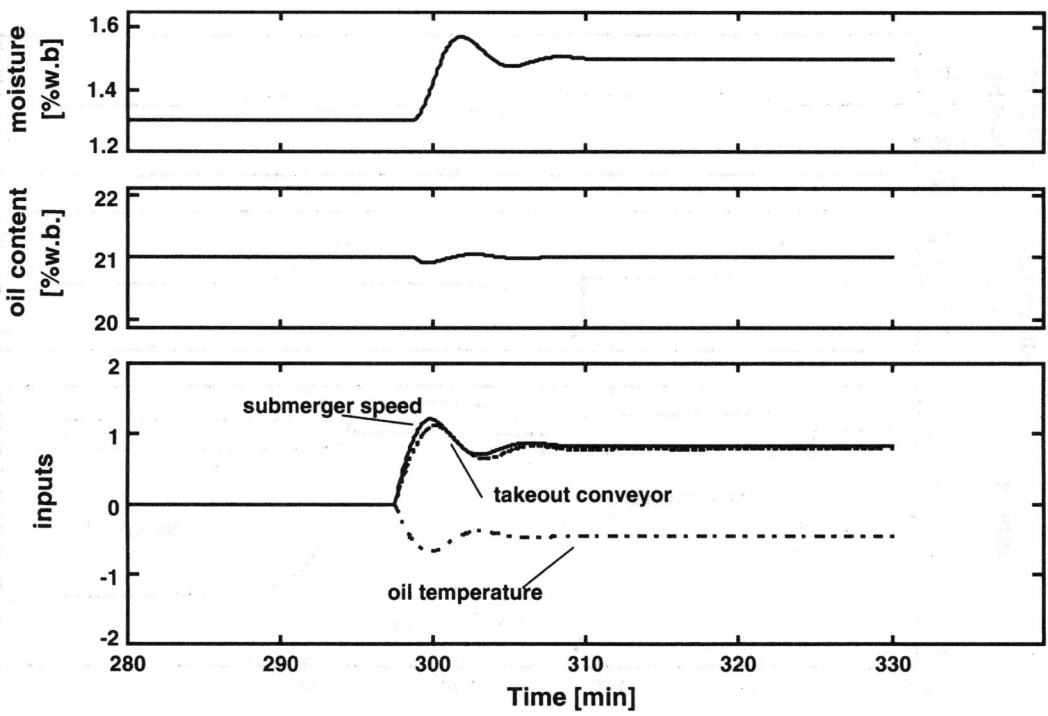

Figure 6–7 Closed-loop response for a step change in the moisture setpoint when the process model has overpredicted gains

where Δ is the operator $(1 - z^{-1})$ so that $\Delta x = x(k) - x(k-1)$. In this model the dead time d is absorbed in B so that its leading $d - 1$ elements are zero (Clarke, 1988).

The GPC algorithm is described in the section "The Control Problem." From Equation 6, the first element of $\tilde{u}$ is $\Delta u(k)$ so that the current control $u(k)$ is given by

$$u(k) = u(k-1) + \bar{g}'(w - p)$$

(21)

where $\bar{g}'$ is the first row of $[G'\,G + RI]^{-1}\,G'$. Therefore, the control includes an integral action that provides zero offset provided that for a constant setpoint $w(k + i) = w$, the vector p involves a unit steady-state gain in the feedback path.

The Control Horizon

The real power of the GPC approach is the assumption made about future control actions. As in the case of DMC, in GPC the projected control elements are assumed to be zero after an interval $NU < N$:

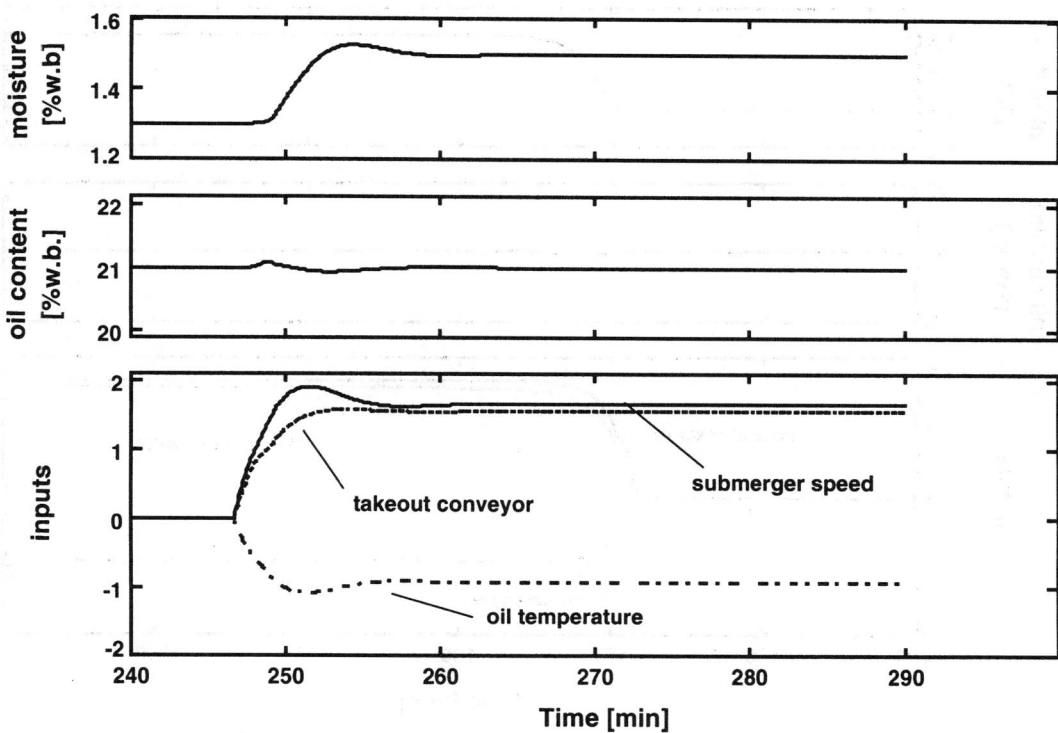

Figure 6–8 Closed-loop response for a step change in the moisture setpoint when the process model has overpredicted time delay in the submerger speed input

$$\Delta u(k + j - 1) = 0 \quad j > NU$$

(22)

The value of NU is called the "control horizon," which is equivalent (in cost-function terms) to placing infinite weights on control changes after some future time. For example, if $NU = 1$, only one control change ($\Delta u(k)$) is considered, after which the control $u(k + j)$ are all taken to be equal to $u(k)$. For this case, suppose that at time k there is a step change in $w(t)$ and that N is large. The choice of $u(t)$ made by GPC is the optimal "mean-level" controller that will place the settled process output at the setpoint w with the same dynamics as for the open-loop process (Clarke et al, 1987).

Selection of the Output and Control Horizons

For the GPC, the primary controller tuning parameters include N and NU. The maximum output horizon, N, in practice should be a large value, corresponding more closely to the rise time of the plant (Clarke et al, 1987).

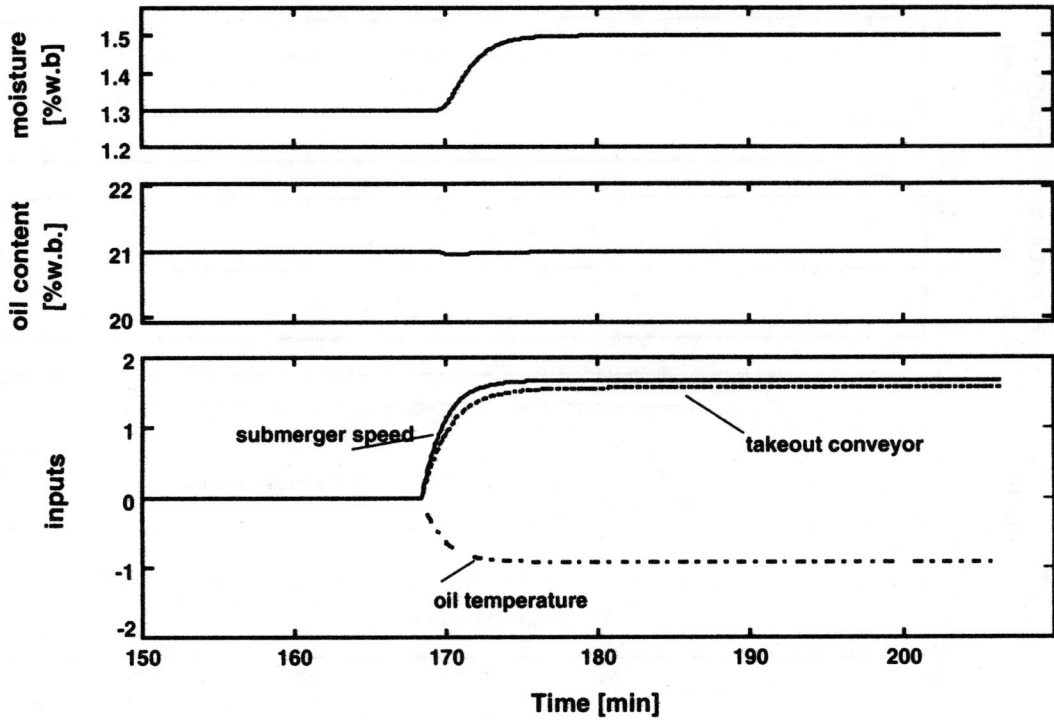

Figure 6–9 Closed-loop response for a step change in the moisture content setpoint

For a simple plant (open-loop stable with possible dead time and non–minimum-phasedness) a value of 1 for the control horizon, *NU*, gives generally acceptable control. For high-performance applications such as the control of coupled oscillators, a large value of *NU* is desired.

Properties for effective control for a plant of order *n* are outlined as follows by Clarke and Mohtadi (1988) (Equation 1):

- $N_1 = n$; the first costing horizon at least exceeds the plant dead time.
- N_2 is large; at least $2n - 1$ but up to the rise time of the plant may be better.
- $NU \leq n$; *NU* determines the degrees of freedom in future control increments. For processing plants *NU* = 1 is generally satisfactory, while difficult plants require *NU* = number of unstable, underdamped poles.
- $R = 0$ is the easiest choice. Sometimes R = small number helps numerical robustness.

Despite the popularity and success of GPC, one major drawback shared by many other MPC methods is the lack of a general theory that predicts closed-loop stability. Although ad hoc methods can be used to stabilize any example, there is no tuning technique that can guarantee stability.

The MIMO GPC

The extension of GPC to the multivariable case is straightforward. The process model is now in vector difference-equation form. With the same control horizon *NU* for each input and the same prediction horizon *N* for each output, the control loops are ordering-independent, which is useful in applications where there is no obvious pairing of inputs and outputs (Clarke, 1988). For detailed discussion on MIMO-GPC, the reader is referred to Shah et al (1987) and Kinnaert (1987).

GPC Applied to a Twin-Screw Food Extrusion Process

An example of application of GPC to a food-processing system is as follows. The ARX-GPC was developed for the food extrusion process described in Chapter 2 in the section "The Food Extrusion Process" (Schonauer & Moreira, 1997). Color-b (CB) was selected as the controlled variable. Feed rate (FR), water rate (WR), and screw speed (SS) were selected as the manipulated variables.

An ARX model with two parameters for each input and output was used in an interactive search procedure as the time delay varied from 1 to 30 sampling intervals (1 to 120 seconds). The delay with lowest AIC (Chapter 4) was chosen; then the model orders varied from 1 to 20 and the delays by 1 or 2. The final model structure chosen for the parsimonious model was two parameters for each input and output. Previous modeling results using only one numerator parameter and two denominator parameters resulted in inconsistent directions in steady-state gains. More consistent values were obtained by increasing the numerator order.

The coefficients were estimated using least-squares algorithms. The loss functions using AIC were calculated. The structure selected for CB-FR, WR, SS ARX model was 2 2 2 2 5 5 1 (where 5 5 1 are the delays associated to FR, WR, and SS) with an AIC (Chapter 4) loss function of 0.0205. The coefficients for this model are shown in Table 6–2. The output can be determined from the model in Table 6–2 according to

$$\Delta CB(k) + \sum_{i=1}^{ny} [a_{CBi+1} \times \Delta CB(k-i+1)] = \sum_{i=1}^{nu} [b_{FRi} \times \Delta FR(k-1) + b_{WRi} \times \Delta WR(k-i) +$$

$$b_{SS} \times \Delta SS(k-i) + \sum_{c=1}^{ny} [c_i \times e(k-i+1)]$$

$$(23)$$

where a_i's, b_i's, and c_i's are the coefficients starting at $i = 1$; n_y and n_u are the numbers of past inputs and outputs; CB is the output; and FR, WR, and SS are the inputs. The polynomial C is equal to 1 for the ARX model (Chapter 4).

The controller to be analyzed is the MISO controller with CB as the output and FR, WR, and SS as the inputs. FR was heavily weighted to create an MIB-based operation (Schonauer & Moreira, 1997). The tuning parameters and weights for the controller were: prediction horizon, $N = 30$; control horizon; $NU = 10$; input weights or penalties on the

Table 6–2 CB-ARX Parsimonious Model Coefficients (FR, WR, SS)

a_{CBi} ([]/[])	b_{FRi} ([]/kg/h)	b_{WRi} ([]/kg/h)	b_{SSi} ([]/rpm)
1	0.0000	0.0000	−0.0010
−1.7836	0.0000	0.0000	−0.0005
0.8121	0.0000	0.0000	0.0000
	0.0000	0.0000	0.0000
	0.0114	0.0609	0.0000
	−0.0161	0.0416	0.0000

Note: CB(k) − 1.7836 CB(k − 1) + 0.8121 CB(k − 2) = 0.0114 FR(k − 5) − 0.0161 FR(k − 6) + 0.0609 WR(k − 5) + 0.0416 CB(k − 6) − 0.0010 SS(k − 1) − 0.0005 SS(k − 2).
Source: Schonauer and Moreira (1997)

control actions, FR: 50, WR: 20, SS: 1. The performance of the controller was analyzed by tracking and regulation tests. These are the implementation results.

Tracking

Figure 6–10 shows the controller response to a +4 CB step-change request using the ARX-based controller (Schonauer & Moreira, 1995). The WR increased approximately 0.5 kg/h, and the SS decreased 20 rpm to reduce the energy input. This caused CB to increase. The response time was adequately short (1.5 to 2 minutes). The input variations were not as smooth, exhibiting noisy response created by the controller using the noisy past outputs in the control input calculation. The controller has a higher dependence on the past outputs than the past inputs, so any deviation that occurred in the past outputs was responded to strongly.

Regulation

The controller was also compared for its ability to compensate for disturbances in such things as cornmeal moisture and screw wear.

Cornmeal Moisture Content Disturbance. The controller was disturbed by adding regular yellow cornmeal (YCM) that had been adjusted to higher moisture content (+1.8%). The wetter YCM would result in increased CB due to lower energy input as a result of the higher moisture content as per Figure 6–11 (no control) (Schonauer & Moreira, 1995). As expected, the controller reacted by requesting lower WR and higher SS inputs to compensate for the wetter cornmeal.

Figure 6–12 shows the reaction of the GPC controller. The WR reduced approximately 1 kg/h when the wet meal was introduced at 15 hours. The SS increased approximately 30 rpm. The large shifts in control inputs were necessary to compensate for the large CB shift (5 units). The WR and the SS tended to oscillate over wider ranges, resulting in a cyclical CB response. Because the ARX-based control places high importance on past outputs compared to past inputs, the controller attempts to adjust the process based on false error. When it attempts to correct the process based on this false error, it automatically

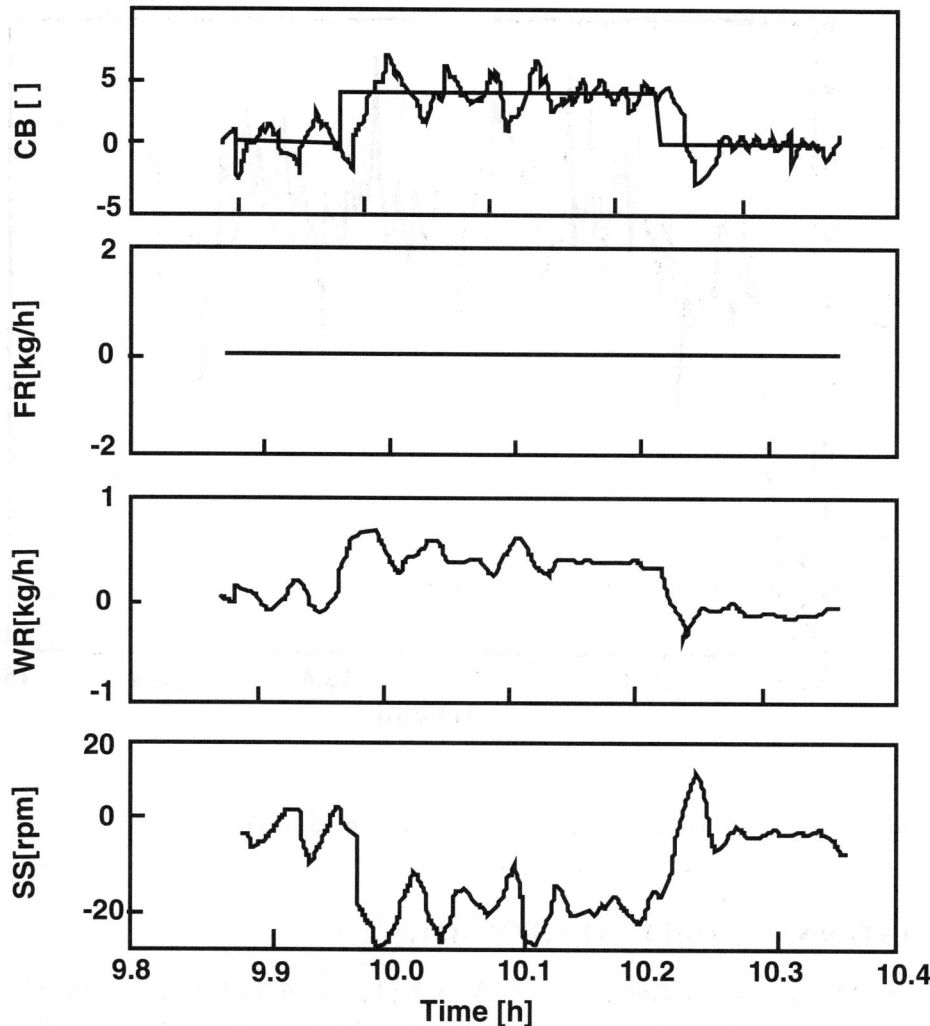

Figure 6–10 Tracking performance of the ARX-based GPC controller.

introduces change, which creates a real variation in CB, which it then has to correct, ultimately setting up the cycles so evident in Figure 6–12.

Simulated Wear. In an attempt to determine the response of GPC controllers to a condition of extruder wear, the die resistance was increased by removing one shim, 0.254 mm (0.010 inch) from the die configuration. The process was started and equilibrated to the same state points as with the previous die configuration. As expected, CB decreased (4 units) due to the higher energy driven by the greater die resistance (Figure 6–13). The controller corrected CB to the setpoint with a cycle in the response.

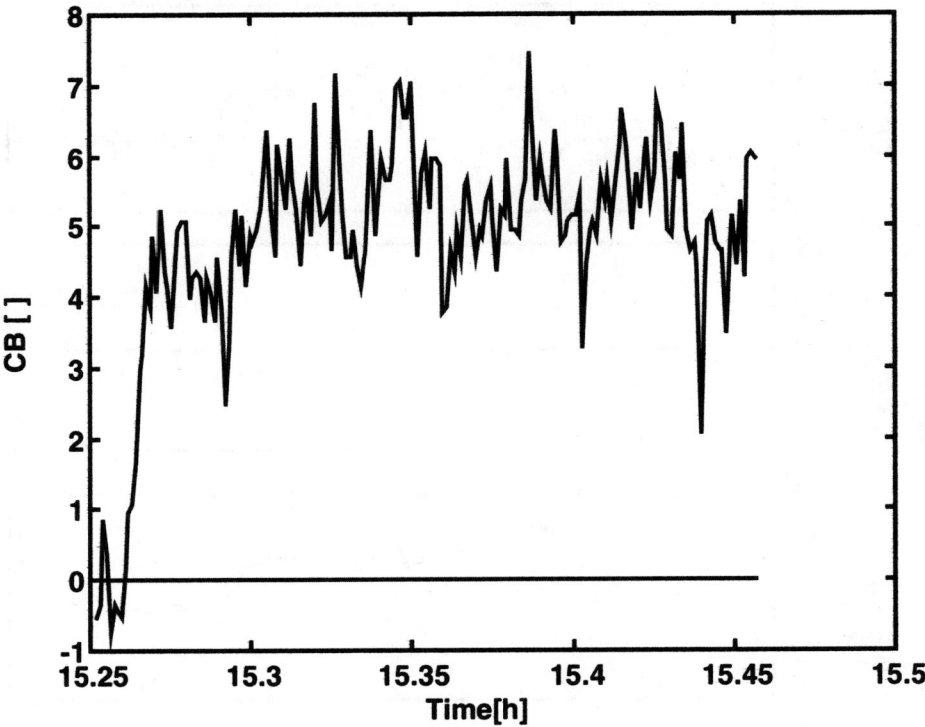

Figure 6–11 CB response to disturbance in cornmeal moisture content (no control).

THE INTERNAL MODEL CONTROL (IMC)

The IMC is an unconstrained form of MPC and uses a model in transfer function form. A disadvantage of this method is that for MIMO systems a matrix with transfer functions has to be inverted, becoming more complicated if the number of model parameters is large.

The IMC Structure

In IMCs, the controller design is based on the inverse of the process model; and the error between the plant and the model outputs is used as a feedback signal.

The IMC structure is shown in Figure 6–14. In this figure, G_p is a discrete transfer function of the process, G^*_p a model transfer function of the process, G_c the controller transfer function, v the disturbance, and w the setpoint. The IMC is a model predictive control in nature because it employs the process model directly for prediction of the output behavior (Morari & Zafirou, 1988). This structure is equivalent to a classical feedback loop with controller (Figure 6–15):

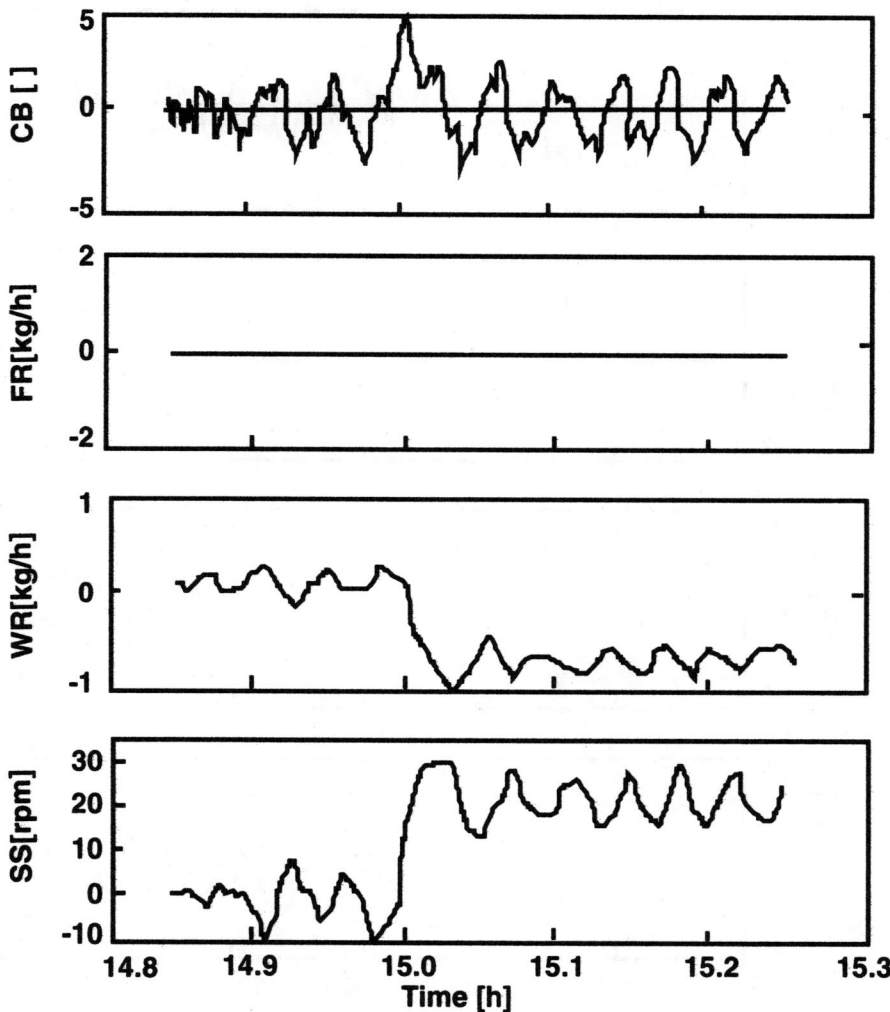

Figure 6–12 GPC controller response to cornmeal moisture disturbance.

$$C = G_c \left[I - G_p^* G_c \right]^{-1} G_c$$

(24)

On the other hand, any feedback loop can be put into an IMC structure:

$$G_c = C \left[I + G_p^* C \right]^{-1} C$$

(25)

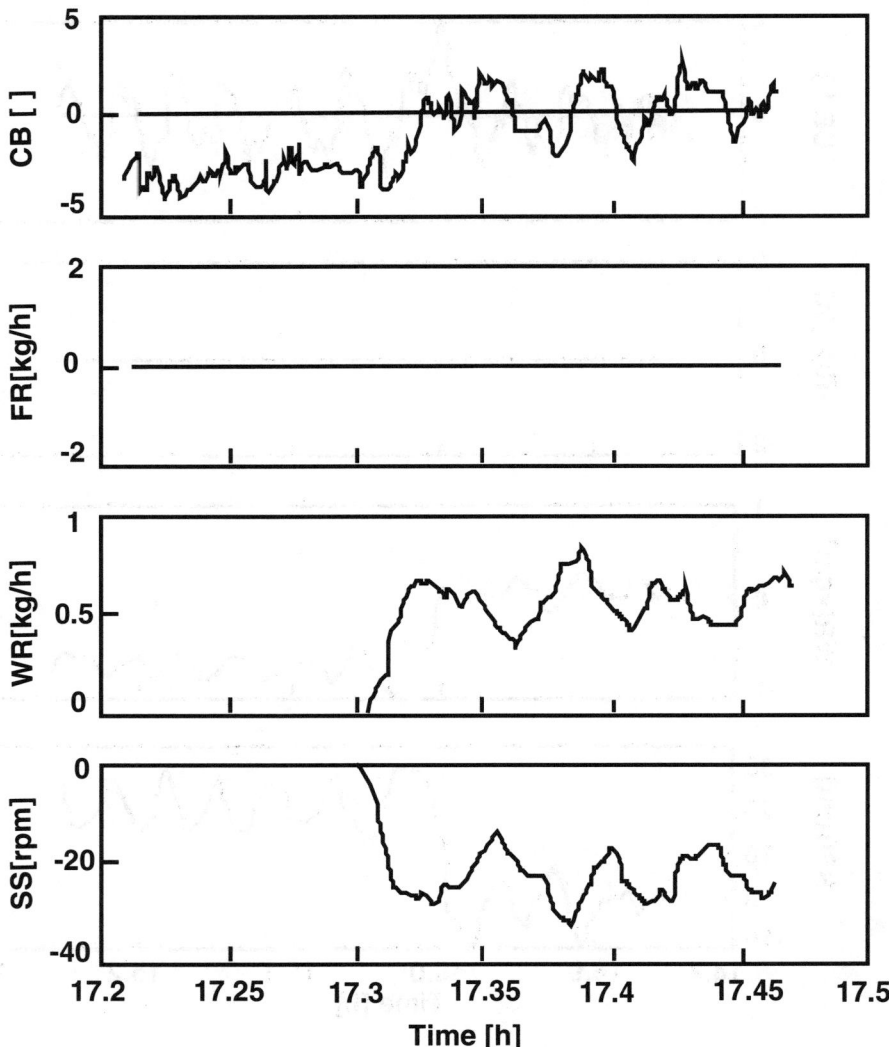

Figure 6–13 GPC controller response to wear disturbance.

However, the IMC structure has certain properties that make the design simpler than the classical feedback control. One advantage of the IMC is that generally, it acts like a feedforward controller (can be designed like one) and can become a feedback controller only if necessary. Consider, for instance, the feedback signal (v^*) of Figure 6–14:

$$v^* = v + \left[G_p + G_p^* \right] m$$

(26)

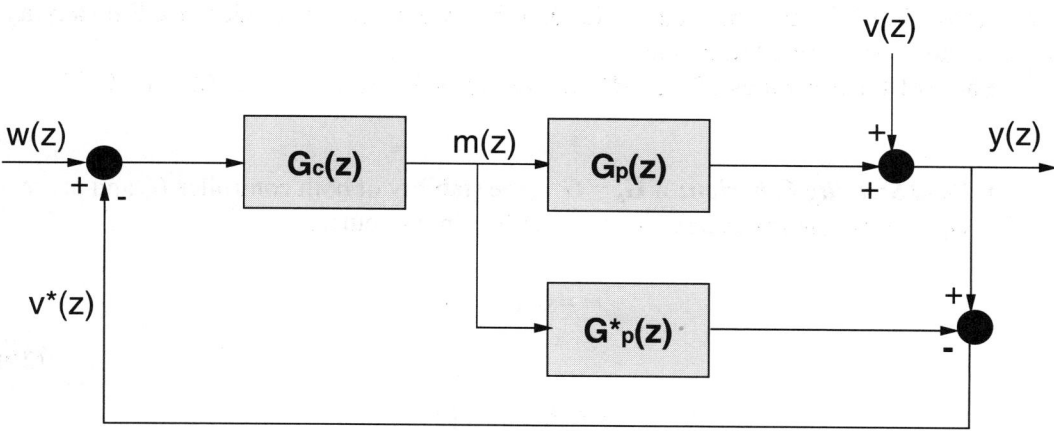

Figure 6–14 The IMC structure

If no disturbances affect the system and if $G_p = G^*_p$ (ie, the model describes the process exactly), $v^* = 0$. From Figure 6–14, the input and output transfer functions are

$$m(z) = \left[I + G_c(z)\left(G_p(z) - G^*_p(z)\right)\right]^{-1} G_c(z)[w(z) - v(z)]$$

(27)

$$y(z) = v(z) + G_p(z)\left[I + G_c(z)\left(G_p(z) - G^*_p(z)\right)\right]^{-1} G_c(z)(w(z) - v(z))$$

(28)

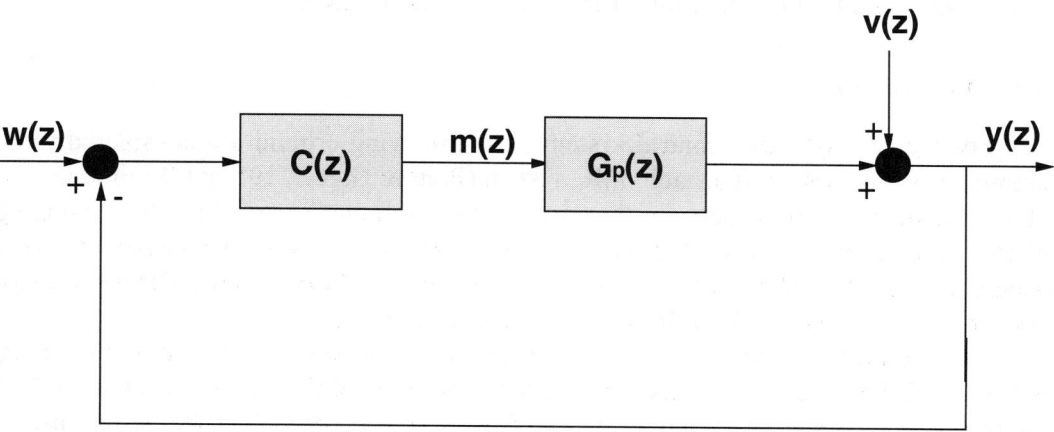

Figure 6–15 The classical feedback structure

Equations 27 and 28 are obtained by standard block diagram manipulations. For stability, both m and y are required to be stable.

Some of the properties of the IMC are discussed below (Garcia & Morari, 1985).

- **Dual Stability Criterion:** If $G_p = G^*_p$, the stability of both controller G_c and process G_p is sufficient for overall system stability. In particular:

$$m = G_c[w - v]$$

$$(29)$$

$$y = G_p G_c[w - v] + v$$

$$(30)$$

Therefore, the system poles and the controller poles must lie inside the unit circle for stability. The design of the classical feedback controller C requires the roots of the det $[I + G_p C] = 0$ to be checked for stability even when the model is exact. Conversely, if $G_p = G^*_p$ and the process is open-loop stable, the stability of G_c is sufficient for closed-loop stability.

Another important characteristic of IMC is that stability is not affected by adding constraints on the inputs (as long as the constrained input is also fed to the model).

- **Perfect Controller:** If $G_c = [G^*_p]^{-1}$, the closed loop is stable, so the IMC will achieve perfect setpoint satisfaction, regardless of any disturbances and model and process mismatch.

- **Zero Offset:** Any control G_c that equals $[G^*_p]^{-1}$ and that produces a stable IMC will yield zero offset. Therefore, integral action is achieved automatically by making the steady-state controller gain the inverse of the model gain.

Design Procedure

For the case of SISO control systems, the following criteria are considered to be important for the design of a stable IMC system (Prett & Garcia, 1988): (1) minimization of the sum of square errors between the controlled variable and its setpoint, (2) no swinging of the manipulation variable, (3) finite settling time, (4) no excessive overshoot and/or undershoot, and (5) offset-free tracking of steps and ramps. In the case of MIMO systems, decoupled response could be a desired performance criterion.

These criteria will be met by a stable controller G_c as long as the model of the plant is perfect. Otherwise, the designed controller must be modified to reach some level of robustness, thus affecting the controller performance as compared to the perfect model case. Thus, the IMC design consists of two steps:

1. Design a "predicted" $\hat{G}_c$ so that a good controlled variable response is achieved regardless of manipulated variable response and robustness;
2. The IMC controller G_c is found by augmenting $\hat{G}_c$ with a low-pass filter F so that $G_c = F \, \hat{G}_c$. Then $\hat{G}_c$ can be calculated as

$$\hat{G}_c = [G_M^*]^{-1}$$

(31)

where G_M^* has a stable and realizable inverse (Prett & Garcia, 1988).

The low-pass filter F for a SISO commonly used is first order:

$$F(z) = \frac{(1-\alpha)s}{z-\alpha}$$

(32)

where $0 < \alpha < 1$ is the filter time constant. The closed-loop response for the perfect model case can be written as

$$y(z) = G_A^*(z)F(z)w(z) + [I - G_A^*(z)F(z)]v(z)$$

(33)

with the important requirements:

$$G_A^*(1) = I; \quad F(1) = I$$

(34)

In the case of model errors, the closed-loop response is shown by Equation 28.

IMC Applied to a Cross-Flow Grain Dryer

Liu and Bakker-Arkema (1999) developed a controller for a cross-flow grain dryer using the ideas of IMC. The process is described in detail in Chapter 2 in the section "High-Capacity Continuous-Flow Grain-Drying Process." The process model for calculating the moisture removal in one bed element (Figure 3–18) of a cross-flow maize dryer is

$$\overline{M}_f = \overline{M}_i - \frac{a_1(\overline{M}_i - M_{e1})\Delta Y}{\dot{m}_g\left(\dfrac{1}{\rho_g} + a_1 c_1 q \overline{M}_i - a_1 c_2 q M_{e1}\right)}$$

(35)

where M_i is the average inlet moisture content of a bed element, ΔY is the height of the bed element, M_{el} is the equilibrium moisture content at the inlet air conditions to the bed element, M_f is the outlet moisture content of the bed element, $\dot{m}_g$ is the grain flow rate, and a, c, and q are constants, such as

$$a_1 = c \exp\left(\frac{-r}{T_{Rl}}\right)$$

$$c_1 = \frac{1 - \exp\left(\frac{r}{T_{Rl}} - \frac{r}{618}\right)}{T_{Rl} - 618}$$

$$c_2 = \frac{1 - \exp\left[\left(\frac{r-b}{T_{Rl}}\right) - \left(\frac{r-b}{618}\right)\right]}{T_{Rl} - 618}$$

$$q = \frac{h_{fg}(0.9H)}{c_a \dot{m}_a}$$

$$M_{el} = 0.01 \exp\left(\frac{a+b}{T_{Rl}}\right)$$

(36)

For implementation of the control action, an inverse of the process model Equation 35 is used to calculate the ideal grain discharge rate, given the setpoint of the grain moisture content and the current position of the grain in the dryer:

$$\dot{m}_{g1} = \frac{-a_1 \rho_g Y}{c_3 \ln\left(\dfrac{\overline{M}_t - M_{el}}{\overline{M}_{yi} - M_{el}}\right) + c_4(\overline{M}_t - \overline{M}_{yi})}$$

(37)

where M_{yi} is the moisture content of the grain at a distance Y from the outlet of the drying section, M_t the setpoint, and $\dot{m}_{gi}$ the ideal grain flow rate, and

$$c_3 = 1 + a_1(c_1 - c_2)\rho_g q M_{el}$$

$$c_4 = a_1 c_1 \rho_g q$$

(38)

Since the grain discharge rate calculated with Equation 37 is usually different for each bed element, an optimum control was then calculated by minimizing an objective error function so that the same amount of grain is overdried and underdried. If the inlet moisture content of the grain in a bed element i is M_i, and the actual discharge rate of the dryer is $\dot{m}_g$

during the next i sampling periods before the grain reaches the outlet of the drying section, the final moisture content of the grain currently in bed element i can be estimated by

$$\overline{M}_{fi} = \overline{M}_i - a \frac{\rho_g (i\Delta Y)}{\dot{m}_g}$$

$$\underbrace{\qquad\qquad}$$
time for grain moving down from
bed element i to the end of drying section

(39)

where α is a constant of the dryer depending upon the dryer model and operating conditions. If the target moisture content is $\overline{M}_t$, the ideal discharge rate, $\dot{m}_{gi}$, for the grain currently in bed element i is calculated using Equation 38, with $\overline{M}_{fi}$ and replaced by $\overline{M}_t$ and $\dot{m}_{gi}$, respectively. When the actual discharge rate is different from the ideal discharge rate during the next i sampling periods, an error term is calculated as $e_i = \overline{M}_t - \overline{M}_{fi}$. The actual error caused by using $\dot{m}_g$ instead of $\dot{m}_{gi}$ for grain bed element i is ith of the e_i values. The average error caused by using $\dot{m}_g$ during the next sampling period for all grain in the n bed element is

$$e = \frac{1}{n} \sum_{i=1}^{n} \frac{e_i}{i}$$

(40)

So by minimizing the error, $e = 0$, the optimum grain flow rate can be calculated as

$$\dot{m}_g = \frac{n}{\sum\limits_{i=1}^{n} \dfrac{1}{\dot{m}_{gi}}}$$

(41)

The controller structure is shown in Figure 6–16. If the feedback signal, the error of the process model defined as the difference between the measured and the predicted outlet moisture contents, is not zero, the drying constant, a_1 in Equations 36 and 37, is modified as

$$a_1^* = \beta a_1$$

(42)

with

$$\beta = \beta_1 + \gamma \ln \frac{\hat{M}_f}{M_f}$$

(43)

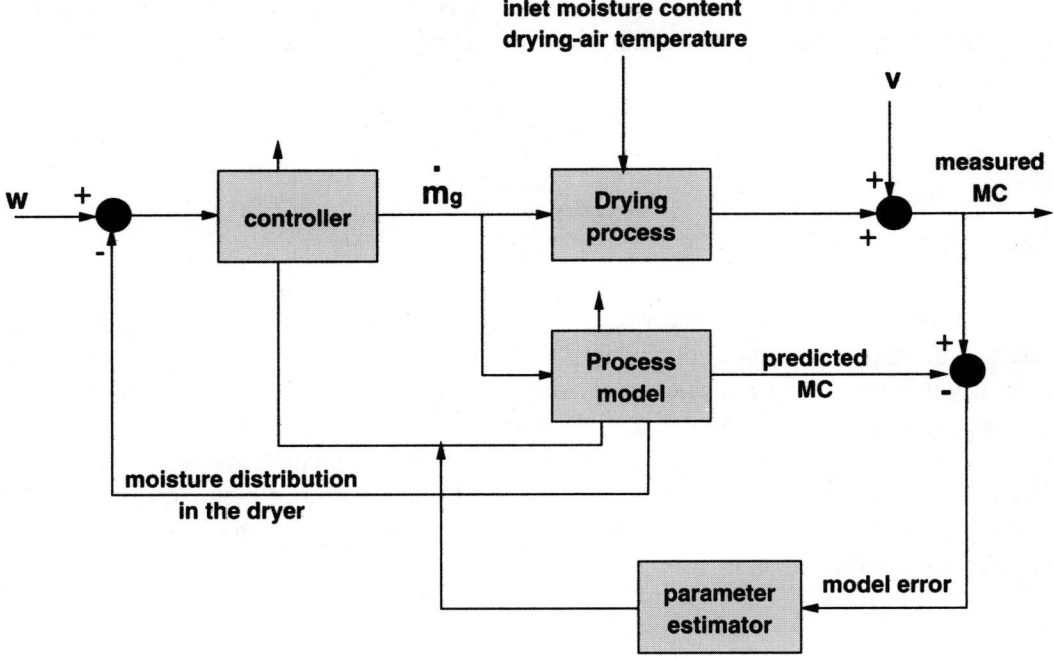

Figure 6–16 IMC of a cross-flow grain dryer

where γ is the filter factor, M_f the measured final moisture content, $\hat{M}_f$ the predicted final moisture content, and β_1 a multiplier used in last sampling period (initially equal to 1.0).

Simulation Results

The cross-flow dryer was modeled in LabView (National Instruments, Austin, TX) using the fundamental partial differential equation model (Brooker et al, 1992). The IMC described above controlled the operation of the virtual dryer. The dryer specifications include:

- Overall dryer height = 17.01 m
- Outside dryer diameter = 3.56 m
- Columns' width = 0.32 m
- Drying section
  —Hold capacity = 28.44 m³
  —Length = 8.76 m
  —Airflow rate = 97,222 m³/h
- Cooling section
  —Hold capacity = 9.9 m³
  —Length = 3.04 m
  —Airflow rate = 33,685 m³/h
- Drying-air temperature = 93.3°C

• Rate capacity (5% moisture removal for dry maize) = 30×10^3 kg/h

Perfect-Process Model Controller. It was first assumed that the model predicted the process exactly ($M_f = \hat{M}_f$). The responses of the controller to step changes in the inlet moisture content and drying-air temperature are shown in Figures 6–17 and 6–18 (Liu, 1998).

Figure 6–17 shows the control response to a series of 5% w.b step changes in the inlet grain moisture content. In response to the step up in the inlet moisture content, the final moisture content starts to decrease before the step change has reached the outlet of the dryer; when the step change reaches the outlet of the dryer, the final moisture content jumps from a condition of overdrying to one of the underdrying before quickly returning to the setpoint. This transition period lasts about two residence times of the grain in the drying section. During this period, the same amount of grain is underdried as is overdried (average control error is equal to zero).

In Figure 6–18, the response of the controller to step changes in the drying-air temperature shows that there is an immediate response in the discharge rate to each step change in the drying temperature. The change in final moisture content is very small (less than 0.3% w.b.) for inlet air temperature changes ranging from 80 to 120°C.

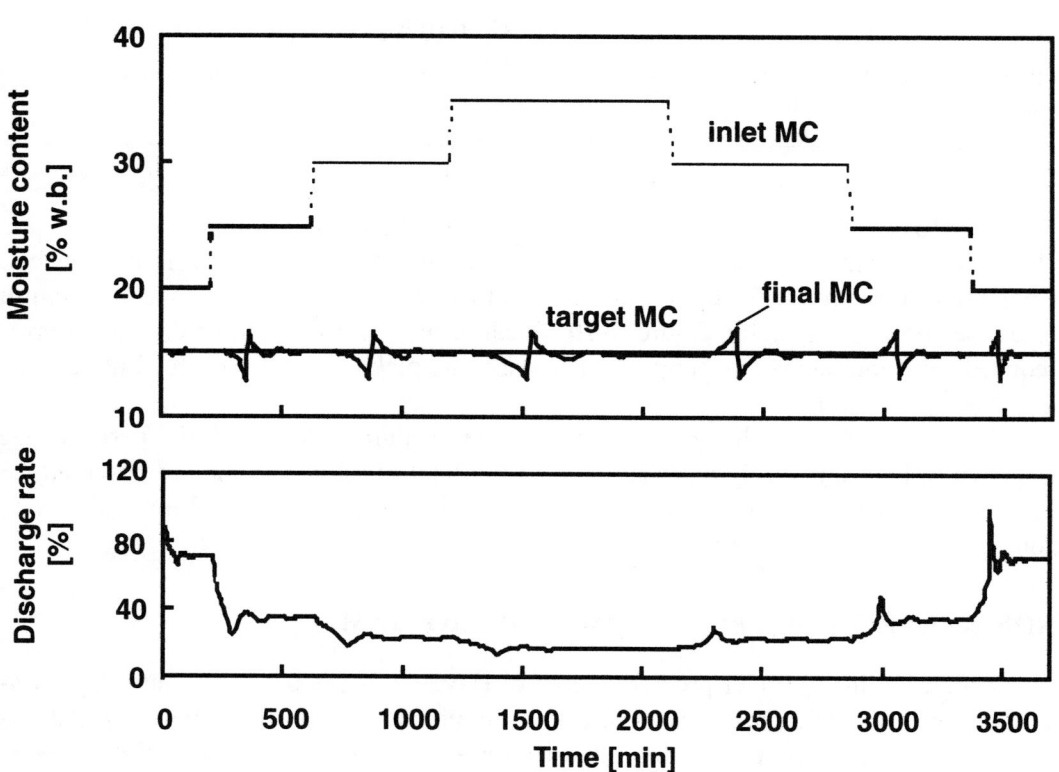

Figure 6–17 Simulated response of feedforward to step changes in the grain inlet moisture content (drying-air temperature = 85°C).

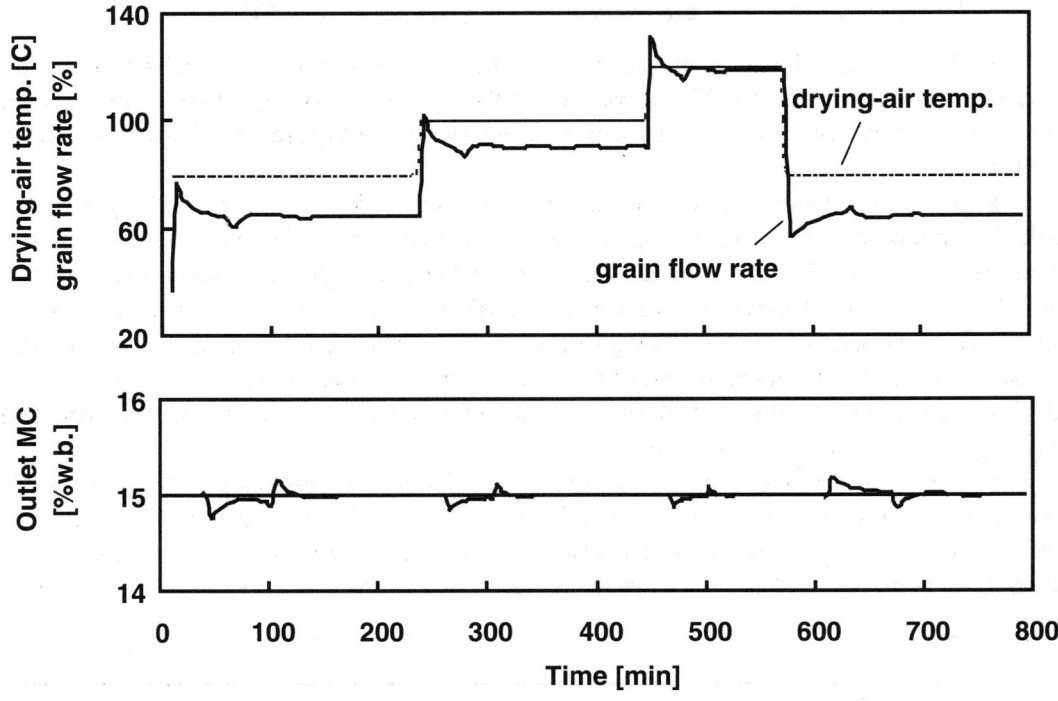

Figure 6–18 Simulated response of feedforward to step changes in the drying-air temperature (inlet grain moisture content = 20% w.b.).

Feedback Activated Controller. Figure 6–19 shows the controller response to step changes in the inlet moisture content when the model of the process is not perfect. Compared to Figure 6–17, the transition period for the outlet moisture content to return to the setpoint is longer for Figure 6–19. Because the compensation of the model error requires time and delays the controller response, the pick error is larger for Figure 6–19 than for Figure 6–17.

The effect of a step change in the inlet air temperature on the controller performance is shown in Figure 6–20. Larger variations in the final moisture content are observed for Figure 6–20 than for Figure 6–18. Again, the discharge rate is adjusted immediately to compensate for changes in the inlet air temperature.

NONLINEAR MODEL PREDICTIVE CONTROL (NMPC)

Although linear model predictive control (LMPC) is standard for controlling constrained multivariable processes in industrial applications, one of its major limitations is that the process behavior is described by linear dynamic models. Therefore, LMPC is inadequate for controlling highly to moderately nonlinear processes (Henson, 1998).

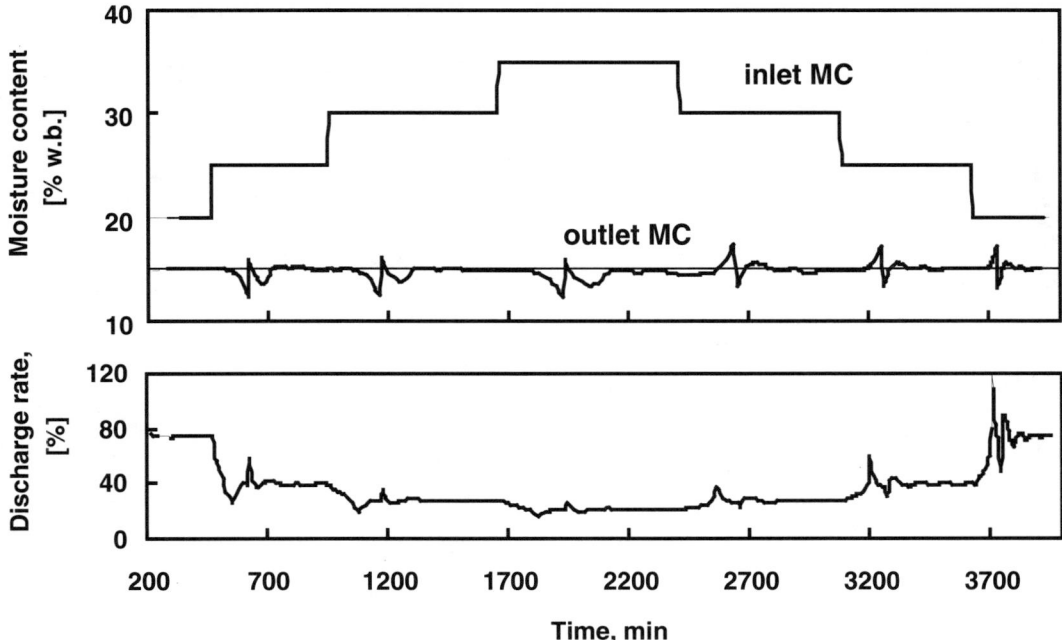

Figure 6–19 Simulated response of the feedforward/feedback controller to step changes in the inlet grain moisture content (drying-air temperature = 85°C).

The increased strict demands on throughput and product quality have encouraged the development of nonlinear model predictive control (NMPC).

NMPC is appropriate for controlling multivariable nonlinear processes with constraints. Conceptually, NMPC is similar to MPC, with the exception that nonlinear models are used for process prediction and optimization. However, the theoretical and practical problems of those NMPC are more challenging than those of LMPC.

Optimization Problem

The nonlinear process model is assumed to have the following discrete-time representation:

$$x(k+1) = f[x(k), u(k)]$$

(44)

$$y(k) = h[x(k)]$$

(45)

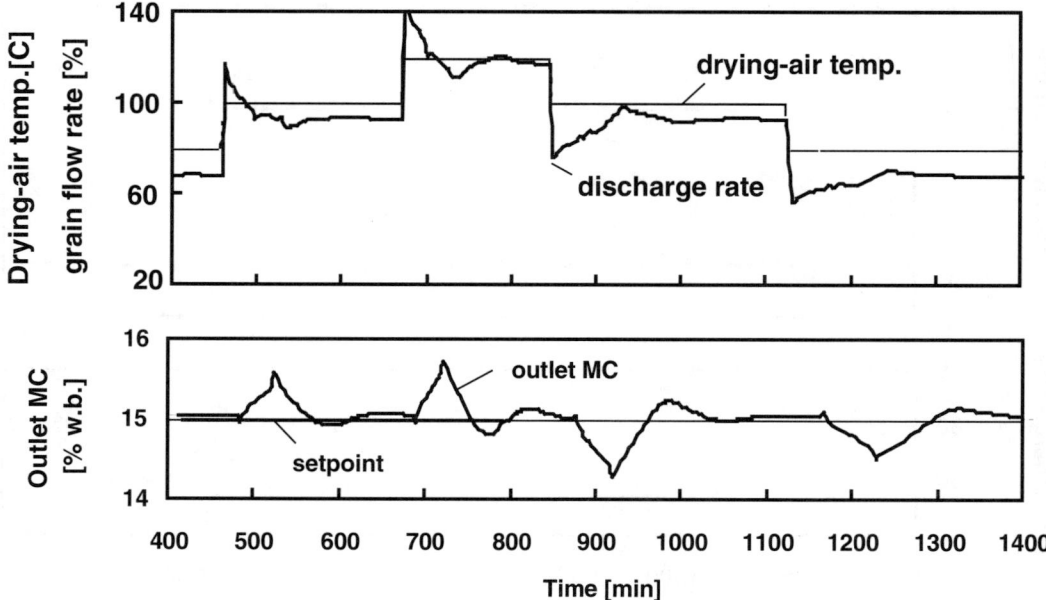

Figure 6–20 Simulated response of the feedforward/feedback controller to step changes in the drying-air temperature (inlet grain moisture content = 20% w.b.).

where f and h are nonlinear mapping, x is an n-dimensional vector of state variables, u is an s-dimensional vector of manipulated input variables, and y is an r-dimensional vector of controlled output variables.

The optimization problem for NMPC formulation is (Meadows & Rawling, 1997):

$$\frac{\min}{u(k/k),u(k+1/k),\ldots,u(t+NU-1/k)} J = P[y(k+N/k]\sum_{j=0}^{N-1} L[y(k+j/k),u(k+j/k),\Delta u(k+j)/k]$$

$$(46)$$

where $u(k+j/k)$ and $y(k+j/k)$ are the input $u(k+j)$ and output $y(k+j)$, respectively, calculated from information available at time k; NU is the control horizon; N is the prediction horizon; and P and L are possibly nonlinear functions of their arguments.

The optimization problem can be solved by, for example, considering a quadratic function as

$$L = [y(k+j/k)-w(k)]'Q[y(k+j/k)-w(k)]+[u(k+j/k)-u_s(k)]'$$
$$R[u(k+j/k)-u_s(k)]+\Delta u'(k+j/k)S\Delta u(k+j/k)$$

$$(47)$$

$$P = [y(k + N/k) - w(k)]'Q[y(k + N/k) - w(k)]$$

(48)

where $u_s(k)$ is the steady-state target for u and Q, R, and S are weighting matrices. The principal controller tuning parameters are NU, N, Q, R, S, and the sampling period Δt.

CONCLUSION

MPC has been applied to a wide variety of simulated processes. Multivariable MPC has been applied to a growing number of food manufacturing processes. The industrial success of MPC is largely attributable to the availability of commercial software packages that can be used to develop linear dynamic models directly from process data. MPCs have been predicting process conditions and product-quality characteristics for dairy ingredients, beer, fruit juices, corn starch, snack foods, bakery products, and waste water.

One form of MPC, the generalized predictive control (GPC) algorithm, is capable of stable control of processes with variable parameters, with variable dead time, and with a model order that changes instantaneously. It is effective with a plant that is simultaneously non–minimum phase and open-loop unstable and whose model is over- or underparameterized by the estimation scheme without special precautions being taken. Essentially, GPC tends to overcome these issues and create a robust control scenario.

Another form of MPC, dynamic matrix control (DMC), has been used extensively in the petrochemical industry for controlling complex processes that are multivariable and have operating constraints. Like GPC, DMC calculates moves on manipulated variables that minimize future projections of controlled variable errors and constraint violations in the least-squares sense.

A main difference between DMC and GPC is the underlying plant model. DMC is based on coefficients of the step response; GPC is based on ARX or ARMAX models, but the algorithm also simplifies if an X model is available.

REFERENCES

Brooker, D.B., Bakker-Arkema, F.W. & Hall, C.W. (1992). *Drying and Storage of Grains and Oilseeds*. New York, NY: Van Nostrand Reinhold.

Byun, D.G. & Kwon, W.H. (1988). Predictive control: a review and some new stability results. In *Proceedings of IFAC: Model Based Process Control*, pp 81–87. Edited by T.J. McAvoy, Y. Arkum, E. Zatiriou. Oxford, England: Pergamon Press.

Clarke, D.W. (1988). *Application of Generalized Predictive Control. IFAC Adaptive Control of Chemical Processes*. Copenhagen, Denmark, August 17–19.

Clarke, D.W. & Mohtadi, C. (1988). Properties of generalized predictive control. In *IFAC Proceedings Series*, 15, pp 65–76. Edited by Rolf Isermann. New York, NY: Pergamon Press.

Clarke, D.W., Mohtadi, C. & Tuffs, P.S. (1987). Generalized predictive control. Part I: the basic algorithm. *Automatic* 23, 137–148.

Garcia, C.E. & Morari, M. (1985). Internal model control. 2: design procedure for multivariable systems. *Ind Eng Chem Process Design Dev*, 24, 472–484.

Garcia, C. & Morshedi, A. (1986). Quadratic programming solution of dynamic matrix control (QDMC). *Chem Eng Commun* 46,73–87.

Haarsma, G.J. (1994). *Development of a Dynamic Matrix Controller for a Frying Process.* College Station, TX: Department of Agricultural Engineering, Texas A&M University. Internal report.

Henson, M.A. (1998). Nonlinear model predictive control: current status and future directions. *Comput Chem Eng* 23, 187–202.

Kinnaert, M. (1987). Generalized predictive control of multivariable linear systems. *Proceedings of the 26th Conference on Decision and Control*, pp 515–610. Los Angeles, CA: Institute of Electrical and Electronic Engineers.

Liu, Q. (1998). Stochastic modeling and automatic control of grain dryers: optimizing grain quality. East Lansing, MI: Michigan State University. PhD dissertation.

Liu, Q. & Bakker-Arkema, F.W. (1999). Automatic control of crossflow grain drying: Part 1: design of a model predictive controller. Presented at IFT Annual Meeting, Chicago, IL, July 24–28.

Meadows, E.S. & Rawling, J.B. (1997). Model predictive control. In *Nonlinear Process Control*. Edited by M.A. Henson & D.E. Seborg. Upper Saddle River, NJ: Prentice Hall.

Morari, M., Garcia, C., & Prett, D. (1988). Model predictive control: theory and practice. In *Proceedings of IFAC: Model Based Process Control*, pp 1–12. Atlanta, GA: IFAC.

Morari, M. & Zafiriou, E.(1988). *Robust Process Control*. Englewood Cliffs, NJ: Prentice Hall.

Prett, D.M. & Garcia C.E. (1988). *Fundamental Process Control*. Boston, MA: Butterworth- Heinemann.

Qin, S.J. & Badgwell, T.A. (1997). An overview of industrial model predictive control technology. In *Chemical Process Control*, pp 232–256. Edited by V. Kantor, C. Garcia, & B. Carnahan. Tahoe, CA: American Institute of Chemical Engineers.

Schonauer, S.L. & Moreira. R.G. (1995). Development of a fixed-GPC controller for a food extruder based on PQA. Part II: control development, implementation and analysis. *Trans Inst Chem Eng* 73(C), 200–210.

Schonauer, S. & Moreira, R.G. (1997). Dynamics analysis of on-line product quality attributes for automation of food extruders. *Food Sci Technol Int* 3, 413–421.

Seborg, D.E., Edgar, T.F., & Mellichamp, D.A. (1989). *Process Dynamics and Control*. New York, NY: John Wiley.

Shah, S.L., Mohtadi, C. & Clark, D.W. (1987). Multivariable adaptive control without a prior knowledge of the delay matrix. *Syst Control Lett* 9, 295–306.

APPENDIX 6–A

CONVOLUTION MODEL

To develop a convolution model, consider, for example, the open-loop step response shown in Figure 6–A–1. The values of the unit-step response are given by $a_0, a_1, a_2, \ldots, a_T$, with $a_i = 0$ for $i \le 0$ and using the sampling period Δt. The variable $T\Delta t$ is the settling time of the process, and the integer T is defined as the *model horizon*.

For a change Δu in the input, the resulting step response can be described by the following convolution model as:

$$\hat{y}_{n+1} = y_0 + \sum_{i=1}^{T} a_i \Delta u_{n+1-i}$$

(a.1)

where $\hat{y}_n$ is the predicted value of the output variable, u_c are the values of the manipulated variable at the nth sampling time, and $\Delta u_i = u_i - m_{i-1}$. The actual input y_n is equal to $\hat{y}_n$ if no modeling error or disturbances are presented. Both y and u are expressed as deviation variables. Equation a.1 can be interpreted as the sum of a series of step changes Δu_i.

From Figure 6–A–1, we can suppose that the system is initially at a value y_0 and that a step change in the input Δu_0 is made with no consecutive input changes. Thus, $\hat{y}$ can be calculated as:

$$\hat{y}_1 = y_0 + a_1 \Delta u_0$$
$$\hat{y}_2 = y_0 + a_2 \Delta u_0$$
$$\cdots$$
$$\hat{y}_T = y_0 + a_T \Delta u_0$$

(a.2)

From Equation a.2, you can see that the predicted values of c simply follow the step-response coefficients $a_1, \ldots a_T$ (Figure 6–A–1) multiplied by the magnitude of the input change. If y_0 is the normal operating point, $y_0 = 0$ since it is a deviation variable.

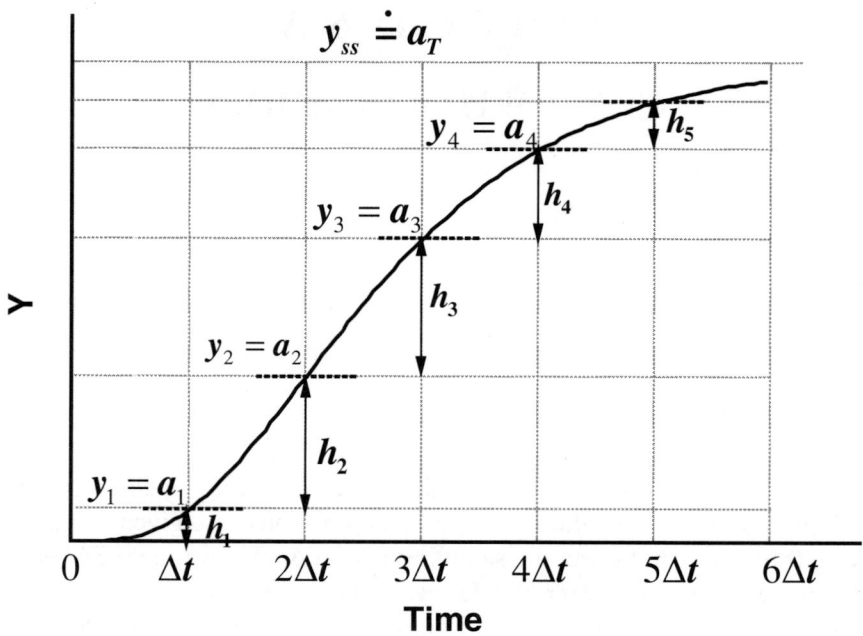

Figure 6–A–1 Coefficient of step-response (a_1) and convolution (l_1) models

If two sequential input changes, Δu_0 and Δu_1, are made at times $i = 0$ and $i = 1$, the predicted value of c can be calculated using the principle of superposition as

$$\hat{y}_1 = y_0 + a_1 \Delta u_0$$
$$\hat{y}_2 = y_0 + a_2 \Delta u_0 + a_1 \Delta u_1$$
$$\ldots$$
$$\hat{y}_T = y_0 = a_T \Delta u_0 + a_{T-1} \Delta u_1$$

(a.3)

Equation a.3 is expressed in a generalized form by the convolution model of Equation a.1.

The impulse response (Figure 6–A–1) can be expressed as the first derivative of the step response. In the case of a digital system with a zero-order hold, the impulse response can be found by taking the first backward difference of the step response:

$$h_i = a_i - a_{i-1} \quad i = 1, 2, \ldots, T$$
$$h_0 = 0$$

(a.4)

where h_1, h_2, . . ., h_T are the unit impulse-response coefficients. The discrete convolution model using the impulse-response coefficients is then

$$\hat{y}_{n+1} = y_0 + \sum_{i=1}^{T} h_i \Delta u_{n+1-i}$$

(a.5)

The z transform of Equation a.5 results in

$$\frac{\hat{Y}(z)}{U(z)} = HG(z) = \sum_{i=1}^{T} h_i z^{-1}$$

(a.6)

CONTROL AND PREDICTION HORIZONS

A central idea in predictive control is the use of horizons. Previously, we defined the model horizon T as the process settling time: that is, the time for the open-loop step response to reach 99% completion.

Two other horizons are used in MPC: (1) the control horizon NU and (2) the prediction horizon N. The control horizon NU is the number of calculated control actions that affect the predicted outputs over the prediction horizon N. Therefore, at time-step n, the next NU values of the manipulated variable u are calculated (u_n, u_{n+1}, . . . $n + NU^{-2}$) as well as the next N output predictions ($\hat{y}_{n+1} + \hat{y}_{n+1}$, . . . $\hat{y}_{n+N}$). The following matrix equation can be used to calculate the response of an arbitrary sequence of NU input changes and an initial steady state $y_0 = 0$:

$$\begin{bmatrix} \hat{y}_1 \\ \hat{y}_2 \\ \cdots \\ \hat{y}_N \end{bmatrix} = \begin{bmatrix} a_1 & 0 & \cdots & 0 \\ a_2 & a_1 & & 0 \\ \cdots & & & \cdots \\ a_N & a_{N-1} & \cdots & a_{N-NU+1} \end{bmatrix} \begin{bmatrix} \Delta u_0 \\ \Delta u_1 \\ \cdots \\ \Delta_{NU-1} \end{bmatrix}$$

(a.7)

Equation a.7 is based on Equation 1 and a prediction horizon of NU sampling periods. The dimension of the matrix in Equation a.7 is $N \times NU$.

One-Step Ahead Prediction

Equation a.5 can be expressed in recursive form by shifting the model back one time step as

$$\hat{y}_{n+1} = y_n + \sum_{i=1}^{T} h_i \Delta u_{n+1-i}$$

(a.8)

Equation a.8 does not provide any corrections for the model errors or unmeasurable load changes, since it is an open-loop prediction. To compensate for these shortcomings, the DMC uses a corrected prediction of $\hat{y}_{n+1}$, defined as

$$y_{n+1}^* = \hat{y}_{n+1} + (y_n - \hat{y}_n) = y_n + \sum_{i=1}^{T} h_i \Delta u_{n+1-i}$$

(a.9)

This shift offsets for model errors and unmeasured disturbances during the previous steps and acts like a feedback control correction (Seborg et al, 1989).

Infinite-Step Ahead Prediction

The prediction horizon N is a design parameter that affects control system performance. A N-step predictor can be expressed in terms of incremental changes in the manipulated variable as

$$\hat{y}_{n+j} = y_{n+j-1} + \sum_{i=1}^{T} h_i \Delta u_{n+j-i} \qquad (j = 1,2,3,\dots,N)$$

(a.10)

The recursive version of Equation a.9 is

$$y_{n+j}^* = y_{n+j-1}^* + \sum_{i=1}^{T} h_i \Delta u_{n+j-1}$$

(a.11)

For $u = 1, 2, \dots, N$ and $c_n^* = c_n$. Equation a.11 can also be written in a vector-matrix form by taking the future incremental input changes Δu_{n+j} out of the summations and rearranging (Marchetti et al, 1983).

$$\begin{bmatrix} \hat{y}_{n+1}^* \\ \hat{y}_{n-2}^* \\ \dots \\ \hat{y}_{n+N}^* \end{bmatrix} = \begin{bmatrix} a_1 & 0 & \dots & 0 \\ a_2 & a_1 & & \\ \dots & & & NU \\ a_N & a_{N-1} & \dots & a_{N-NU+1} \end{bmatrix} \times \begin{bmatrix} \Delta u_n \\ \Delta u_{n+1} \\ \dots \\ \Delta u_{N+NU-1} \end{bmatrix} + \begin{bmatrix} y_n + P_1 \\ y_n + P_2 \\ \dots \\ y_n + P_v \end{bmatrix}$$

(a.12)

where

$$a_i = \sum_{j=1}^{i} h_j$$

(a.13)

$$P_i = \sum_{j=1}^{i} S_j \quad \text{for } i = 1,2,\ldots,N \quad S_i = \sum_{j=1}^{i} h_i \Delta m_{n+j-i} \quad \text{for } j = 1,2,\ldots,N$$

(a.14)

REFERENCES

Marchetti, J.L., Mellichamp, D.A. & Seborg, D.E. (1983). Predictive control based on discrete convolution models. *IEC Research* 27, 956–962.

Seborg, D.E., Edgar, T.F., & Mellichamp, D.A. (1989). *Process Dynamics and Control*. New York, NY: John Wiley.

CHAPTER 7

Adaptive Control

Adaptive control is usually defined as a control system that automatically adjusts controller settings to accommodate changes in the process or its environment. Simply stated, adaptive control is a control with on-line parameter estimation.

Unlike fixed control, adaptive control adjusts (adapts) its behavior to the changing properties of controlled processes and their signals. Adaptive controllers can be divided into two main groups: self-tuning controllers and model reference adaptive controllers.

In this chapter, a short overview of the most important basic structures of adaptive control systems will be presented. Examples of applications to food-processing systems will be shown. The emphasis will be given to digital automation systems. Digital systems permit the implementation of complex control algorithms that would be impossible with analog techniques. Having models and controllers in a discrete-time form is advantageous, especially for theoretical development and for computational effort. For detailed discussion on the subject, the reader is referred to Isermann (1989), Åström and Wittenmark (1990), and Clarke and Gawthrop (1979).

INTRODUCTION

Adaptive control offers an alternate way to maintain consistent performance of a system with uncertainty or unknown variation in plant parameters. Robust control has been proposed to replace adaptive control (Åström, 1983); however, the differences between the two really suggest that they should be combined instead. These differences include the following (Slotine & Li, 1991):

- Adaptive control is superior in dealing with uncertainty in constant or slow-moving parameter change. Robust control is superior in dealing with quickly varying parameter change and unmodeled dynamics.
- Adaptive control requires little a priori information about unknown parameters. Robust control requires reasonable a priori information: typically, an estimation of parameter bounds or uncertainty.

Design of a controller incorporating both adaptive control and robust control techniques would accommodate uncertainties on slowly varying parameters by adaptation, while the robust technique would handle other sources of uncertainty, including unmodeled dynamics.

Two types of adaptive control are predominant in the literature (Åström &Wittenmark, 1990):

1. *Self-tuning controllers* (*STCs*), which combine a controller with an on-line recursive parameter estimator
2. *Model reference adaptive controllers* (*MRACs*), which consist of a process with unknown parameters, a reference model for specifying the desired output of the control system, a control law containing adjustable parameters, and an adaptation mechanism for updating the adjustable parameters

The two methods are quite different in terms of design, stability analysis, and implementation. Compared with MRACs, STCs are flexible because of the possibility of separating estimation and control. However, stability and convergence of STCs are very difficult to guarantee because the closed-loop systems obtained with an STC are nonlinear and time varying.

Design of an adaptive control usually involves the following steps:

1. Selection or development of a control law containing variable parameters
2. Selection of an adaptation law for adjusting those parameters
3. Analysis of the convergence properties of the resulting control system

Performing step 3 is extremely difficult due to the inherent nonlinearities and time-variant nature of the problem. Significant advances in stability results of special cases, such as MRAC, have been made using the Lyapunov theory. Many of these stability results assume the following a priori knowledge—model order n, relative degree of plant model, sign of gain, and minimum-phase plant as well as persistently exciting signals (Isermann, 1989). However, many of the assumptions used in the stability proofs are violated in practice. Adaptive control of non–minimum-phase processes is still a topic of active research.

The adaptive control problem can be divided into explicit (indirect) and implicit (direct) algorithms. In an indirect algorithm, the parameters of the process model are estimated and the controller parameters calculated using these estimated model parameters. If the controller parameters are calculated directly, without separate identification of the process model, the algorithm is considered direct. Historically, MRAC has been implemented as direct control algorithms and STC as indirect control algorithms (Isermann, 1989).

SELF-TUNING CONTROLLERS (STCs)

Self-tuning controllers or model identification adaptive controllers (MIACs) are based on the identification of a closed-loop process model. Such a system is shown in Figure 7–1. The controller is composed of two loops. The inner loop consists of the process and an ordinary feedback controller. Here, the input signal $u(k)$ and the output signal $y(k)$ are

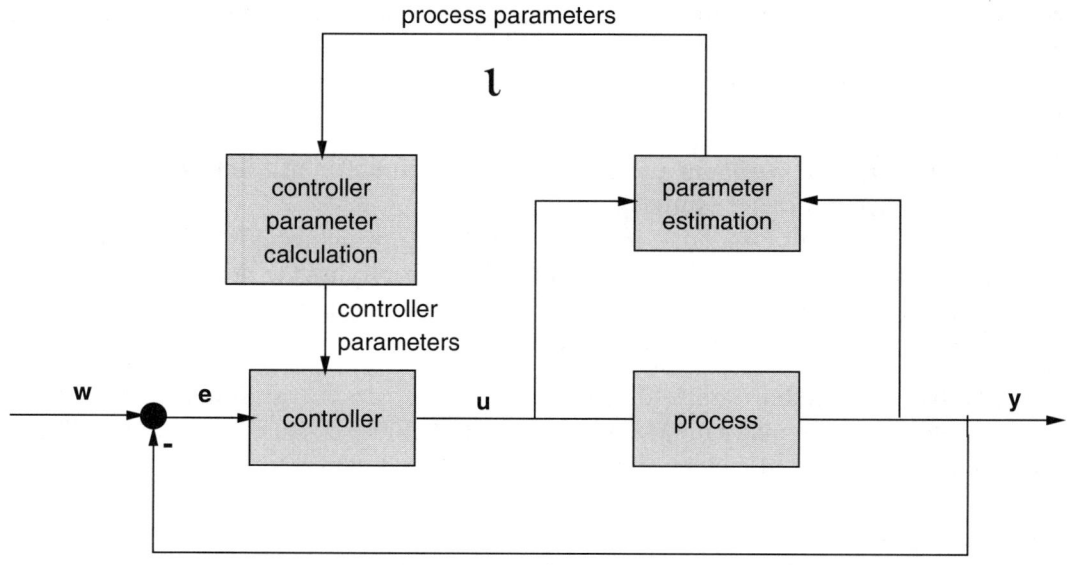

Figure 7–1 Block diagram of a self-tuning controller (explicit combination)

measured. A recursive identification method, for example, determines the process model and may be, the noise model, on-line and in real time. On the basis of the model(s), the controller parameters are calculated using an appropriate controller design method.

The self-tuning controller is very flexible with respect to design methods. Nearly any design technique can be accommodated, including phase and amplitude margins, pole placement, minimum-variance control, linear quadratic Gaussian (LQG) control, and model predictive control (MPC). Many different parameter estimation schemes may be used: for example, stochastic approximation, least squares, extended and generalized least squares, instrumental variables, extended Kalman filtering, and maximum likelihood (Isermann, 1989; Åström & Wittenmark, 1990).

The self-tuning algorithms can be divided into two main classes: explicit self-tuners (Figure 7–1) and implicit self-tuners (Figure 7–2).

Explicit self-tuners explicitly estimate the process model parameters and then calculate the controller parameters. Implicit self-tuners directly estimate the controller parameters. The model parameters are implicitly integrated in the self-tuning algorithm.

An explicit self-tuner converges if the parameter estimates converge. This requires that the model structure used in the estimation be correct and that the input signal be sufficiently rich in frequencies—one frequency for each parameter (Isermann & Lachmann, 1985). However, good regulation using an adaptive control provides poor excitation for the estimation algorithms. Therefore, adaptation under closed-loop adaptive control may require introduction of perturbation signals or switching off the adaptation when excitation is poor (Wittenmark & Åström, 1984).

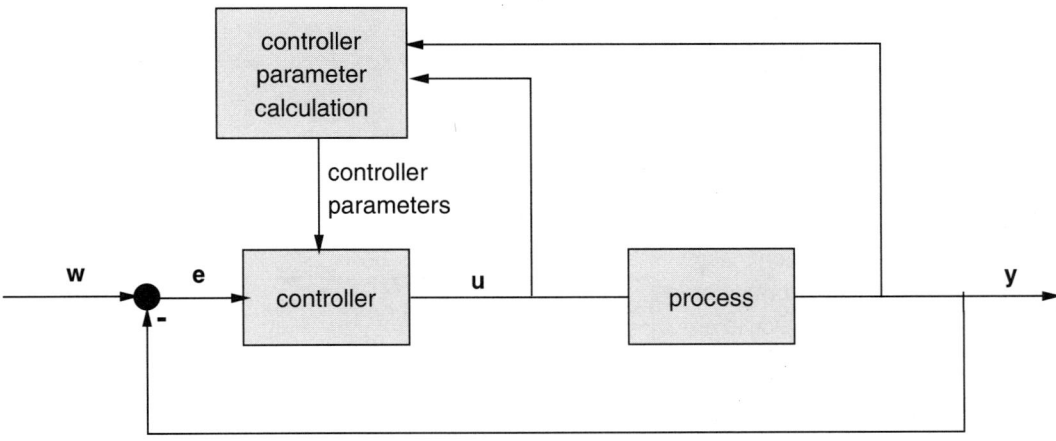

Figure 7–2 Block diagram of a self-tuning controller (implicit combination)

MODEL REFERENCE ADAPTIVE CONTROLLERS (MRACs)

The block diagram of an MRAC is shown in Figure 7–3. Here, the controller tries to obtain a closed-loop response close to that of a given reference model for a given input signal. An external signal—for example, the reference signal—is measured, and a difference signal is created using the signals of the control loop and the reference model. An adaptation method then uses that difference signal to change the controller parameters.

As the STC, the MRAC controller has two loops. The inner loop is an ordinary control loop composed of the process and the controller. The parameters of the controller are adjusted by the outer loop in such a way that the error $e(k)$ between the model input $y_m(k)$ and the process output $y(k)$ becomes small. The outer loop thus also looks like a controller loop. The key problem is to find the adjustment mechanisms so that a stable system is obtained. This problem is not trivial, since it cannot be solved with a simple linear feedback from the error to the controller parameter (Åström & Wittenmark, 1990).

Most of the emphasis in this chapter will be given to STCs. Detailed information dealing with MRACs can be found in Landau (1974, 1979) and Parks (1981).

PARAMETER ADAPTIVE CONTROLLER DESIGN

It is assumed that the process is linear and has either constant or time-varying parameters. Self-tuning controllers or MIACs can be classified according to

- process model
- process identification method
- information about the process
- control design criterion
- control algorithm

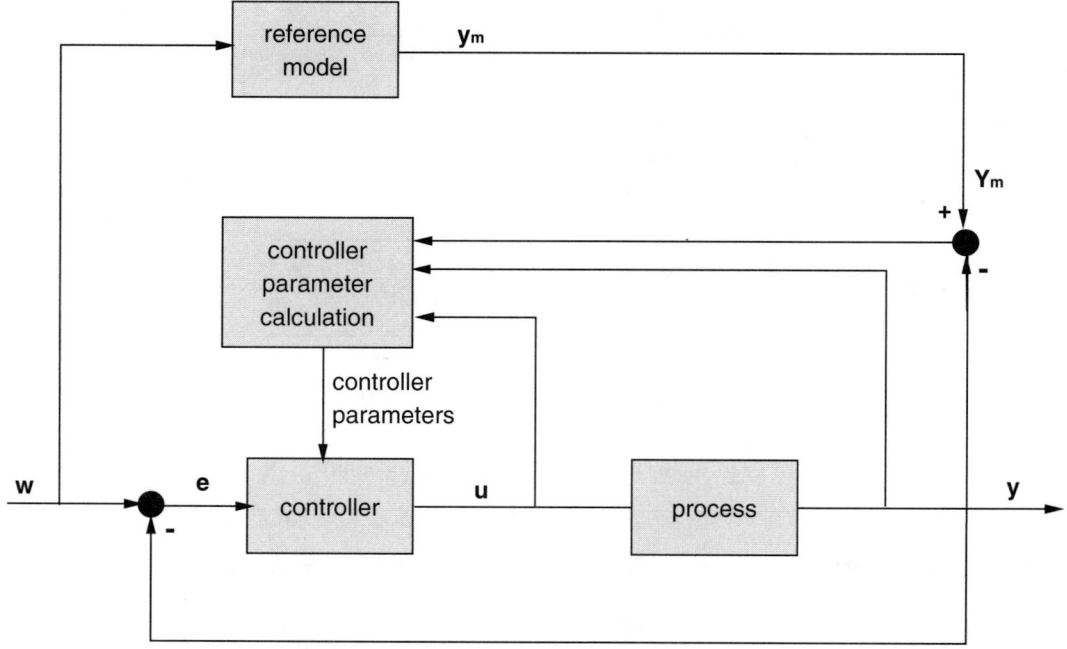

Figure 7–3 Adaptive controller with reference model

Process Model

Only parametric process models are of interest in this chapter. Chapter 4 presents a classification of mathematical process models used in process control.

The *input/output model* is in difference equation or z-transfer function model forms, such as the ARMAX model (Chapter 4):

$$A(z^{-1})y(z) = B(z^{-1})z^{-d}u(z) + D(z^{-1})v_1(z)$$

(1)

where $v_1(k)$ is a statistically independent stochastic signal with $E\{v_1(k)\} = 0$ and variance equal to σ^2. The parameters $\theta' = [a_1 \ldots a_n; b_1 \ldots b_m; d_1 \ldots d_p]$ are assumed to be constant. The state representation model form is also valid, but only the difference equation model form will be considered here. Isermann (1989) presented results for the state model form.

Process Identification Methods

For the identification of parametric process models, the use of parameter estimation methods is straightforward. Recursive parameter estimation methods are best suited for on-line identification in real time. Isermann (1989) described different recursive parameter estimation methods developed for identification in an open loop, where the input signal can be chosen freely. However, special conditions for closed-loop identification have to be satisfied (Isermann, 1989).

For *on-line identification in a closed loop*, consider the linear time-invariant process with z transform as shown in Figure 7–4, which is to be identified in a closed loop:

$$G_p(z) = \frac{y_u(z)}{u(z)} = \frac{B(z^{-1})}{A(z^{-1})} z^{-d} = \frac{b_1 z^{-1} + \ldots + b_m z^{-m}}{1 + a_1 z^{-1} + \ldots + a_n z^{-n}} z^{-d}$$

(2)

and the noise filter:

$$G_{pv}(z) = \frac{v_2(z)}{v_1(z)} = \frac{D(z^{-1})}{A(z^{-1})} = \frac{1 + d_1 z^{-1} + \ldots + d_p z^{-p}}{1 + a_1 z^{-1} + \ldots + a_n z^{-n}}$$

(3)

so that the process and noise are described as

$$y(z) = \frac{B(z^{-1})}{A(z^{-1})} z^{-d} u(z) + \frac{D(z^{-1})}{A(z^{-1})} v_1(z)$$

(4)

The controller transfer function is

$$G_c(z) = \frac{u(z)}{e(z)} = \frac{Q(z^{-1})}{P(z^{-1})} = \frac{q_0 + q_1 z^{-1} + \ldots + q_r z^{-\eta}}{1 + p_1 z^{-1} + \ldots + p_j z^{-j}}$$

(5)

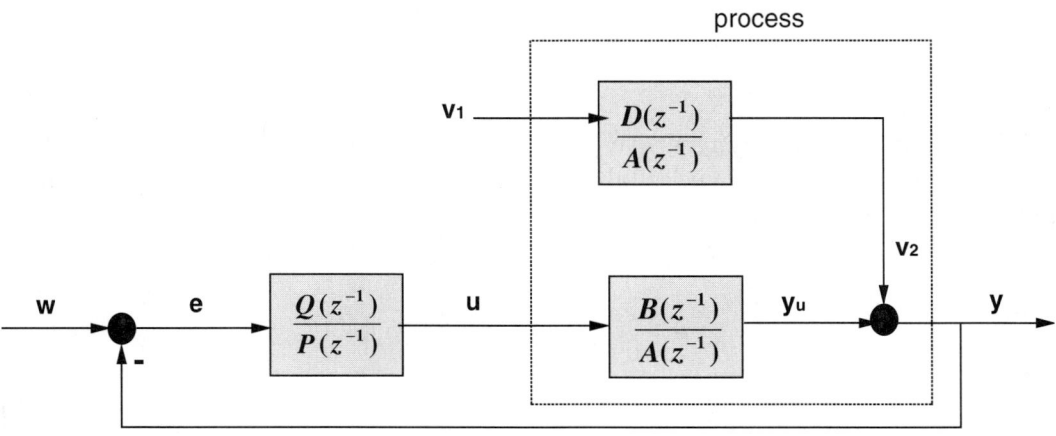

Figure 7–4 Block diagram of the process to be identified in a closed loop

where $y(z) = y_u(z) + v_2(z)$. It is assumed that $w(z) = 0$, so $e(z) = -y(z)$.

The closed loop can be identified by several ways; for example

- *Indirect process identification:* A model of the closed loop is identified; only the output signal has to be measured; the controller has to be known; the process model is calculated on the basis of the closed-loop model.
- *Direct process identification:* The process is directly identified (does not use a closed-loop model) by measuring $u(k)$ and $y(k)$; the controller does not need to be known.

Other considerations include whether the closed-loop identification is performed without a perturbation or with a perturbation signal, in addition to the acting noise $v_1(k)$.

Consider the case of *indirect process identification without a perturbation signal.* The closed loop with the noise as input is

$$\frac{y(z)}{v_1(z)} = \frac{G_{pv}(z)}{1 + G_R(z)G_p(z)} = \frac{D(z^{-1})P(z^{-1})}{A(z^{-1})P(z^{-1}) + B(z^{-1})z^{-d}Q(z^{-1})}$$

(6)

$$\frac{y(z)}{v_1(z)} = \frac{1 + \beta_1 z^{-1} + \ldots + \beta_r z^{-r}}{1 + \alpha_1 z^{-1} + \ldots + \alpha_p z^{-\ell}} = \frac{\mathcal{B}(z^{-1})}{\mathcal{A}(z^{-1})}$$

(7)

where the polynomial orders are $r = p + j$ and $\ell = \max [n + j, m + \eta + d]$. The controlled variable $y(k)$ in Equation 7 is an ARMA (autoregressive moving average) process generated by $v_1(k)$ and the closed loop as a noise filter. If only the output $y(k)$ is known, the parameter of the ARMA model

$$\hat{\theta}'_{a,\beta} = \left[\hat{a}_1 \ldots \hat{a}_\ell, \hat{\beta}_1 \ldots \hat{\beta}_r\right]$$

(8)

can be estimated using any recursive parameter estimation method (see Appendix 4–C) as long as the poles of $\mathcal{A}(z^{-1})$ lie within the unit circle of the z plane and the polynomials $D(z^{-1})$ and $\mathcal{A}(z^{-1})$ have no common roots.

To determine the unknown process parameters (given Equation 8):

$$\hat{\theta}' = \left[\hat{a}_1 \ldots \hat{a}_n, \hat{b}_1 \ldots \hat{b}_m, \hat{d}_1 \ldots \hat{d}_p\right]$$

(9)

certain identifiability conditions must be satisfied (Isermann, 1989):

1. The orders of the process model must be exactly known.
2. If the polynomials D and A have no common roots, a unique determination of the process parameters requires that $\ell \geq n + m$. Hence, the orders of the linear controller have to be

$$\eta > j - d + n - m \rightarrow \eta \geq n - d$$
$$\text{or } \eta < j - d + n - m \rightarrow j \geq m$$

(10)

If the process dead time $d = 0$, the orders of the controller polynomials must be

$$\eta \geq n \text{ or } j \geq m$$

(11)

If $d > 0$, then

$$\eta \geq n - d \text{ or } j \geq m$$

(12)

Now consider the case of a *direct process identification without a perturbation signal*. If the undisturbed process output $y_u(k) = y(k) - v_2(k)$ is known, the process

$$G_p(z) = \frac{y_u(z)}{u(z)} = \frac{y(z) - v_2(z)}{u(z)} = \frac{\dfrac{y(z)}{v_1(z)} - \dfrac{v_2(z)}{v_1(z)}}{\dfrac{u(z)}{v_1(z)}}$$

(13)

can be identified, indicating that for closed-loop identification, the noise filter $v_2(z)/v_1(z)$ must be known (Isermann, 1982).

Therefore, if a stochastic process can be described by Equation 4, direct process identification in a closed loop, without the use of an external perturbation signal, can be performed by measuring $y(k)$ and $u(k)$ and applying a proper parameter estimation method as in open loop, as long as the identifiability conditions 1 and 2 defined above are satisfied. If the controller has too low an order or does not satisfy the identifiability condition 2, then identifiability can be obtained by (1) switching between two controllers with different parameters, (2) introducing a dead time in the feedback, and (3) using nonlinear or time-varying controllers (Isermann, 1989).

Both direct and indirect algorithms need a recursive estimation scheme. To track slowly varying process parameters, a forgetting factor is sometimes incorporated to discount old data (Åström, 1983). The weighted algorithm then counts recent data more heavily than old data. In cases where the input is not sufficiently exciting and extra perturbations cannot be introduced to obtain good excitation, a straightforward implementation of an exponential forgetting can result in estimator wind-up or bursting (Wittenmark, 1988). Various techniques such as covariance resetting (Goodwin & Sin, 1984) or directional forgetting (Hägglund, 1983) can be used to prevent these problems.

Information about the Process

The information (Figure 7–1) about the process (ι), obtained by parameter estimation, forms the basis of controller design calculation of the controller parameters. It contains the parameter estimates $\hat{\theta}$ and their $\Delta\hat{\theta}$ uncertainty. If the parameter estimates are assumed to be identical to the real process parameters, the resulting controllers are referred to as *certainty equivalence controllers*. If, in addition to the parameter estimates, the uncertainties of the parameter estimates are taken into consideration for controller calculation, the resulting controllers are called *cautions controllers* (Isermann, 1982).

Controller Design Criteria

The performance of adaptive controllers depends mainly on the quality of the identified process model and thus on the process input signal. Two controllers can result when using design control in adaptive control systems:

1. *Dual controllers:* the input signal has to be determined so that good compensation of current disturbances and good future process identification are achieved.
2. *Nondual controllers:* these use only the present and past signals and the current information concerning the process in their design criteria.

Control Algorithms

The actual design of control algorithms is done before implementation in a digital computer. However, the calculation of the controller parameters (as a function of process parameters) is done on line. Control algorithms for adaptive control should satisfy the following (Isermann, 1982):

- Closed-loop identifiability conditions
- Small storage and computation effort for controller parameter calculations
- Application to a wide range of processes and signals

Within the class of self-optimizing adaptive controllers based on process identification, nondual methods based on the certainty equivalence principle and recursive parameter estimation have been shown to be successful in both theory and practice. Controllers using these methods are called self-tuning controllers.

Some examples of suitable control algorithms for self-tuning controllers are minimum-variance control, pole placement control, quadratic control, and model predictive control.

Minimum Variance Controller

These controllers are designed for stochastic disturbances $v_1(k)$ by minimizing a quadratic performance function:

$$J(k+d+1) = E\left\{y^2(k+d+1) + ru^2(k)\right\}$$

(14)

The variance of the predicted output is minimized for $w(k) = 0$, taking into consideration the weighted variance of $u(k)$. The variance of the process input can be reduced with an increase in output variance (Clarke & Hasting-James, 1971). This minimum-variance controller for the process (Equation 4) is designed by

$$G_{c_{MV}} = \frac{L(z^{-1})}{zB(z^{-1})F(z^{-1}) + \dfrac{r}{b_1}C(z^{-1})}$$

(15)

where

$$C(z^{-1}) = A(z^{-1})F(z^{-1}) + z^{d+1}L(z^{-1})$$

(16)

and

$$F(z^{-1}) = 1 + f_1 z^{-1} + \ldots + f_d z^{-x}$$
$$L(z^{-1}) = l_0 + l_1 z^{-1} + \ldots + l_{m-1} z^{-(m-1)}$$

(17)

So that the noise filter becomes

$$\frac{v_2(z)}{v_1(z)} = \frac{D(z^{-1})}{A(z^{-1})} = F(z^{-1}) + \frac{L(z^{-1})}{A(z^{-1})} z^{-(d+1)}$$

(18)

And the orders of the controller polynomials (see Equation 5) are

$$\text{for } d = 0$$
$$\eta = \max[p, n] - 1$$
$$j = \max[m - 1, p]$$

(19)

$$\text{for } d \geq 0$$
$$\eta = \max[p - d - 1, n - 1] - 1$$
$$j = \max[m + d - 1, p]$$

(20)

Minimum-variance controllers are designed for $E\{v(k)\} = 0$ and $w(k) = 0$, so they do not have integral action. Therefore, offsets may occur for other disturbances.

Pole Placement Controller

The ARMAX model generally assumed in self-tuning control generally leads to an offset between the measuring and the desired value when regulating a process subjected to constant load disturbances about a nonzero setpoint. Inserting an integrator ad hoc solves the problem, but convergence is difficult (Tuffs & Clarke, 1985). One approach is to use the ARIMAX (autoregressive integrated moving average with exogenous input) process model (Clarke & Gawthrop, 1979):

$$A(z^{-1})y(z) = z^{-d}B(z^{-1})u(z) + D(z^{-1})v_1(z) / \Delta$$

(21)

where Δ is the differential operator $(1 - z^{-1})$. Multiplying both sides of Equation 21 by Δ results in a model working in differential data

$$A(z^{-1})\Delta y(z) = z^{-d}B(z^{-1})\Delta u(z) + D(z^{-1})v_1(z)$$

(22)

Unlike the minimum-variance controller, the pole placement controller is based on the desired characteristic equation of the closed loop and not on the criterion minimization.

Consider the case where $D = 1$ in Equations 21 and 22. Let a general integrating control law be defined as (Tuffs & Clarke, 1985):

$$G(z^{-1})\Delta u(z) + F(z^{-1})y(z) - H(z^{-1})w(z) = 0$$

(23)

where G, F, and H are polynomials. The closed-loop system is given by substituting Equation 23 into Equation 24 (the argument z^{-1} is omitted for simplicity):

$$[\Delta AGZ + z^{-d}BF]y(z) = z^{-d}BHw(z) + G_{v_1}(z)$$

(24)

Let the desired characteristic equation of the closed loop be given by the polynomial P:

$$P \equiv \Delta AG + z^{-d}BF$$

(25)

where the degree of F = degree A, and the degree of G = degree $B + d - 1$. Since a unit static gain of the closed loop is desired, it follows from Equation 24 and Equation 25 that

$$P(1) = B(1)H(1)$$

(26)

and a simple choice of H is

$$H(z^{-1}) = \frac{P(1)}{B(1)}$$

(27)

With the controller polynomials given by Equation 25 and Equation 27, the closed loop becomes

$$y(z) = z^{-d}\frac{BP(1)}{PB(1)}w(z) + \frac{G}{P}v_1(z)$$

(28)

An explicit algorithm can be derived to determine the controller parameters. Equation 22 can be rewritten ($D = 1$) as

$$\Delta y(z) = z^{-d}B\Delta u(z) + z(1 - A)\Delta y + v_1(z)$$

(29)

Since $v_1(k)$ is a white noise, the coefficients A and B can be estimated each sampling time using RLS, with Δy regarded as the current measurement and the previous values of Δu and Δy as the regressors. The controller can be calculated using Equations 25 and 27. Generally, these calculations require an algorithm for polynomial multiplication and a numerical algorithm for solving the polynomial identity Equation 25.

Generalized LQG Controller

The explicit pole placement self-tuner described above, although robust against variations of the time delay, is very sensitive to the assumptions made about the model order of the process, which can result in poor performance.

A robust, yet simple self-tuning algorithm based on a linear quadratic performance criterion can control an unstable non–minimum-phase plant, irrespective of incorrect parametrization, and wrong assumption of the time delay (Clarke et al, 1985). The self-tuner is regarded as a linear quadratic Gaussian (LQG) controller.

An ARIMAX model of the process is considered

$$A(z^{-1})\Delta y(z) = z^{-d}B(z^{-1})\Delta u(z) + C(z^{-1})v_2(z) + D(z^{-1})v_1(z)$$

(30)

and the cost function minimized is

$$J = [\psi(k+N) - w(k+N)]^2 + \sum_{i=k}^{k+N-1} \left\{ [\psi(i) - w(i)]^2 + \xi[Qu(i)]^2 \right\}$$

(31)

where d^* is the feedforward delay and the auxiliary output $\psi(k)$ is defined as

$$\psi(k) \hat{=} Py(k) \quad \text{where} \quad P(z^{-1}) = \frac{P_n(z^{-1})}{P_d(z^{-1})} \quad \text{and} \quad P(1) = 1$$

(32)

The LQG self-tuner involves four steps:

1. process parameter estimation
2. state estimation
3. solution of the equation
4. control law calculation

The process model Equation 30 can be (omitting the argument z^{-1}) expressed as

$$\tilde{A}\varepsilon(k) = \tilde{B}\Delta u^*(k-1) + \tilde{C}\Delta v_2(k) - \tilde{F}\Delta\omega(k) + \tilde{D}v_1(k)$$

(33)

where

$$\tilde{A} = \Delta A P_d; \quad \tilde{B} = z^{-d+1} P_n B \frac{Q_d}{P_n(0)}; \quad \tilde{D} = D \frac{P_n}{P_n(0)}; \quad \tilde{C} = z^{-d^*} \frac{P_n}{P_n(0)}; \quad \tilde{F} = A P_d$$

(34)

and the variables are defined as

$$\varepsilon(k) = \frac{[\psi(k) - w(k)]}{P_n(0)}; \quad u^*(k) = \frac{u(k-1)}{Q_d}; \quad \omega(k) = \frac{w(k)}{P_n(0)}$$

(35)

So the cost function Equation 31 can be scaled by $P_n(0)^2$ and written as

$$J = [\varepsilon(k+N)]^2 + \sum_{i=k}^{k+N-1} [\varepsilon^2(i) + \xi^* \Delta u^*(i)^2]$$

(36)

where

$$\xi^* = \frac{\xi}{P_n(0)^2}$$

(37)

The process described by Equation 33 can be expressed in the state-space observable canonical structure (Clarke et al, 1985):

$$x(k+1) = \tilde{A}x(k) + \tilde{b}\Delta u^*(k) + \tilde{e}\xi(k) + \tilde{c}\Delta v_2(k+1) + \tilde{f}\Delta\omega(k+1)$$

(38)

$$\varepsilon(k) = d'x(k) + v_1(k)$$

(39)

The optimal control law $u(k)$ is given by

$$\Delta u^*(k) = -K\hat{x}(k/k)$$

(40)

where K, the Kalman control gain vector, can be obtained from the iterative solution of the Riccati equation:

$$K(k) = \left[\xi^* + \tilde{b}'P(k+1)\tilde{b} \right]^{-1} \tilde{b}'P(k+1)\tilde{A}$$

(41.a)

$$P^*(k) = P(k+1) - P(k+1)\tilde{b}\left[\xi^* + \tilde{b}'P(k+1)\tilde{b} \right]^{-1} \tilde{b}'P(k+1)$$

(41.b)

$$P(k) = Q + \tilde{A}'P^*(k)\tilde{A}$$

(41.c)

where $Q = dd'$ and the terminal condition $P(k + N) = T$.

At every sampling time the covariance $P(k)$ is updated and the control gain $K(k)$ is evaluated. Generally, U-D factorization is used for numerical stability and computation efficiency (Clarke et al, 1985).

Generalized Predictive Controller

Pole-placement and LQG self-tuners perform badly if the order of the plant is overestimated because of the pole/zero cancellations in the identified model. Although significant progress has been made over the last decade in theoretical understanding and application of adaptive control, no single adaptive control scheme has been suitable as a general-purpose algorithm for the majority of real processes. Clarke et al (1987a, 1987b) proposed that the generalized predictive control (GPC) algorithm is capable of stable control of processes with variable parameters, with variable dead time, and with a model order that changes instantaneously, provided that the input/output data are sufficiently rich to allow reasonable plant identification. It is effective with a process that is simultaneously non–minimum phase and open-loop unstable and whose model is over- or underparameterized by the estimation scheme without special precautions being taken. Essentially, GPC tends to overcome these issues and create a robust control scenario.

Clarke (1988) described implementations of the GPC self-tuner using a standard recursive least-squares parameter estimator with a variable forgetting factor and supervisory software that estimated at appropriate times, either by user command or as signals fall within dead bands. Usual practice is injection of a test signal during open-loop operation to collect adequate data to find a reasonable initial plant model. Then the user can either switch off the estimator or choose to maintain its function for tracking subsequent changes in plant dynamics.

Chapter 6 gives some information on predictive controllers, especially the GPC. To apply GPC, a receding horizon strategy is adopted where at each sample instant the following steps are taken following the acquisition of data $y(k)$ (see Chapter 6):

1. An RLS or other algorithm updates estimates of the parameters of the ARIMAX model (eg, Equation 21).
2. The free response y_1 is computed on the basis of the estimated model and known data, where y_1 is defined as

$$y_1 = y(k) + F_j(z^{-1})\Delta y(k) + G_1(z^{-1})\Delta u(k-1)$$

(42)

3. The vector w is formed from preprogrammed setpoints or with constant elements.
4. The step responses $\{g_i\}$ are evaluated from the model as elements of the matrix **G.**
5. The controller

$$\tilde{u} = [G'G + RI]^{-1} G'[w - p]$$

(43)

is computed given the user-chosen values $[N, NU, R]$ to provide the future control increment vector $\tilde{u}$ (see Equation 4 in Chapter 6).
6. The first element of $\tilde{u}$ is extracted, and the control $u(k) = u(k-1) +$ is applied to the process.
7. Data vectors are shifted in preparation for the next sample.

For small values of NU, the above procedure become simple. The proper choice of horizons leads to fast self-tuning control for realistic applications. Moreover, with the finite horizon approach, it is possible to include constraints such as control amplitude limits (something that is not possible for most self-tuning designs).

PERFORMANCE MONITORING

The name *self-tuner* was derived from the controller's automatically tuning the parameters. The controller can be initialized using open-loop estimation until parameter convergence is achieved (identification of process model), the estimator turned off, and the controller implemented using these tuned or fixed parameters (Isermann & Kofahl, 1985). To bypass many of the remaining theoretical difficulties currently associated with adaptive control in a closed loop (parameter convergence and stability), a supervisory level can be implemented to operate the self-tuning control in a closed loop until the performance is adequate, turn off the adaptation, and continue with the current parameters as a fixed

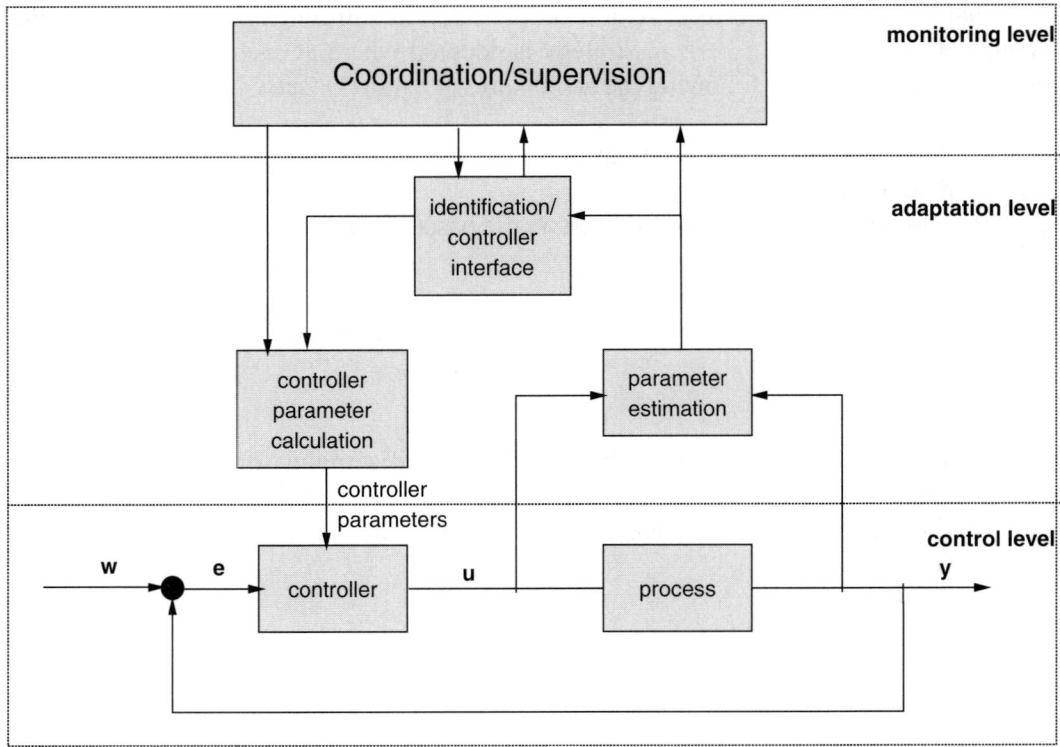

Figure 7–5 Parameter adaptive control loop with monitoring level

controller or continuously monitor the adaptation (Isermann & Lachmann, 1985). Figure 7–5 illustrates the addition of a supervisory level to the adaptive control system.

In supervisory loops, the lower level is the ordinary feedback loop with the process and the controller. The adaptation loop, or level 2, is the level that adjusts the parameters. The third layer, the supervisory level, is used to monitor the performance of the adaptation and subsequent controller modification. It also reviews the overall stability of the system and takes appropriate action to improve the controller's functionality by monitoring such things as the covariance matrix, predictions using new estimates compared to old estimates, and changing of the forgetting factor (Åström & Wittenmark, 1990). Other techniques have been researched to add stability and improve the control. Filtering the estimates, along with prefiltering and antialiasing filtration of the data for improvement of convergence was proposed by Wittenmark (1988). Rohrs et al (1984) recommended the following steps for improving adaptive control performance: keep the adaptation gain small and let it proceed slowly, design the nominal control loop robustly, and sample the system slowly to remove effects of unmodeled dynamics.

One of the problems with adaptive control is that inevitable estimation transients, due to disturbances and parameter tracking, are passed to the controller. Therefore, it would be preferable to update the controller only when the system has changed and parameter estimates have converged with sufficient confidence. The performance of the controller will then be monitored using a supervisory level. A number of performance-monitoring methods have been proposed. Harris (1989), for example, estimated the best possible output variance of the controlled system and monitored the difference from the actual variance. For linear systems with stochastic disturbances, the best possible output variance corresponded to minimum-variance control. It depended solely on the process delay and could be determined by estimating a whitening filter for the output.

Desborough and Harris (1993) proposed the use of a performance index (the ratio of the best variance to the variance of the controlled variable) and the use of ANOVA (analysis of variance) for feedback/feedforward control loops to indicate whether a poor performance is due to the feedforward or feedback controller.

The use of cross-correlation analysis was discussed by Stanfelj et al (1993) to diagnose the cause of poor performances in feedforward/feedback control loops. Qin (1998) has summarized some recent work in this area.

The main tasks of the supervisory level are (1) fault behavior detection, (2) diagnosis of causes, and (3) corrective measures (Isermann, 1989).

The suitable actions required to improve the behavior of self-tuning control systems are described below (Isermann & Lachmann, 1985).

Start-up Procedure

To avoid large and unacceptable process input and output during the start-up phase, a sufficiently exciting perturbation signal must be used to estimate the parameter to obtain a reasonable starting model in an open loop. Comparison of the real process output signal and model output signal must be checked after sufficient identification time of the model is verified. The preidentification also must include an on-line search for structure of the process model (Chapter 4).

Before closing the loop, a backup PID controller is calculated for the working point of the preidentification. Figure 7–6 illustrates the steps required during the start-up.

Parameter Estimation Supervision

The main objective of this step is to check whether the identified process model agrees with the real process. Violations of convergence assumptions could be caused by (Isermann, 1989):

- No persistent excitation
- Noise signals nonstationary (step changes or outliers)
- Rapid changes of process static or dynamic behavior
- Wrong model structure parameters (order and dead time), wrong sample time, wrong forgetting factor, λ (see Appendix 4–C)

For on-line supervision in real time, various measures can be examined, such as

- Equation error (a priori) $e(k)$ and variance value of $e(k)$
- Autocorrelation and cross-correlation functions
- Parameter estimation values: variance of estimated parameters

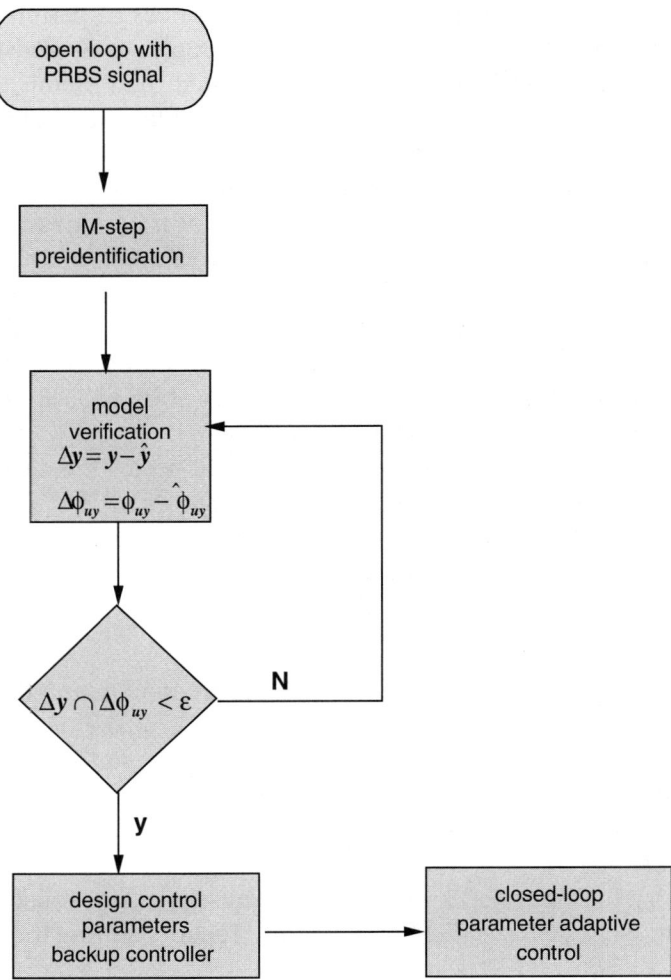

Figure 7–6 Block diagram of start-up procedure with preidentification

- Information matrix $H(k) = P^{-1}(k)$: tr $H(k)$ (Appendix 4–C)
- Control variable: mean $y(k)$ and variance of control signal
- Eigenvalue of the parameter estimator

If these quantities surpass certain limits, some of the measures include stopping the dynamic parameter estimation for M steps, starting parameter estimation again, automatically searching for model order and dead time, and filtering the process signals and parameters estimates (using low-pass filters).

Controller Design

To avoid the application of controllers in cases where the preconditions for controller design are violated, these preconditions should be checked for the actual process model (eg, the cancellation of zeros close or outside the unit circle for minimum-variance controller design). The process model poles and zeros are then calculated based on the parameter estimates, and a new set of controller parameters is determined only if the preconditions are satisfied.

Closed Control Loop

Here, the behavior of the control signal and the manipulated variable is analyzed. If the control deviation is monotonously increasing or the manipulated variable is at the bound for M steps, the self-tuning controller is switched to a fixed backup controller designed during the preidentification phase.

APPLICATIONS OF ADAPTIVE CONTROL

The first example is the use of a pole placement self-tuning controller for a cross-flow grain dryer. The objective of the control system was to maintain the exhaust air temperature

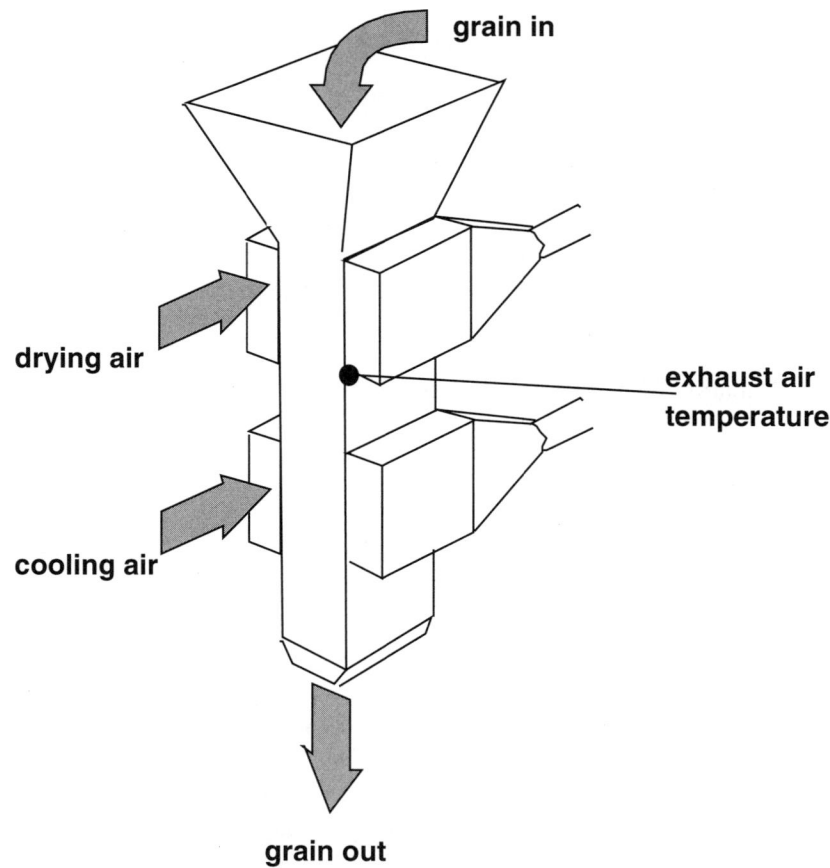

Figure 7–7 Schematic of the cross-flow grain-drying system.

(near the lower edge of the drying section) closer to the setpoint despite changes in the inlet grain moisture content. Figure 7–7 shows the schematic of the drying process. The results presented here are based on the work of Nybrant (Nybrant, 1986).

The model selected to describe the dynamics of the process was the following:

$$y = z^{-1}bSu + \frac{v_1}{\Delta}$$

(44)

where

$$S = 1 + z^{-1} + \ldots + z^{-9}$$

(45)

and y was the exhaust air temperature, u the grain flow rate, and v_1 the white noise.

The closed-loop system was obtained according to Equation 28 as

$$y = z^{-1}\frac{P(1)S}{S(1)P}w + \frac{G}{P}v_1$$

(46)

where P was the desired closed-loop characteristic polynomial and G was given by

$$G = \frac{1}{\Delta}\left[P - z^{-1}\frac{P(1)S}{S(1)}\right]$$

(47)

Since S in Equation 45 had zeros on the unit circle (Nybrant, 1986), the characteristics equation chosen was

$$P = S + \beta$$

(48)

where β was the pole placement parameter that gave desired properties to the closed loop. A small value of β gave poles that almost canceled the zeros of S, a fast closed-loop system response, and large input values. A large β gave poles that were closer to the open-loop poles in $z = 0$. The zeros of S in this case were not canceled given a slower closed-loop response and smaller input values.

The control was implemented in the pilot-size dryer. The value of β was equal to 7, and the forgetting factor in the RLS algorithm (see Appendix 4–C) was equal to 0.98.

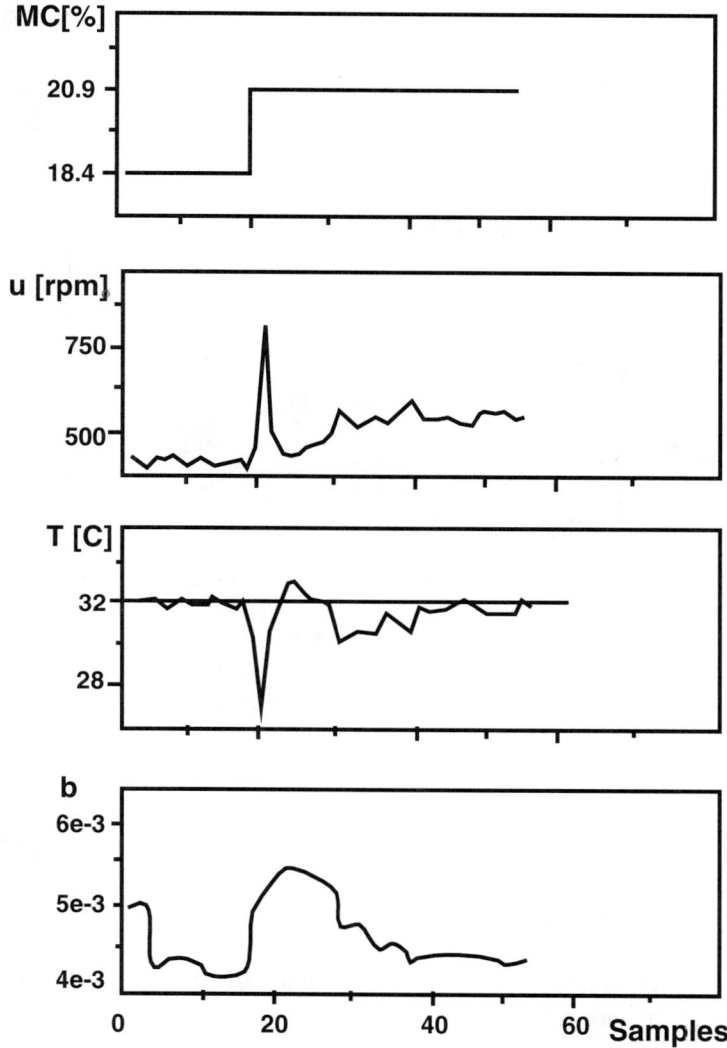

Figure 7–8 Control of exhaust air temperature in a cross-flow grain dryer using a self-tuning pole placement algorithm

The dryer started in steady-state operation, with the only parameter to be estimated, b, equal to 6×10^{-3} and $P = 0.1$. The inlet moisture content of the grain (wheat) was first 18.4% w.b. and then changed to 20.9% w.b. The results of the control performance are shown in Figure 7–8.

The estimated parameter b converged very fast, and it took only few samples for the system to reach good control. Before the step change in the inlet moisture content took place, the control standard deviation error was 0.17 with a mean value of -7.8×10^{-2}. The controller was able to keep the output closer to the setpoint even after a step in the

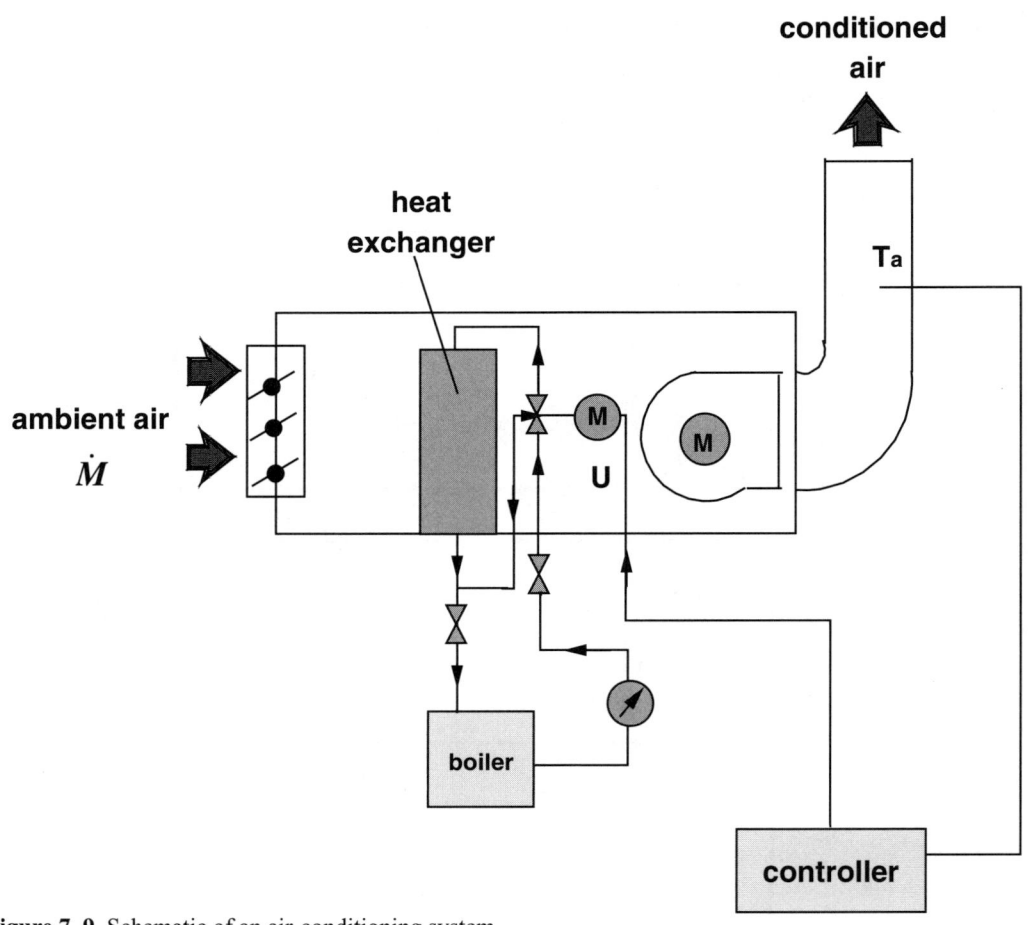

Figure 7–9 Schematic of an air-conditioning system

inlet moisture content occurred. The control responded quickly and accurately to the step change in the inlet moisture content.

Isermann and Lachmann (1985) implemented a self-tuning controller in an air-conditioning unit. Figure 7–9 illustrates the basic structure of the system. Ambient air flows through the plant, where the air flow rate, $\dot{m}$ can be varied from 0 to 1000 m³/h. The air is heated by a water-heater type of heat exchanger whose water inlet temperature can be varied by the three-way valve actuated by the control input U. The temperature of the air, T_a, is measured at the air outlet of the unit. The static and dynamic behaviors of the system are strongly influenced by the characteristics of the valve, the heat transfer, and the air flow rate. In addition, the system is nonlinear.

Figure 7–10 shows the experimental results of the self-tuning controller without additional supervision functions. Large variations of the input and output signal are observed. This is also the case for subsequent setpoint changes for the temperature T_a and for the air flow rate $\dot{m}$. Considerable improvement of the overall control performance is achieved when using supervision functions (Figure 7–11). The signal variations are much

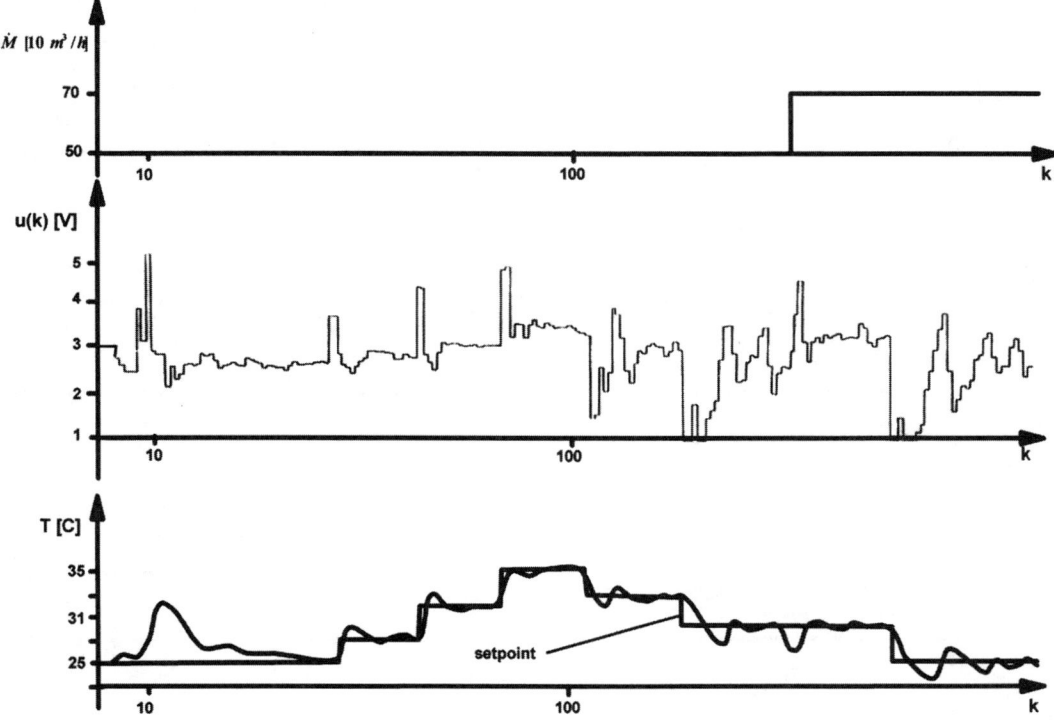

Figure 7–10 Control of an air-conditioning unit without performance monitoring.

smaller. After a brief preidentification, the loop is closed and stabilized. The controller parameters are adapted for each step change of the setpoint.

Chang and Tan (1992) used the generalized predictive control (GPC) in a food extruder. A SISO system was considered with the screw speed (SS) as the manipulated variable and the product pressure in the die (PD) as the controlled variable. The model structure, determined using black box system identification, was

$$[1 - a_1 z^{-1} + a_2 z^{-2}]PD(k) = b_o SS(k - d) + v_1(k)$$

(49)

where d, the time delay, was set to be equal to 10 sampling periods. In the food extrusion process, the time delay varies; therefore, it needs to be estimated at every sample time. One approach used was to expand the polynomial B by adding extra b terms into the $B(z^{-1})$ polynomial. Thus, the b_i terms that corresponded to the correct time delay would have

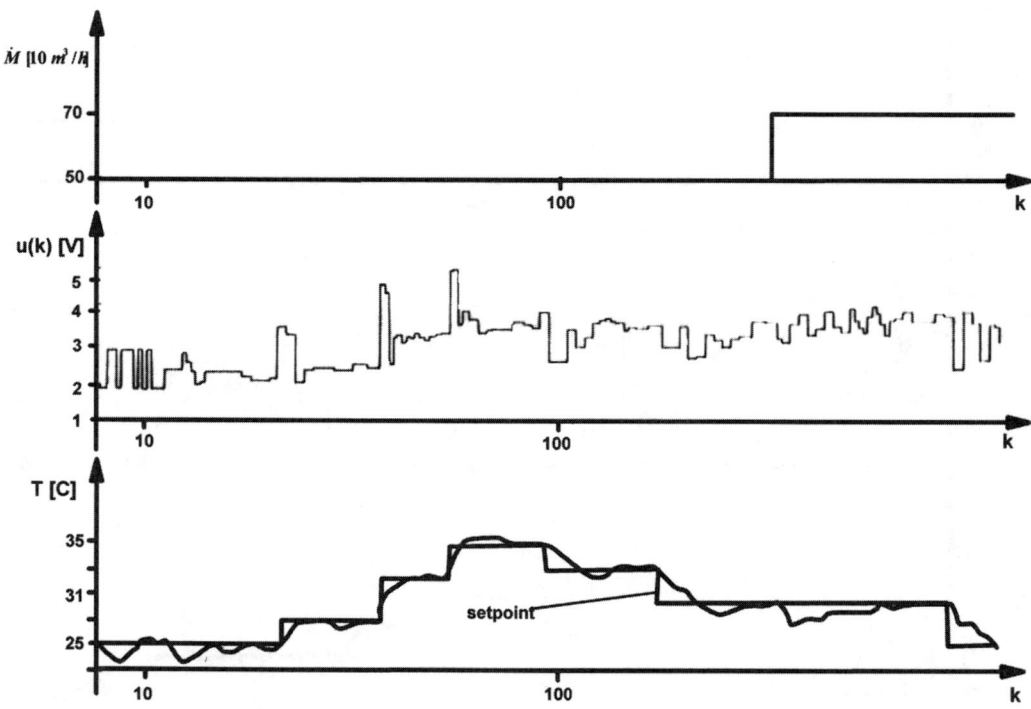

Figure 7–11 Control of an air-conditioning unit with performance monitoring.

more significant values than the others, while the redundant ones would have near-zero values when the estimated process parameters converged.

The GPC and on-line RLS algorithms were implemented in the extrusion process. A square wave disturbance in the moisture content (Figure 7–12) was used to simulate feed material variability. The feed rate was kept constant during the experiments.

Figure 7–12 shows the PD response with and without control under feed moisture content disturbance. When the moisture changed, the product pressure at the die was off the setpoint for the uncontrolled case, while the controlled response always returned to the setpoint. However, the overshoot created by the moisture disturbances on the output response was not significantly reduced with the controller due to the large phase lag. A feedforward loop from the feed moisture to the output would improve the controller performance.

Track performance tests with the controller (Figure 7–13) showed that after an initial learning period the controller was able to track the changes in the setpoint well.

COMMERCIALLY AVAILABLE ADAPTIVE CONTROLLERS

Several commercial adaptive controllers are available in the United States and abroad. In 1983, ASEA (Stockholm, Sweden) introduced an adaptive controller called Novatune,

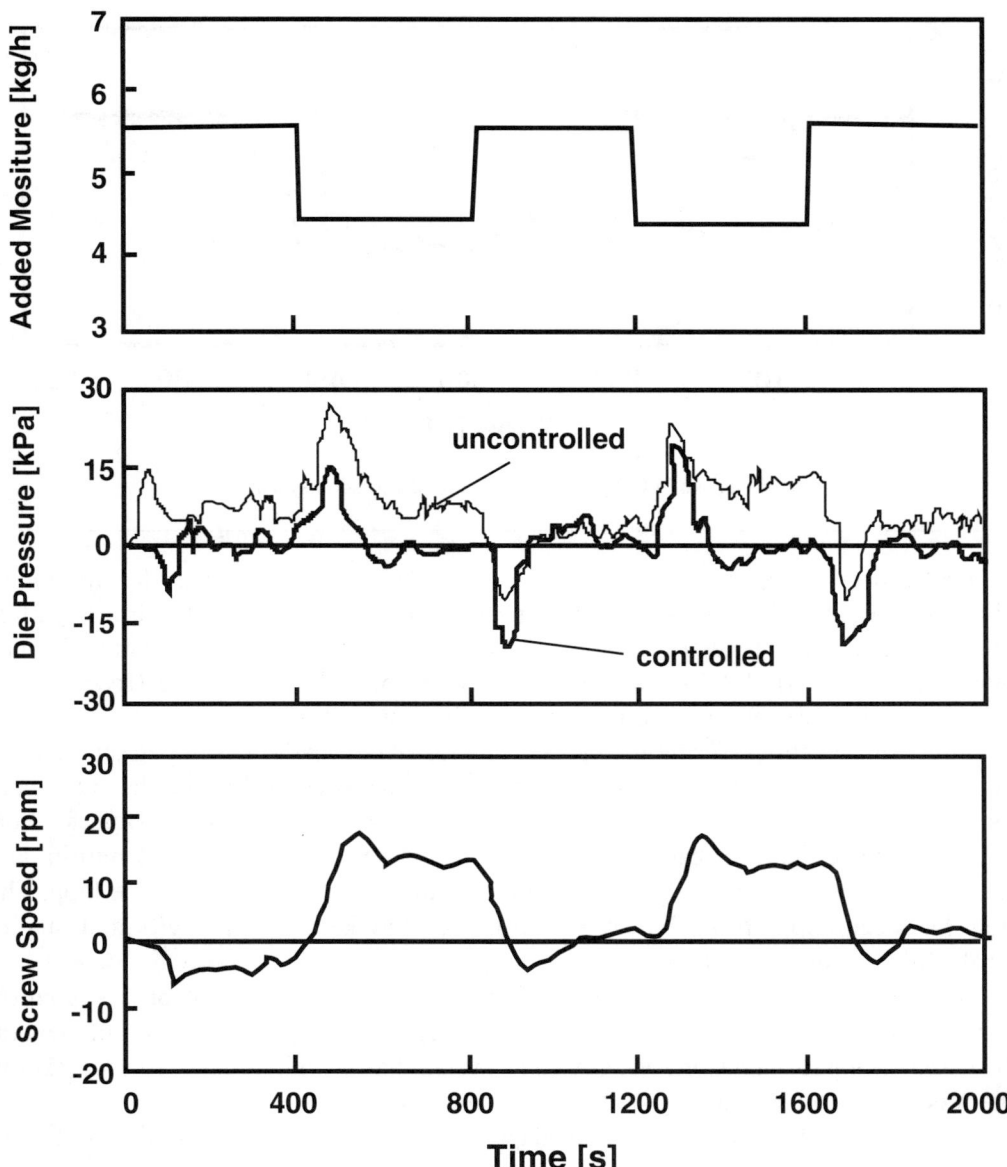

Figure 7–12 Regulation performance of a GPC controller due to changes in the feed moisture content of a food extrusion system.

based on the minimum-variance controller. The Foxboro Company (Foxborough, MA) developed a self-tuning PID controller based on the "expert system" approach.

The Predicted Control (Norwich, Cheschire, UK), formed in 1988, developed the Connoisseur, an adaptive process control software based on the model predictive control approach. The system has been successfully applied to a dairy plant in New Zealand

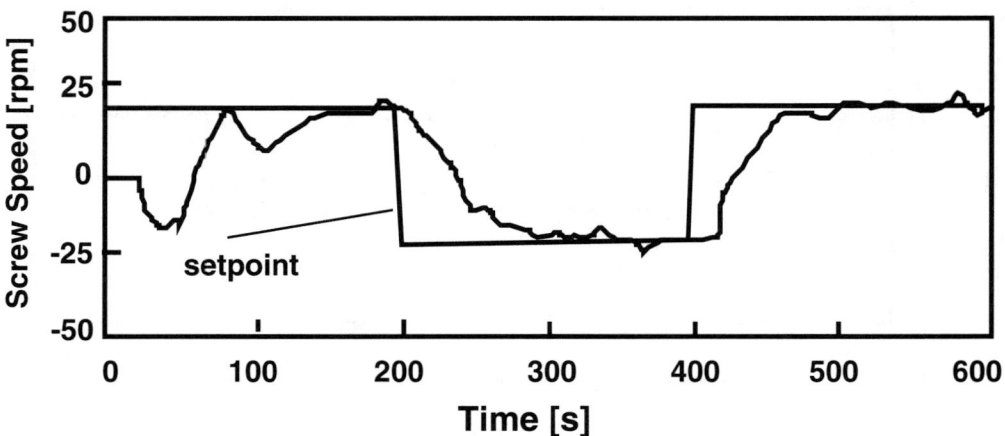

Figure 7–13 Tracking performance of a GPC controller in a food extrusion system.

to control process temperature and pressure and product flow rate and density. Other applications include coffee roasting and sugar refining. The Predictive Control is part of the Siebe group, which, together with other company members such as Simulation Science (SIMSCI (Brea, CA)), APV (Rosemont, IL), Foxboro, and Wonderware (Irvine, CA), creates a potent combination for food-manufacturing applications.

Universal Dynamics Technologies (Richmond, BC, Canada) developed an adaptive controller called BrainWave. It is an adaptive model–based controller that has two main parts: the model-building component and the predictive adaptive controller. Both feedforward and feedback control capabilities reside in the software. It also has multimodel capability. The system can apply up to 10 process models to control the same loop. The system has been applied to a brewery to improve brew-kettle productivity and quality control while reducing steam costs.

CyboTune is another commercially available software product used in process control. It is defined as a model-free adaptive controller developed by CyboSoft, a unit of General Cybenation Group (Ranch Cordova, CA). The system has been implemented successfully in the waste water treatment of a food-processing plant.

The structure of a 2×2 model-free adaptive (MFA) controller is described in Figure 7–14. Basically, it consists of two controllers with decoupling compensators. Generally, each single-variable MFA controller is designed to have only one tuning parameter, the controller gain (Cheng et al, 1998). The details of the technique, however, are proprietary.

CONCLUSION

The automatic tuning of controllers for process plants is of increasing industrial relevance. The first generation of self-tuners concentrated on automating classical control systems by adjustment of PID parameters. It was found that more consistent performance was obtained with auto-tuned PID controllers than with manually tuned PID controllers.

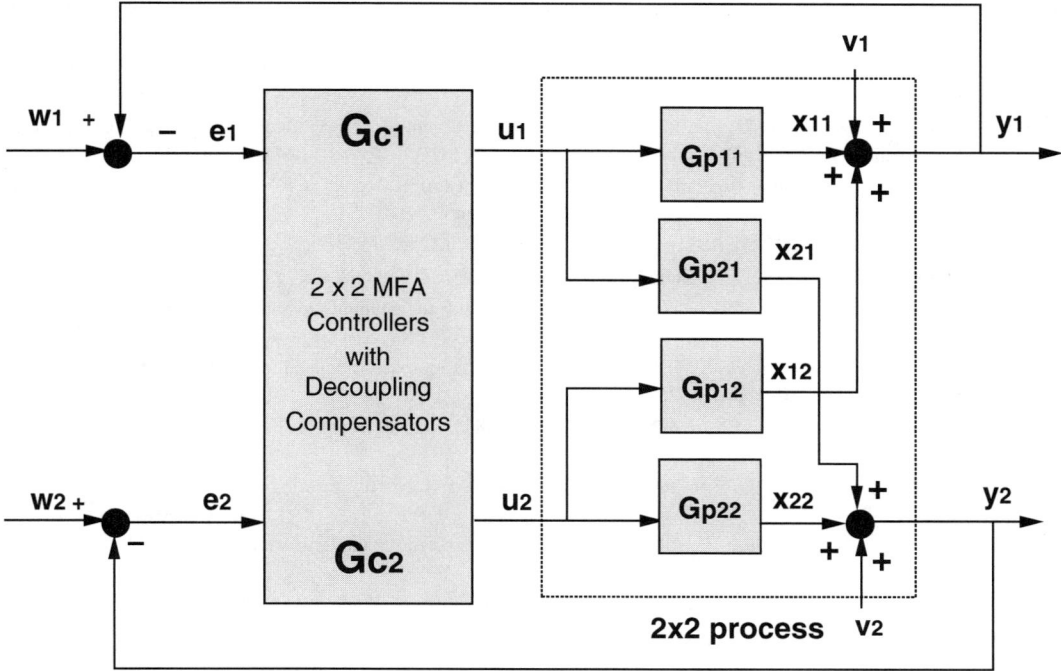

Figure 7–14 A 2 × 2 model-free adaptive (MFA) control system

However, the basic PID design is not very effective for plants with dead times or where feedforward signals and multiloop interactions are present. Fortunately, the self-tuning controller is very flexible, being able to accommodate a variety of estimation procedures and controller design techniques.

The self-tuning controller bypasses many of the theoretical difficulties associated with adaptive control. Its main limitation is identification of the system under closed-loop conditions. Both feedback and lack of persistence of excitation can interfere with accurate identification. The controller's objective is to regulate the process smoothly, while the closed-loop identification algorithm requires process perturbation to operate efficiently. Consequently, the better the controller performs, the worse the adaptation scheme predicts the model parameters. One solution to this problem is supervisory control that monitors the process inputs/outputs and the identification results and adapts only when the supervisory criteria indicate that adaptation is needed. Self-tuning controllers often adapt only in the initial controller start-up phase until parameter convergence occurs; then adaptation is switched off, and the controller is implemented as a fixed controller using the new parameters.

REFERENCES

Åström, K.J. (1983). Theory and applications of adaptive control: a survey. *Automatica* 19, 471–486.

Åström, K.L. & Wittenmark, B. (1990). *Computer Controlled Systems: Theory and Design*. Englewood Cliffs, NJ: Prentice Hall.

Chang, Z. & Tan, J. (1992). *Adaptive Control of Food Processes with Time Delay*. St. Joseph, MI: American Society of Agricultural Engineers. Paper No. 92-6544.

Cheng, G., Wang, J. & Smialkowski, S. (1998). Model-free adaptive control of evaporators. Processing IEEE Dynamic Control Applications for Industry Workshop. Vancouver, Canada, March.

Clarke, D.W. (1988). Application of generalized predictive control. In *Proceedings of IFAC: Adaptive Control of Chemical Processes*, pp 1–7. Edited by M. Kummel. New York, NY: Pergamon Press.

Clarke, D.W. & Gawthrop, P.J. (1975). Self-tuning control. *Proc IEEE* 122, 929–934.

Clarke, D.W. & Gawthrop, P.J. (1979). Self-tuning control. *Proc IEEE* 122, 929–934.

Clarke, D.W. & Hasting-James, R. (1971). Design of digital controller for randomly disturbed systems. *Proc IEEE*, 118, 1503–1508.

Clarke, D.W., Kanjilak, P.P. & Mohtadi, C. (1985). A generalized LQG approach to self-tuning control. Part II: implementation and simulation. *Int J Control* 41, 1525–1544.

Clarke, D.W., Mohtadi, C. & Tuffs, P.S. (1987a). Generalized predictive control: part I. The basic algorithm. *Automatica* 23, 137–148.

Clarke, D.W., Mohtadi, C. & Tuffs, P.S. (1987b). Generalized predictive control: part II. Extensions and interpretations. *Automatica* 23, 149–160.

Desborough, L. & Harris, T.J. (1993). Performance assessment measures for univariate feedback control. *Can J Chem Eng* 71:605–616.

Goodwin, G.C. & Sin, K.S. (1984). *Adaptive Filtering Prediction and Control*. Englewood Cliffs, NJ: Prentice Hall.

Hägglund, T. (1983). The problem of forgetting old data in recursive estimation. *In Proceedings of IFAC: Adaptive Systems in Control and Signal Processing*, pp 213–214. San Francisco, CA: IFAC.

Harris, T.J. (1989). Assessment of control loop performance. *Can J Chem Eng* 67, 856–861.

Isermann, R. (1982). Parameter adaptive control algorithms: a tutorial. *Automatica* 18, 513–528.

Isermann, R. (1989). *Digital Control Systems. Vol 2. Stochastic Control, Multivariable Control, Adaptive Control, Applications*. New York, NY: Springer-Verlag.

Isermann, R. & Kofahl R. (1985). On the application of parameter adaptive control systems for industrial processes. In *Proceedings of IFAC: Adaptive Control of Chemical Processes*, pp 121–126. Frankfurt, Germany: IFAC.

Isermann, R. & Lachmann, K.H. (1985). Parameter-adaptive control with configuration aids and supervision functions. *Automatica* 21, 625–638.

Landau, I.D. (1974). A survey of model reference adaptive techniques: theory and applications. *Automatica* 10, 353–379.

Landau, I.D. (1979). *Adaptive Control: The Model Reference Approach*. New York, NY: M. Dekker.

Nybrant, T. (1986). *Modeling and Control of Grain Dryers*. Uppsala, Sweden: Institute of Technology, Uppsala University. UPTEC 8625 R.

Parks, P.C. (1981). Stability and convergence of adaptive controllers: continuous systems. In *Self-tuning and Adaptive Control: Theory and Applications*. Edited by Harris & Billings. London, England: P. Peregrinus.

Qin, S.J. (1998). Control performance monitoring: a review and assessment. *Comput Chem Eng* 23, 173–186.

Rohrs, C.E., Athans, M., Valavani, L. & Stein, G. (1984). Some design guidelines for discrete-time adaptive controllers. *Automatica* 20, 653–660.

Slotine, J.J.E. & Li, W. (1991). *Applied Nonlinear Control*, pp 311–389. Englewood Cliffs, NJ: Prentice Hall.

Stanfelj, N., Marlin, T.E. & MacGregor, J.F. (1993). Monitoring and diagnosis of process control performance: the single loop case. *Ind Eng Chem Res* 67, 856–861.

Tuffs, P.S & Clarke, D.W. (1985). Self-tuning control of offset: a unified approach. *Proc IEEE* 132, 100–108.

Wittenmark, B. (1988). Implementation and application of adaptive control. In *Proceedings of IFAC: Adaptive Control of Chemical Processes*, pp 9–14. Edited by M. Kummel. Copenhagen, Denmark: IFAC.

Wittenmark, B. & Åström, K.J. (1984). Practical issues in the implementation of self-tuning control. *Automatica* 20, 595–605.

CHAPTER 8

Statistical Process Control and Intelligent Controllers

Intelligent methods in control are becoming part of mainstream control approaches. They integrate concepts and methods from such areas as control, identification, estimation, communication theory, and artificial intelligence. Intelligent control paradigms can endow control systems with functionality that traditional methods could hardly make available. They can also improve the analysis, design, operation, and maintainance characteristics of control systems.

In the design of controllers for complex dynamic systems, the existing conventional control theory cannot address all the needs required to successfully control these systems. The controller should include a series of functions to make the controller intelligent/autonomous to effectively control complex processes. For example: (1) heuristic methods may be needed to tune the parameters of an adaptive control law; (2) new control laws to perform novel control functions to meet new objectives should be designed while the system is in operation; (3) learning from past experience and planning control actions may be necessary; and (4) failure detection and identification are needed. All these functions have been performed by human operators in the past. An intelligent controller, if well designed, can increase the speed of response, relieve operators from mundane tasks, and protect operators from hazards, thus creating a high degree of autonomy.

Examples of methods for using computers to automate a number of process control tasks that would not be practical using traditional tools include fuzzy logic and neural networks. Neither technique is a replacement for conventional control systems. They are additional tools that can be added to, or used beside, more traditional methods.

Fuzzy logic and neural networks both aim at mimicking the operations of the human brain to some extent. A major goal of fuzzy control is to model the control actions of the operator. A control law described by fuzzy rules is more understandable by a human operator who knows the process but who is not an expert control engineer. It facilitates the integration of heuristic knowledge about the system and theoretical control laws from mathematical models.

Neural networks go further by aiming at building models of the way a human being thinks and reaches conclusions. In general, neural networks are useful for representing and approximating nonlinear relationships. They are applicable for multivariable systems and require less restrictive assumptions for the process than conventional modeling and control techniques. The reader is referred to Verbruggen et al (1999) for detailed information on fuzzy control and to Hunt et al (1992) and Bhat and McAvoy (1990) for information on neural networks.

Another important technique that only recently has become widely applied is statistical process control (SPC). It involves the application of statistical concepts to determine whether a process is operating satisfactorily. A process is said to be operating satisfactorily—that is, to be in control—if the product quality variations fall within acceptable bounds (minimum and maximum values of a specific attribute). The subject of SPC is very broad, and only a simple introduction will be presented in this chapter. For detailed information on control charts, sampling, probability distribution, and so forth, the reader is referred to Kear (1998), Grant and Leavenworth (1988), and Montgomery (1985).

STATISTICAL PROCESS CONTROL

For several decades, SPC and quality control have been synonymous in most high manufacturing environments. However, little is written on the successful application of SPC in the food industry. SPC provides many benefits to the manufacturing industries, including economic, predictive, and systematically documented process control (Gaafar & Keats, 1992).

In continuous processes where automatic feedback control has been implemented, the feedback mechanisms theoretically ensure that product quality is at or near the setpoint, regardless of process disturbances. Nevertheless, even under feedback control, there may be variations in product quality due to disturbances or equipment or sensor failures. The concepts of SPC can be used to analyze these incidents.

The development of SPC started about 50 years ago (Deming, 1982). Basically, SPC focuses on four key points relating to the manufacturing process:

1. Understanding the process fully and completely
2. Making sure that the process is entirely capable of meeting its goal
3. Making sure that the process is operating optimally at all times
4. Continuously improving the process

Take, for example, the food-manufacturing process. A change in personnel involved in the processing line, raw material variability, or changes in sensors and equipment will have a potential effect on the final product quality. To maintain profitability and stay in the market, the company will have to minimize the effect of these changes on the final product characteristics.

To know what variations there are in processed food products, we must have some method of measuring the product quality attributes (PQAs) and of determining how these PQAs change from product to product. Examples of PQAs are

- appearance
- color
- texture
- bulk density

- dimensions
- taste
- flavor

Other product attributes that are useful for measuring manufacturing processes are

- function
- cost
- competitiveness and affordability
- availability
- on-time delivery

Basically, SPC aims at managing the connection between the above attributes and the manufacturing process.

SPC in manufacturing practice is concerned with examining the products manufactured to determine the extent of variation that they exhibit. These variations can be quantified by measuring various aspects of the products. A number of SPC tools then may be used to mathematically describe the manufacturing process. The main objective of these tools is to design, correct, and manage the process to eliminate unacceptable products.

When measuring any attributes or characteristics of the products being processed, the measured results are distributed, with most of the measurements near a mean value. Figure 8–1 illustrates the variation of the controlled variable, in this case a PQA, represented in the form of a histogram, that may be expected to occur under typical steady-state operating conditions. The mean value ($\bar{y}$) and the standard deviation σ are identified in Figure 8–1 and can be calculated as

$$\bar{y} = \frac{1}{n}\sum_{i=1}^{n} y_i$$

(1)

$$\sigma = \left[\frac{1}{n}\sum_{i=1}^{n}(y_i - \bar{y})^2\right]^{1/2}$$

(2)

The standard deviation, σ, is a measure of the spread of observations around the mean. A large value of σ indicates that wide variations in y (or PQA) are occurring. The probability that the controlled variable lies between two arbitrary values, y_1 and y_2, is given by the area under the histogram between y_1 and y_2. If the histogram follows a normal probability distribution (the curve in Figure 8–1), then 99.7% of all observations should lie within $\pm 3\sigma$ of the mean. These upper and lower control limits are used to check if the process is operating satisfactorily. The target value or setpoint of the controlled variable should be selected near the mean $\bar{y}$ so that both upper and lower control limits lie inside the operating constraints. In other words, violations of the product specification are very unlikely (ie, they have a low probability).

SPC is a diagnostic tool: it is used as an indicator of quality problems, but it does not recognize the source of the problem or the corrective action to be taken. Consider, for example, the process control chart for lethality data over a time period of 0.6 hours from a high-temperature, short-time (HTST) pasteurization system (Figure 8–2). Assume that the mean value and the standard deviation have been calculated on the basis of early observations. If the new data lie within the lethality limits ($\pm 3\sigma$), the process is *in control*. However, if repeated violations of the $\pm 3\sigma$ limits occur, the process is *out of control*. This situation occurs in Figure 8–2 at $t = 0.27$ hour. This may be caused by changes in raw material, equipment, or instrument malfunction and should be corrected in real time. In Figure 8–2, the lethality rate could be adjusted by increasing the residence time of the product in the holding tube, for example.

Quality Control Charts

Quality control charts are used to monitor and control manufacturing processes. The idea simply consists of plotting the data as soon as they become available and observing

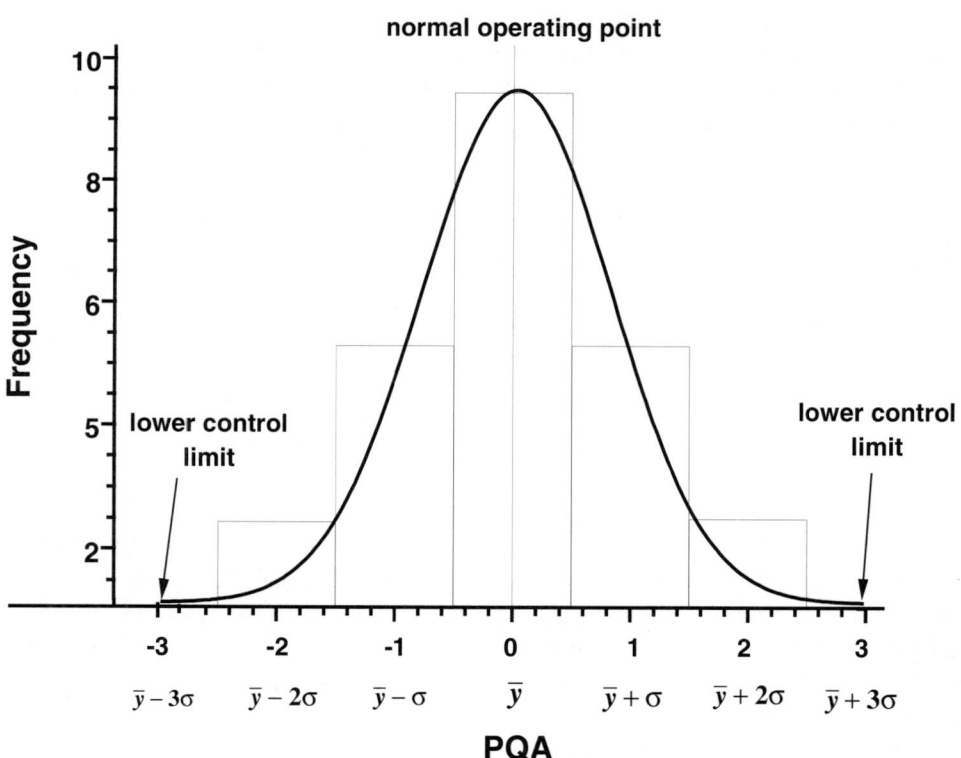

Figure 8–1 A histogram and a normal probability distribution fit to the data

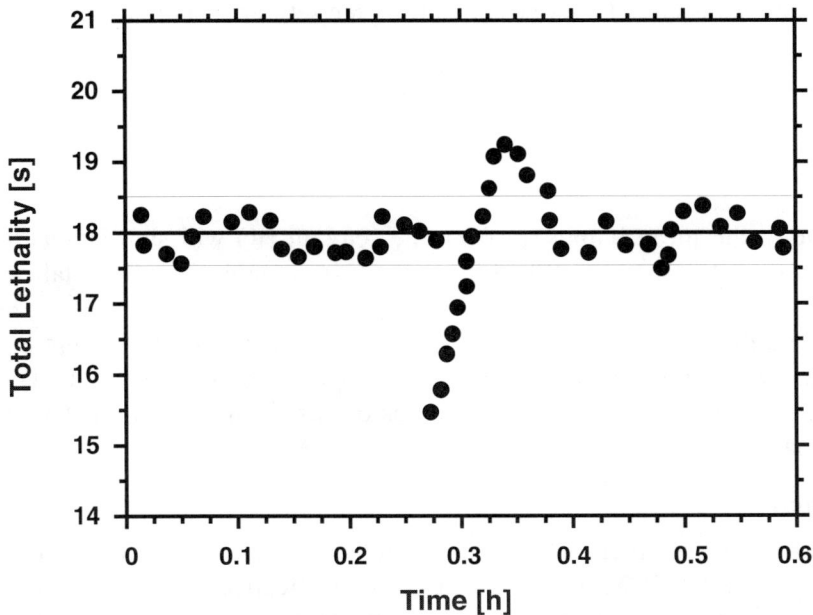

Figure 8–2 Process control chart for the average lethality readings

trends and changes. Some of these charts are the Shewhart, the CUSUM, and the EWMA charts.

Shewhart Chart

To help in assessing whether changes have occurred in the process, Shewhart (1931) suggested plotting the data sequentially in time on a chart containing the target (or setpoint) value and upper and lower action limits. This chart (Figure 8–2) is still the most used control chart.

The objective of Shewhart's procedure is to indicate when an out-of-control situation has occurred so that by looking carefully at the process data around that period of time one can find an assignable cause and improve the process. It does not provide a decision mechanism for when to take feedback control action.

A major disadvantage of the Shewhart control chart is that it uses only the information about the process contained in the last plotted point and ignores any information given by the entire sequence of points.

CUSUM Chart

The CUSUM (cumulative sum) chart directly incorporates all of the information in the sequence of sample values by plotting the cumulative sums of the deviations of the sample values from a target value. The chart is a sequential likelihood ratio test (Prins, 1993). It checks the hypothesis that the process mean is equal to the setpoint value against the alternative hypothesis that it deviates by some specified amount.

The cumulative sum of the deviations from the setpoint is defined as

$$S = \sum_{i=1}^{n} [y_i - w]$$

(3)

Any change in the mean output (y) from the setpoint (w) will show up as a change in the slope of the CUSUM plot. When the process in on target, a horizontal trend will be obtained.

To assess the statistical significance of any change, a V mask (Figure 8–3) is often used (Prins, 1993). When either leg of the V mask crosses the plotted CUSUM values, a statistically significant change in the mean has occurred. In general, the CUSUM is able to detect smaller changes than the Shewhart procedure.

EWMA Chart

The standard Shewhart control chart is relatively insensitive to small shifts in the process mean. The CUSUM control chart is more effective than the Shewhart chart in detecting small shifts. Control charts based on the moving average are also very effective in detecting small process deviations.

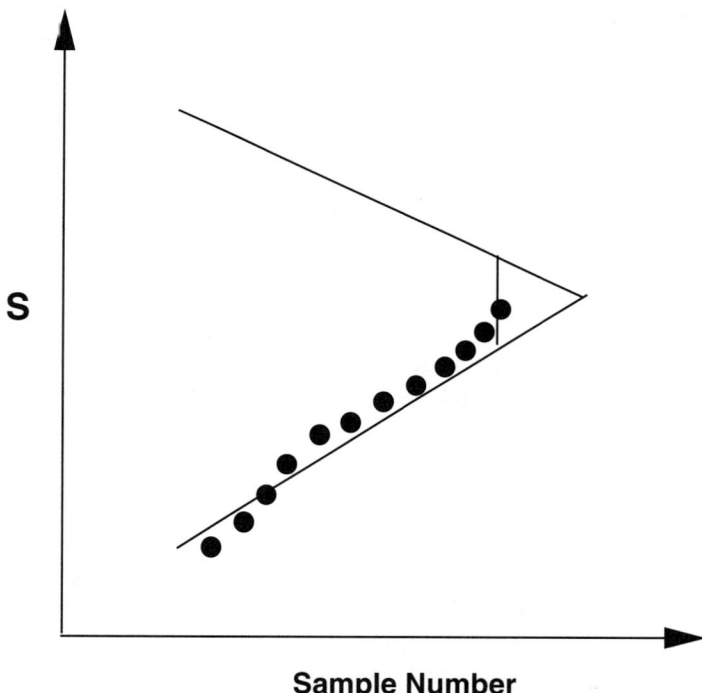

Sample Number

Figure 8–3 CUSUM plot with V mask

The exponentially weighted moving average (EWMA) control chart plots EWMA [$\bar{y}(i)$] of the observations, where

$$\bar{y}(i) = \lambda\bar{y}(i-1) + (1-\lambda)y(i)$$

(4)

and λ is the EWMA parameter ($0 < \lambda < 1$), a filter to discount past data (Prins, 1993). Whenever the EWMA exceeds some upper or lower control limit, an adjustment is made. For $\lambda = 0$ (ie, only the current value is weighted), the EWMA chart is equivalent to the Shewhart chart; for $\lambda = 1$, the EWMA chart is equal to the CUSUM chart. The test procedures have been developed as tests of the hypothesis that $\bar{y} = w$ against $\bar{y} \neq w$.

SPC of Multivariable Processes

When several related variables are of interest, a multivariable quality control approach is advised. If the process measurements do not have significant autocorrelation, the principal components are available for monitoring the variability of the process (Kresta et al, 1991). Traditionally, control charts are developed assuming that the process is operating at steady state, and the covariance between lagged values of the variables is not considered. The zero-lag covariance matrix used to calculate the principal components does not include the significant variations between the current value of a process variable with respect to its past values.

An SPC method based on a state-space model is useful for autocorrelated processes. The general form of the state-space model is

$$x(k+1) = Ax(k) + be(k)$$
$$y(k) = Cx(k) + e(k)$$

(5)

where x is the $n \times$ state vector, y the $1 \times p$ output vector, e the stochastic input vector, A the $n \times n$ time-invariant matrix, and C the $p \times n$ observation matrix. The state variables are autocorrelated and cross-correlated except at zero lag.

For stationary linear systems, where the mean of the state vector is zero and its covariance matrix is a constant diagonal matrix, a scalar statistic called Hotelling's T^2 can be defined as (Negiz et al, 1998b):

$$T^2(k) = x'(k)P^{-1}x(k)$$

(6)

where P is the $n \times n$ covariance matrix. Equation 6 is true if all n variables are included in the statistics. The critical value for T^2 is obtained from an F distribution:

$$T_{cr}^2 \sim \frac{n(N^2 - 1)}{N(N - n)} F_{n,N-n}$$

(7)

where T_{cr} is the critical value of T^2, N the number of samples, and $F_{n, N-n}$ the F distribution.

Applications of SPC

The main assumption behind these charts is that when the process is in control, the observations $y(i)$ are independently and normally distributed about a mean ($\bar{y}$) with constant variance (σ^2). If this assumption is valid, it is not possible to predict the value of a specific measurement based on the values of previous measurements of the same variable. For most continuous processes, especially if the measurements are collected at high frequency, this assumption is not correct (Negiz et al, 1998a). Harris and Ross (1991) observed that if the autocorrelation is on the same order of magnitude as the noise in the process data, the use of these control charts on measurements is acceptable.

Another assumption behind the control chart philosophy is the idea that a hypothesis-testing procedure is appropriate. In other words, the hypothesis testing is appropriate when changes in the manipulated variable are infrequent because of their large cost (eg, a process shutdown), and the process is at steady-state. So in this case the use of control charts is optimal. However, when (1) the cost of corrections in the manipulated variable is small compared to the cost of quality deviations, (2) the process dynamics are important, and (3) corrections are frequently made by feedback control, SPC shows that a PID controller is optimal. Therefore, process control charts should not be a substitute for PID regulatory control, and vice versa (Seborg et al, 1989).

Statistical Monitoring of Product Lethality

Negiz et al (1998b) used SPC to obtain information on how close an HTST dairy pasteurization process was to noncompliance with product safety limits and to carry out periodic checks of sensor accuracy at high frequency. The process is described in Chapter 2 (in the section "Milk Pasteurization Process Control").

To reduce the correlations in the data collected (total lethality) from the continuous process, Negiz et al (1998b) suggested using new independent variables from process information by building mathematical models from process data collected when the process was known to operate properly. They found that because of the significant autocorrelation in the total lethality value data, SPC charts based on residuals would perform better: that is, the residuals would be expected to be independent and normally distributed. If the autocorrelation was positive and very high (>0.9), they recommended that monitoring changes in the model parameter values would be more effective.

The feedback control system for the HTST pasteurization process developed by Negiz et al (1998a) maintains total lethality at the setpoint (see Figure 8–4). Variations in total lethality around the setpoint (18 seconds) is caused by the presence of inherent process

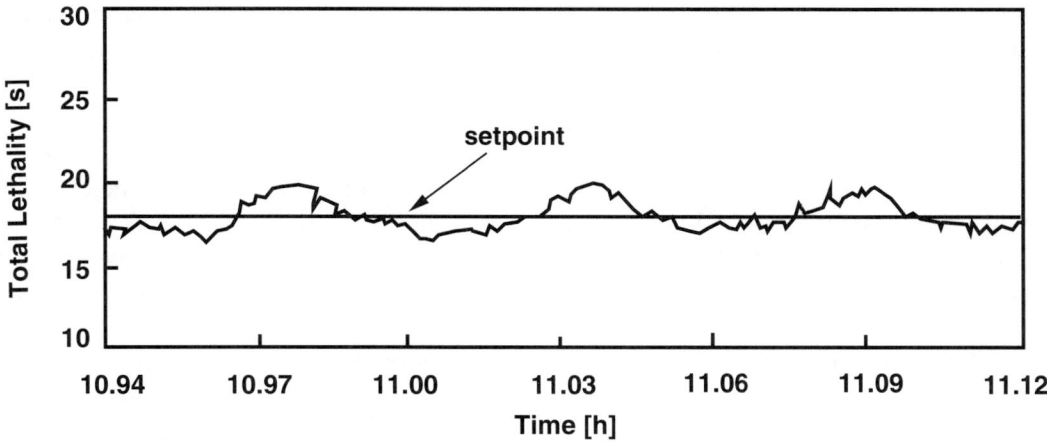

Figure 8–4 Total lethality data from an HTST pasteurization process.

dynamics and random disturbances. As long as the lethality value is always equal or greater than the legal limit (15 seconds), these small variations do not cause a significant process deviation (so the process is in control). Note that the 18-second setpoint corresponds to a 20% overprocessing. However, for safety, it is important to statistically monitor these variations in lethality so that undesirable trends that may lead to noncompliance are detected and corrected quickly by the operator (Negiz et al, 1998b).

Residuals were calculated by subtracting the actual lethality measurements (L) from the model prediction for the measured lethality ($\hat{L}$). The model selected to describe total lethality was a fifth-order autoregressive (AR) model (see Chapter 4). Shewhart charts could then be used for the residuals, and the HTST pasteurization process could be statistically monitored.

In addition, statistical monitoring of product flow rate and temperature and hot water outlet temperature was performed by generating residuals and then constructing Shewhart charts following the procedure used in the lethality monitoring. These charts were used to assess the status of the temperature and flow sensors.

Monitoring the residuals is not suitable for tight monitoring of process variables that are strongly positively correlated (Negiz et al, 1998b). Processes that use feedback control, especially with a controller that has integral action, will have variables that are strongly autocorrelated. So residual-based SPC methods are not suitable, since they generate too many false alarms and detect actual alarms too late (Harris & Ross, 1991). In this case, Negiz and Çinar (1995a, 1997) proposed a technique based on the monitoring of the parameters of the dynamic input-output model. The procedure consists of monitoring the parameters that are updated recursively every sampling time and developing Shewhart charts for these parameters. If the model parameters change beyond their upper and lower limits, the process is out of control. Negiz et al (1998b) applied this method to the HTST pasteurization system and found it to be faster than the residual-based methods.

In the case of multivariable processes yielding highly autocorrelated and cross-correlated measurements, Negiz et al (1998b) suggested the canonical variates state-space realization methodology. The procedure can be implemented by computer to carry out the assessment of sensor behavior periodically at high frequency and to report the outcome to plant personnel for inspection and maintenance.

SPC for a Food Extrusion Process

An example of an application of SPC to a food extruder is illustrated below. The twin-screw extruder described in Chapter 2 (Schonauer & Moreira, 1997) was controlled using an ARX-GPC control combination (Chapter 6). Color-b (CB) was selected as the controlled variable. Feed rate (FR), water rate (WR), and screw speed (SS) were selected as the manipulated variables.

The twin-screw extrusion process under study exhibited a high-amplitude, low-frequency output disturbance that was generated in part from the presentation of the collets (extrudates) to the sensor, as well as a low-frequency, process-generated disturbance (Schonauer, 1995). Four filtering techniques were incorporated during simulation and implementation to eliminate the effects of these unwanted inputs. These techniques included simple low-pass digital filtering; a combination of low-pass digital filtering with forward and reverse filtering of the past output vector to smooth it without causing a phase shift; and SPC principles in combination with each of these filtering techniques. The SPC controller was based on properties of the steady-state CB signal. The signal's noise/disturbance amplitude and characteristics were used to impose error bands around the setpoint for upper and lower control limits. Corrective action was taken by the controller only when both the current measurement and the past three measurements were outside these control limits, thus reducing unnecessary corrective action based on non–process-related or false output disturbances.

The controller tested was a combination of SPC logic and the ARX-GPC control scheme with simple low-pass digital filtering. The ARX-based control scheme was selected for SPC incorporation because it appeared to be more sensitive to the low-frequency disturbance that was present in the system. Figure 8–5 shows the controller tracking of a CB step change of +4 units (Schonauer & Moreira, 1995). CB tracks as well or better than the ARX-based GPC controller tested (Figure 6–10). It is also clear that the inputs WR and SS are more stable with less noisy reaction in response to CB "noise." Figure 8–6 shows the simulation results with the ARX-SPC-GPC controller if the combination had been with the smoothing GPC controller rather than the low-pass digital filtering GPC controller. The response to the change is smoother and would be the recommended approach for the implementation of an SPC-GPC type controller.

Other examples of applications of SPC in the food industry include the work of Hayes et al (1997) to monitor and evaluate the hazard analysis critical control point (HACCP) hygiene data; of Grigg et al (1998) on SPC applied to fish product packaging; and of Fitzgerald et al (1998) on cheese manufacturing.

FUZZY CONTROLLERS

Fuzzy control has always been a controversial subject. A lack of mutual understanding between the fuzzy control and the traditional control communities has resulted in exag-

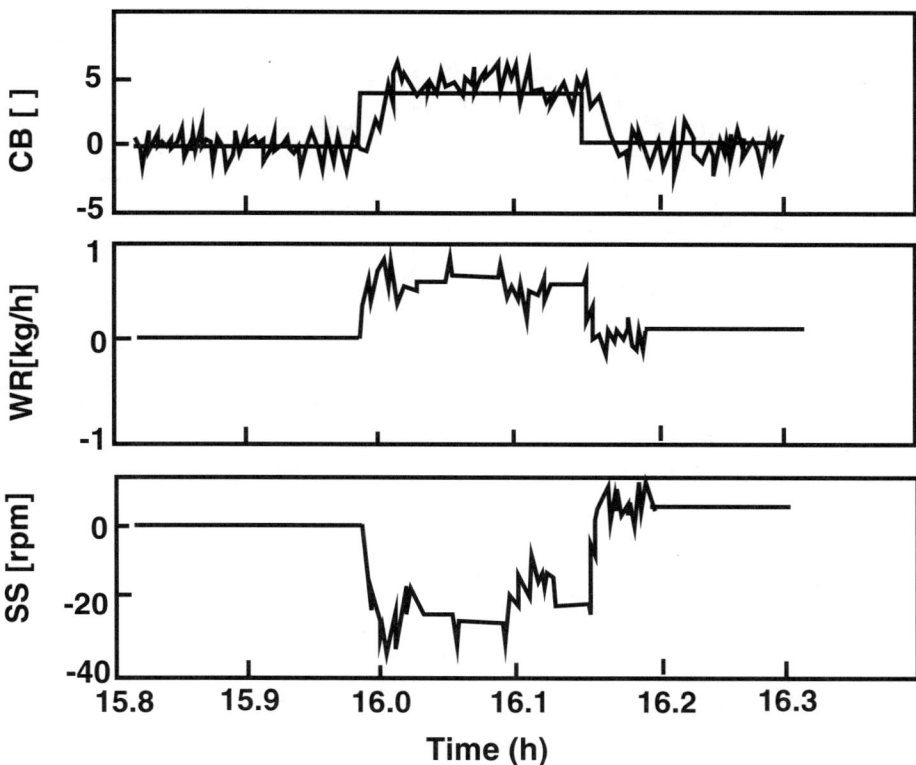

Figure 8–5 The track performance of an ARX-GPC controller with SPC logic in a food extrusion process.

gerated claims in certain fuzzy control publications (Årzén & Åström, 1996). However, fuzzy control has shown indisputable success when compared to linear PID control due to many reasons: (1) fuzzy control is a direct approach to nonlinear control design; (2) the rule-based approach is intuitive and easy to understand for non–control engineers; (3) it makes easy to implement nonlinear control elements, and nonlinear control elements can give better control performance than linear controllers for both linear and nonlinear processes; and (4) CAD environments for fuzzy control have a user-friendly graphical environment and are available in industrial hardware, making fuzzy control easy to apply in the industry.

The roots of fuzzy control are in manual control. The early applications of fuzzy control were based on the idea of mimicking the control actions of the human operator. Fuzzy logic control is an excellent alternative to more traditional control techniques when the development of a mathematical model of the process is not justified technically or economically.

The foundation of fuzzy control, fuzzy sets, was introduced by Zadeh (1965) as a way of expressing nonprobabilistic uncertainties. Since that time, fuzzy set theory has evolved, and today it is applied to database management, decision support systems, signal processing, data classifications, and computer vision.

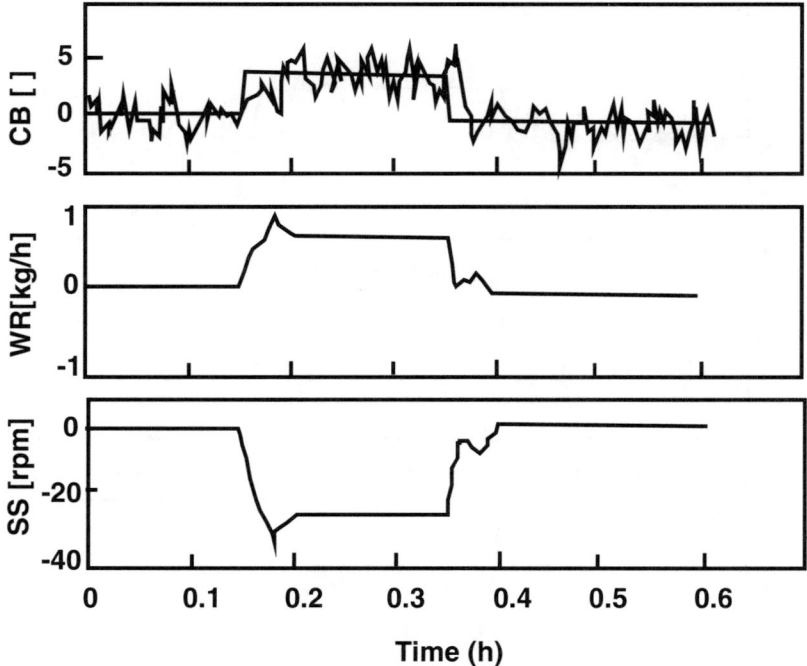

Figure 8–6 Simulation of the track performance of an ARX-GPC controller with SPC logic in a food extrusion process.

A fuzzy controller is a controller that contains a mapping, often nonlinear, that has been defined using fuzzy logic–based rules (Årzén & Åström, 1996). A fuzzy controller consists of a *set of rules*, each stating the control action to be taken in a certain process state.

A fuzzy process control has a number of advantages for many food processes (Davidson et al, 1999): (1) some process variables can be introduced by the operator in a linguistic form that is meaningful in the context of the industry; (2) measured process variables can be introduced directly into the controller as a fuzzy value without any statistical computations; (3) shifting in the process characteristics due to inherent variation in raw materials is easily controlled by fuzzy control systems, since no crisp setpoint exists that can cause instability; and (4) complex biological processes may be adequately described by fuzzy rule-based models, since a heuristic model that trained operators have learned by experience can be easily converted into fuzzy process models.

Fuzzy Logic Principles

To illustrate the fuzzy logic principle, let us consider the quality attribute of "color" used by the USDA standards for grades of french fries (Moreira et al, 1999). The product can be classified into good color, reasonable good color, and standard. The idea of reasonable good color, for example, is difficult to specify in a precise way. Based on sensory evaluation, the U.S. grades state that french fries with reasonable good color are those products that have a grade point between 24 and 26; above 27 the product is classified as having good color, and below 24 as standard. Therefore, we may define these choices as follows (Figure 8–7a):

IF x < 24 OR x > 26 THEN "the french fries have reasonable good color" *is wrong* (= 0)

IF x ≥ 24 OR x ≤ 26 THEN "the french fries have reasonable good color" *is true* (= 1)

(8)

where x is the french fries' color. The problem with these rules is that they do not take into consideration the uncertain aspect of "reasonable good color." It represents only a narrow view of this attribute: french fries having a grade-point of 23 are considered standard, while those with a grade-point of 24 are considered to have reasonable good color.

Fuzzy logic allows a better representation of the uncertain aspect of human reasoning. The idea consists of defining the true value of a proposition by a number varying between 0 and 1. The classical logic, on the other hand, only takes the value 0 or 1. Figure 8–7b illustrates this approach. Here, the color grade-point 25 has reasonable good color, which is defined as being true with a degree 1, the grade-points 24 and 26 have reasonable good color with a coefficient of 0, while 24.5 and 25.5 have reasonable good color with a degree 0.25. The qualifications "good color," "reasonable good color," and "standard" are called linguistic terms.

From concepts defined as fuzzy, we can have the following logical connectives:
IF (condition) *THEN* (conclusion)

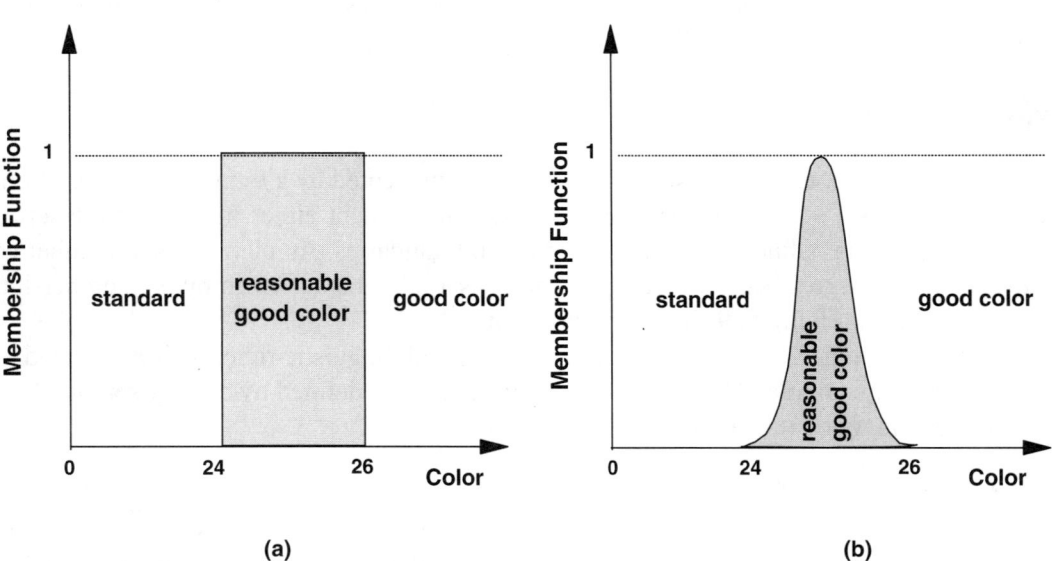

(a)   (b)

Figure 8–7 Description of a quality attribute using (a) classical logic and (b) fuzzy logic

In classical logic, the conclusion is *true* (= *1*) if the condition is *true* (= *1*). In fuzzy logic, the conclusion will be *true with a degree d* if the condition is *true with a degree d*.

To connect different conditions, it is necessary to add the definitions of AND and OR to work with fuzzy logic. The definition of the logical connective AND (conjunction) takes the minimum degree of truth:

IF f_1 is true with a degree d_1, AND f_2 are true with a degree d_2, THEN f_1 and f_2 are true with a degree min(d_1,d_2)

(9)

Similarly, the logical connective *OR* (disjunction) is defined as being the *maximum* degree of truth. All these elements allow the realization of a fuzzy logic controller; that is, they specify the rule of action in function of certain fuzzy concepts that describe the state of a system. Consider temperature adjustment of water flowing in a double pipe heat exchanger:

- *Rule:* If the water temperature is high, adjust the cold fluid.
- *Condition:* The water temperature is a little high.
- *Conclusion:* Adjust the cold fluid a little.

In classical logic we would have

- *Rule:* If the water temperature is high, adjust the cold fluid.
- *Condition:* The water temperature is a little high.
- *Conclusion:*

    –*Case I: T* is classified as high: adjust the water (maximum flow).

    –*Case II: T* is classified as not high: do not do anything.

Therefore, one of the major characteristic of fuzzy logic, as shown in the example above, is its progressive correction to the setpoint (Figure 8–8).

Fuzzy Sets and Fuzzy Logic

In fuzzy logic, each linguistic relationship is represented by a *fuzzy set*. By definition, a classical set is a set with crisp boundaries. So, an element either belongs to the set or not. A fuzzy set is defined as a set with noncrisp boundaries. An element is a member of a fuzzy set to a degree between 0 and 1. A fuzzy set A is characterized by its membership function, $\mu_{A:x[0,1]}$. Figure 8–9 shows this concept.

The fuzzy logic lies on the idea of fuzzy sets. Each linguistic function is associated to a fuzzy set. For example, the linguistic magnitude "hot" is defined by a fuzzy set, that is, a membership function under the set of temperatures.

The Logical Connectives AND and OR

The logical connectives *AND* and *OR* can be defined by fuzzy sets, similarly to the classical set theories. Recall that these connectors are defined by the intersection and union operators, respectively:

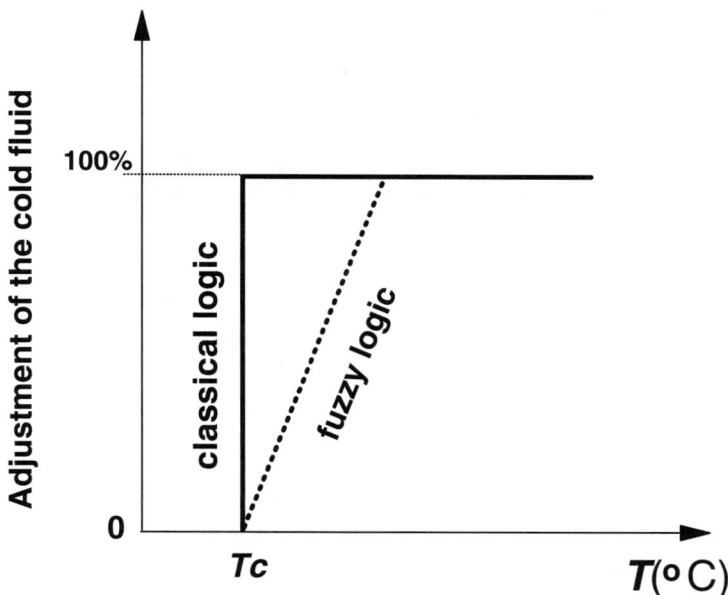

Figure 8–8 Representation of fuzzy and classical logics for the adjustment of hot water in a heat exchanger

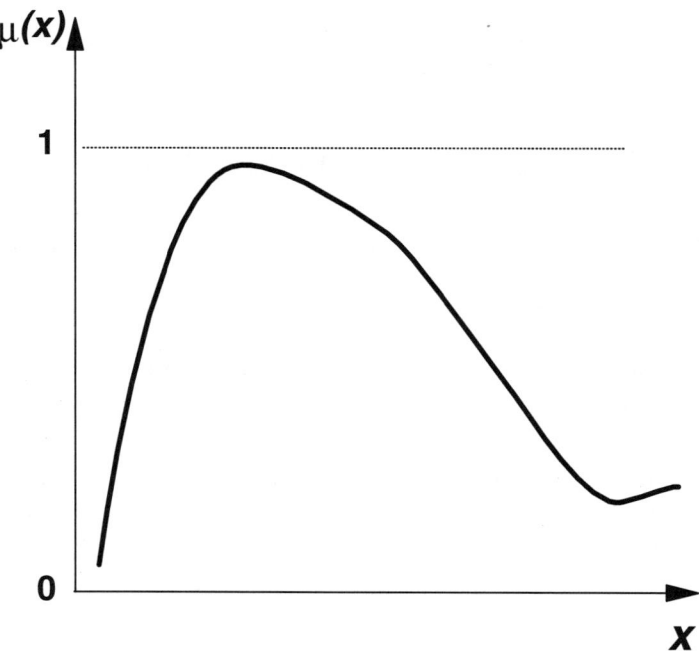

Figure 8–9 Membership function of a fuzzy set.

$$x \text{ is hot AND dry} \Leftrightarrow x \in A_{hot} \cap B_{dry}$$
$$x \text{ is hot OR dry} \Leftrightarrow x \in A_{hot} \cup B_{dry}$$

(10)

In fuzzy logic, union and intersection are defined by the membership functions through the corresponding fuzzy sets as:

$$x \text{ is hot AND dry} \Leftrightarrow \mu_{A_{hot} \cap B_{dry}(x)}$$
$$x \text{ is hot OR dry} \Leftrightarrow \mu_{A_{hot} \cup B_{dry}(x)}$$

(11)

Fuzzy logic uses the Boolean set operators to operate in fuzzy sets. Several definitions for the operators union and intersection are possible. The two operator definitions most often used are the following:

$$\mu_{A \cap B} = \min(\mu_A, \mu_B) \text{ and } \mu_{A \cup B} = \max(\mu_A, \mu_B)$$
$$\mu_{A \cap B} = \mu_A \cdot \mu_B \text{ and } \mu_{A \cup B} = 1 - (1 - \mu_A)(1 - \mu_B)$$

(12)

Other examples of the most common operators are shown in Table 8–1.

Fuzzy Inference

Recall that in classical logic, the rule of modus ponens is a reasoning rule with two statements:

1. *x is A*
2. *IF x is A THEN y is B,* so *y is B.*

These rules provide only the conclusion that the first statement coincides with conditional part of the second statement.

This rule has been studied by Zadeh (1979), under the fuzzy logic context, with the objective of applying it to the case where the coincidence is not exact. This principle is called generalized modus ponens (GMP), a form of approximation reasoning, that is, the process of reasoning with information given by fuzzy sets. The key issue here is that a fuzzy rule, for example, *IF x is A THEN y is B* (where *A* and *B* are linguistic values defined by fuzzy sets on *X* and *Y*, respectively) is defined as a binary fuzzy relation *R* on the space *X* × *Y*. For example, if we have the implication rule *IF* the temperature is low *THEN* the humidity is high and we know that the temperature is slightly low, then we may say that the humidity is slightly high.

Table 8–1 The Most Common Operator Definitions in Fuzzy Logic

Operation	Definition	Name
$\mu_{A \cap B}(x)$	$\mu_A(x) \bullet \mu_B(x)$	Product
	$\min[\mu_A(x), \mu_B(x)]$	Minimum
$\mu_{A \cup B}(x)$	$\min[\mu_A(x) + \mu_B(x)]$	Bounded sum
	$\max[\mu_A(x), \mu_B(x)]$	Maximum
$\mu_{\cup A}(x)$	$1 - \mu_A(x)$	—

Fuzzy Logic Control

An essential feature of conventional control is that the control law can be analytically described using equations, such as

$$u(k) = f[e(k), e(k-1), \ldots, e(k,n), u(k-1), \ldots, u(k-n)]$$

(13)

where u represents the control input, $e = w - y$, the error between the setpoint value w and the output y of the controlled system. The main goal of the conventional control theory is to find the function f in Equation 13.

In fuzzy control, the same dependence can be described using logical formulas in the form of the following rule-base:

$$IF\ e(k)\ is\ A_i^o\ AND \ldots AND\ e(k-n)\ is\ A_i^n\ AND$$

$$u(k-1)\ is\ B_i^l\ AND \ldots AND\ u(k-n)\ is\ B_i^n$$

$$THEN\ u(k)\ is\ B_i^o,\quad l \leq i \leq N$$

(14)

where N is a number of rules; $A^o_i, \ldots, A^n_i, B^o_i, \ldots, B^n_i, 1 \leq i \leq N$, are fuzzy predicate symbols; and *IF–THEN* (implication) and *AND* are logical connectives.

Structure of a Fuzzy Logic Controller

The basic structure of a fuzzy logic controller is based on that of a discrete PI controller. Recall that a discrete PI controller is described by the following equation:

$$u(k) = u(k-1) + K_p \Delta e(k) + K_p \frac{T_e}{T_i} e(k-1)$$

(15)

where K_p is the gain, T_i the integral constant, and T_e the sampling time. The variable $e(k)$ is the error between the measured value, and the setpoint, $\Delta e = e(k) - e(k-1)$ is the error variation between two sampling times, and $u(k)$ is the input. Thus, a PI controller can be seen as a function that associates a variation of the manipulated variable and an error to that variation: $\Delta u = f(e, \Delta e)$.

In the case of a fuzzy logic controller, the mathematical expression may be described by linguistic rules, as in Equation 15. Consider, for example, a temperature controller. The error between the measured value of the temperature and the setpoint as well as the variation of this error can be described by three linguistic values as

$$\{negative,\ zero,\ positive\}\ and\ \{decreasing,\ constant,\ increasing\}$$

The error and its variation are then described by three membership functions as

$$e \rightarrow \mu_{negative}(e),\ \mu_{zero}(e),\ \mu_{positive}(e)$$
$$\Delta e \rightarrow \mu_{decr.}(\Delta e),\ \mu_{cons.}(\Delta e),\ \mu_{incr.}(\Delta e)$$

$$(16)$$

The action that the controller must take is characterized by the following linguistic values: $\{low,\ medium,\ high\}$. This can be written by the following rules:

$$IF\ e\ is\ negative\ AND\ \Delta e\ is\ increasing\ THEN\ set\ \Delta u\ to\ high$$
$$IF\ e\ is\ negative\ AND\ \Delta e\ is\ constant\ THEN\ set\ \Delta u\ to\ medium$$
$$IF\ e\ is\ negative\ AND\ \Delta e\ is\ decreasing\ THEN\ set\ \Delta u\ to\ low$$

$$(17)$$

These rules give a result in fuzzy form, described by its membership function. In fuzzy logic control, the membership function must be converted into crisp values so that it can be applied to the real system. Several approaches exist to convert crisp inputs to noncrisp ones, as will be illustrated in the next section.

The structure of a fuzzy logic controller includes three parts: (1) a prefiltering device for calculating the fuzzy system inputs, (2) the fuzzy system mapping(s), and (3) a postfiltering device for computing the actual control signal (Figure 8–10).

The prefiltering device represents the signal processing performed on the controller inputs to obtain the inputs of the fuzzy systems. It may perform operations such as sampling, signal conditioning, dynamic filtering, etc. The postfiltering device represents the signal processing performed on the fuzzy system output to obtain the actual control signal. It can perform several operations, including precomputing part of the control, signal conditioning, dynamic filtering, and sampling (Årzén et al, 1999).

One of the characteristics of fuzzy controllers is that they contain a static nonlinear mapping defined by a fuzzy inference system. The external input-output view of a fuzzy

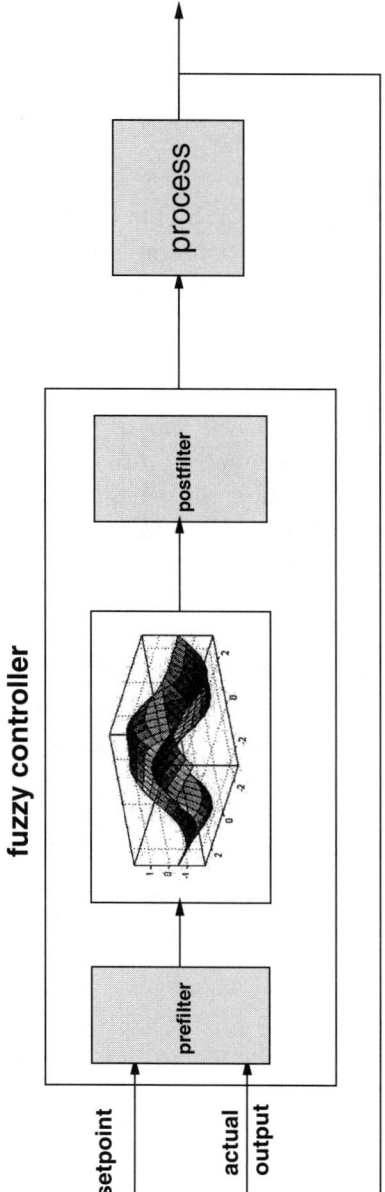

Figure 8–10 The fuzzy controller structure

inference system is shown in Figure 8–11. Therefore, fuzzy control is a subset of nonlinear control. The nonlinear mappings generated by fuzzy systems can generally be written as basis function expansions, that is, as weighted sums of basis functions $g_i(x)$ according to

$$f(x,\theta) = \sum_{i=1}^{M} \overbrace{g_i(x)}^{\text{IF - part}} \underbrace{w_i}_{\text{THEN - part}}$$

(18)

The exact nature of the basis functions is determined by the inferences systems used and how the fuzzy operations are defined. The basis function expansion can be viewed as a canonical form that is common in several nonlinear function approximation methods (sigmoidal neural network, radial basis function, splines, and wavelets).

In fuzzy control, the inputs and outputs are crisp rather than fuzzy. Therefore, fuzzification and defuzzification are needed. Fuzzification transforms the numerical inputs into fuzzy sets, and defuzzification approximates the output fuzzy set to a single crisp number. Therefore, the calculation of a control input by a fuzzy logic controller can be broken up into three steps: (1) fuzzification, (2) inference, and (3) defuzzification.

Figure 8–12 shows the internal configuration of a fuzzy logic controller. Here w is the setpoint, y the controlled variable, x_i the input variable to the inference block, and x_o the output variable from the inference block. The fuzzification block generally contains the preliminary data treatment, for example, the control error calculation ($e = w - y$) or the determination of the deviation of a variable (i.e., the difference between two sampling times). These variables are then handled by the linguistic variables. The fuzzification provides a series of fuzzy variables, collected under the vector x_i.

In the inference block, the values of the linguistic variables are connected by several rules that take into consideration the static and dynamic behavior of the system. In the inference block, a fuzzy information is obtained to the variable x_o. The defuzzification block then transforms the fuzzy values to crisp values.

This configuration was introduced by Mamdani (Årzén et al, 1999), who was one of the first researchers to apply fuzzy logic to control a system.

Fuzzification

Fuzzification basically consists of defining the membership functions of the different variables (i.e., x_i). The crisp variables are transformed to fuzzy variables so they can be handled by the inferences. Usually, the fuzzy predictive symbols are expressed linguistically using terms such as *positive small*, *near zero*, *OR negative big*. For example, the process states and control actions in fuzzy control are expressed in linguistic form as

IF error is large
THEN control action is large

(19)

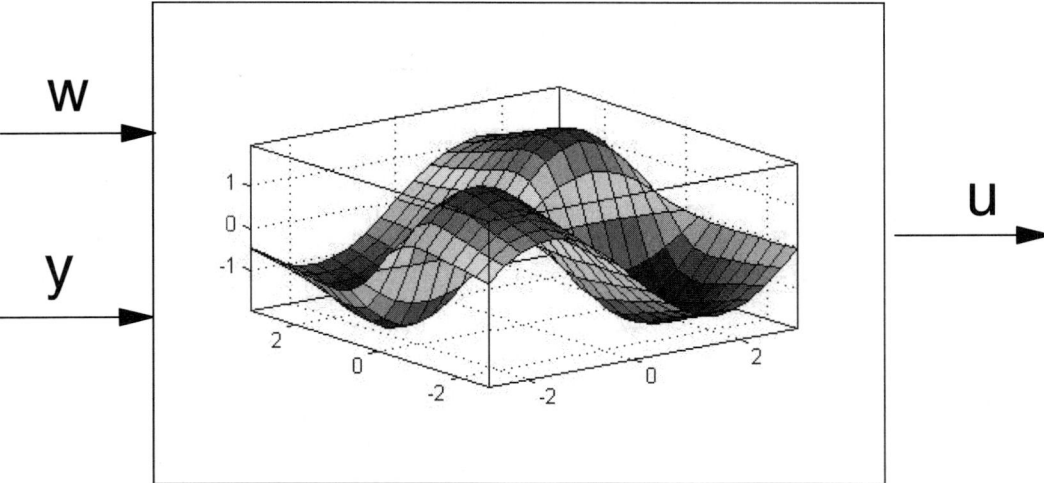

Figure 8–11 The external view of a fuzzy inference system with crisp inputs and output

---

The main idea of fuzzy sets is that statements like "Error is large" are not just true or false but can be satisfied to any degree in [0,1]. For a given observation x, the membership function determines to what extent the corresponding linguistic relations apply. The membership function types used in fuzzy control include triangular functions and trapezoidal functions or the smoother Gaussian functions and bell-shaped functions (Verbruggen et al, 1999). A bell-shaped function

$$f(x) = e^{-\left[\frac{x-x_o}{2a}\right]^2}$$

(20)

is illustrated in Figure 8–13, for example.

In the fuzzification step, a numerical variable X (e.g., the control error) is generally transformed to a normalized variable x. It is supposed that this variable varies between $-1 \leq x \leq 1$. The scale factor between x and X is selected based on the system so that the x-domain is not exceeded. However, it is not necessary to normalize the variables. Defined membership functions maintain the physical magnitude of the variables without need for normalization.

Inference

This step consists of applying fuzzy rules to the fuzzy variables. Consider the following rules applied to the fuzzy variables e and Δe:

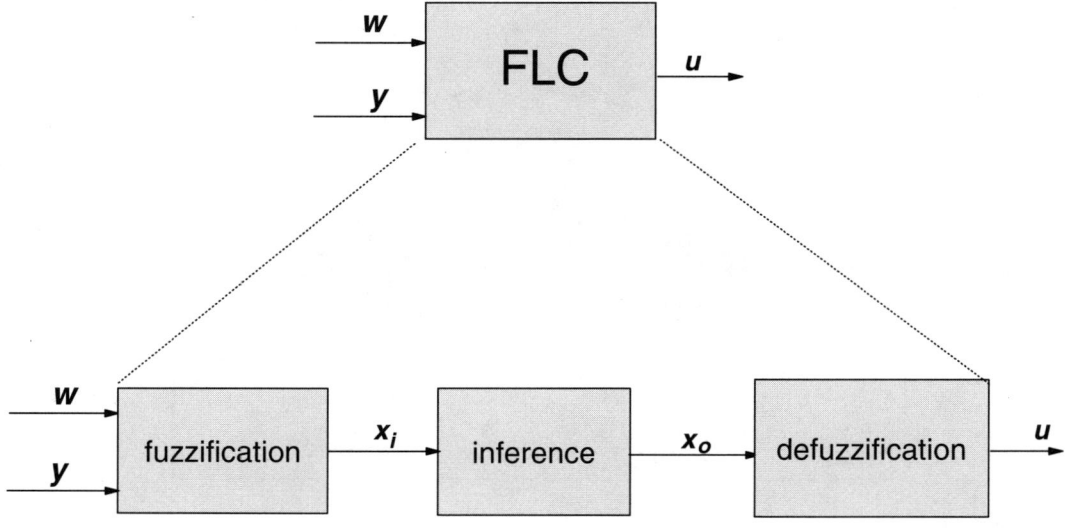

Figure 8–12 Internal configuration of a fuzzy logic controller

$$IF\ (e\ is\ positive\ big)\ AND\ (\Delta e\ is\ near\ zero)\ THEN\ \Delta u\ is\ near\ zero,\ OR$$
$$IF\ (e\ is\ near\ zero)\ OR\ (\Delta e\ is\ negative\ big)\ THEN\ \Delta u\ is\ negative\ big$$

(21)

The inference function in Equation 21 is composed of two rules. The second rule has the operator *OR* in the interior of the condition.

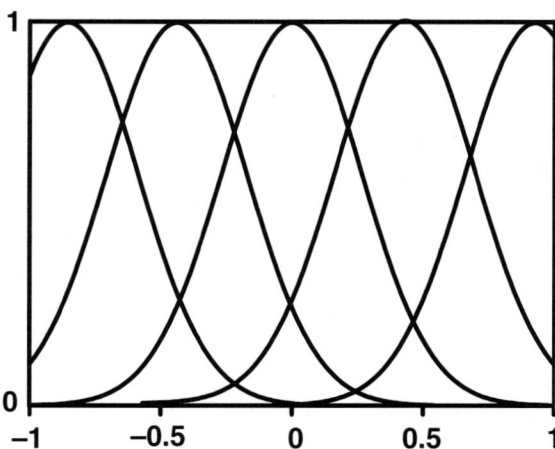

Figure 8–13 A function of bell-shaped type fuzzification

As discussed before, several possible choices can produce these operations as applied to the membership functions. For a fuzzy logic controller, the following methods can be used:

- method of inference max-min
- method of inference max-prod
- methods of inference add-prod

Only the method of inference max-min will be described here. For information on the other methods the reader is referred to Zimmermann (1991).

The method of inference max-min defines the operator *OR* by a function maximum and the operator *AND* by a function minimum (see Equation 12). The conclusion of this rule, introduced by the operator *THEN*, connects the condition membership factor with the membership function by the operator *AND* (defined in this case by a function maximum). The operator *OR* that connects the two rules is defined by the formation of a maximum.

Note that the name of this method is related to the definition of *OR* associated to the rules (max) and the definition of *THEN* (min).

Figure 8–14 represents the principles of the membership method max-min. Suppose that the numerical values for e and Δe are 0.74 and –0.37, respectively. The condition of the first rule (*e is positive big AND e is near zero*) implies the membership factors $\mu_e(e = 0.74) = 0.72$ and $\mu_{\Delta e}(\Delta e = -0.37) = 0.22$. So the condition takes a membership factor of $\mu_{C1} = 0.22$ (the minimum between two values due to the operator *AND*). The membership function $\mu_{\Delta u1}(\Delta u)$ is then written as 0.22 (due to the formation of a minimum connected to *THEN*).

The second rule indicates by its condition (*e is near zero OR Δe is negative big*) the membership factors $\mu_e(\Delta e = 0.74) = 0.24$ and $\mu_{\Delta e}(\Delta e = -0.37) = 0.70$. So the condition has a membership factor $\mu_{C2} = 0.70$ (the maximum between two values due to the operator *OR*). The membership function $\mu_{\Delta u2}(\Delta u)$ is then written as 0.70 (due to the formation of a minimum connected to *THEN*).

The final membership function $\mu_{\Delta u}(\Delta u)$ is obtained by the maximum of two membership function partials, since the two rules are connected by the operator *OR*. The final membership function is then shown at the lower right-hand side of Figure 8–14.

Defuzzification

The transformation of the fuzzy information to crisp information is called defuzzification. The most common defuzzification strategy is the center of gravity defined as:

$$u = \frac{\int\limits_{-\infty}^{\infty} \mu_G(x_o) \cdot x_o dx_o}{\int\limits_{-\infty}^{\infty} \mu_G(x_o) \cdot dx_o}$$

(22)

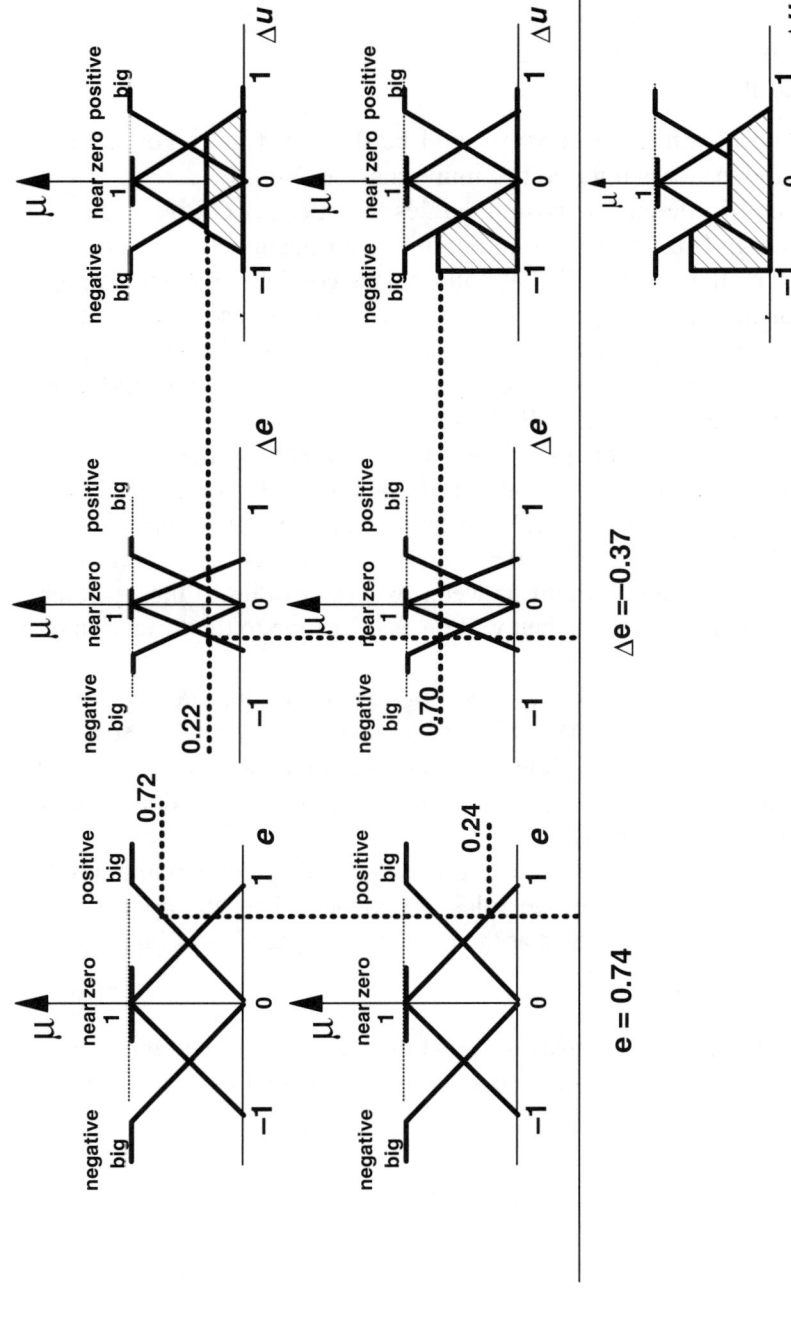

Figure 8–14 Method of inference min-max for two variables x_i and two rules

Thus, the control signal is calculated as the ratio between the moment and the area of its fuzzy set.

Figure 8–15 shows this principle of defuzzification. The abscissa of the center of gravity (described by Equation 22) forms the output signal that must be submitted to a treatment. Note that the output variable x_o (with $-1 \leq x_o \leq 1$) is the argument of the membership function $\mu_G(x_o)$, and the output signal u is a crisp value (determined by defuzzification).

When the membership function is composed by pieces of straight line, as is generally the case of a fuzzy logic controller, it is possible to make the integration analytically. With the coordinates x_j, μ_j containing p points of intersection of straight line segments, the abscissa of the center of gravity can be calculated as

$$u = \frac{\sum_{j=1}^{p}(x_{j+1} - x_j)[(2x_{j+1} + x_j)\mu_{j+1} + (2x_j + x_{j+1})\mu_j]}{3\sum_{j=1}^{p}(x_{j+1} - x_j)(\mu_j + \mu_{j+1})}$$

(23)

However, the determination of the coordinates x_j, μ_j of intersection points, which must be done in real time by using Equation 23, is time consuming. The calculation of

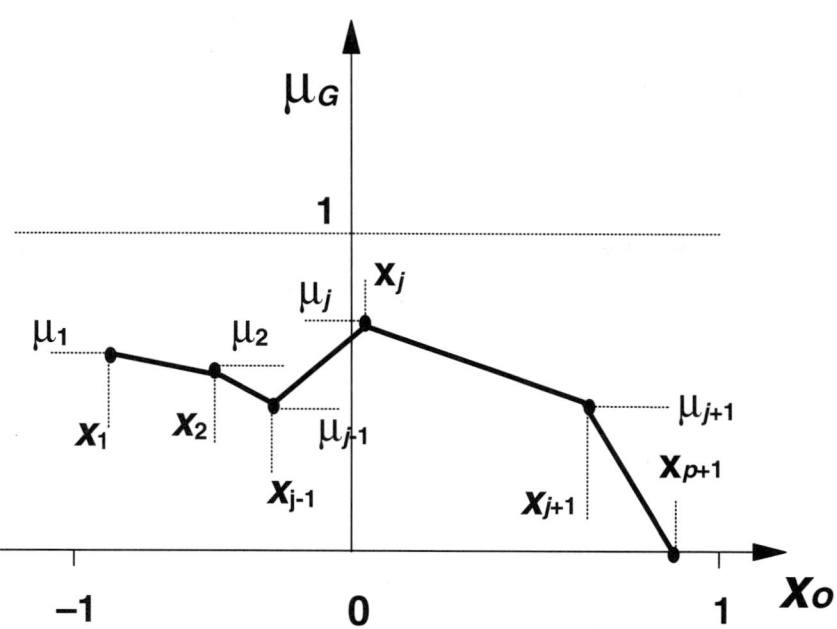

Figure 8–15 Intersection point coordinates of segments of membership function straight lines

the abscissa of the center of gravity can be substantially simplified when the membership function is determined by the inference method add-prod (Büller, 1994).

Application of Fuzzy Control

The most widely used control structure in the industry today is still the linear PID control. The main reason is that it is intuitive and close to the way humans manually control a process. The control action depends on the current control error (the *P*-term), the time history of the error (the *I*-term), and the prediction of the future value of the error (the *D*-term). In addition, PID controllers can be used in most industrial environments as long as the demands on performance are not high; and they perform reasonably well even when no detailed information about the process is available (Årzén et al, 1999).

In cases in which the process is nonlinear or transient performance improvement is desired, it would be advantageous to use nonlinear PID controllers. The combination of the advantages of an ordinary linear PID with the possibility of introducing nonlinearities in the control law by developing a fuzzy PID control has resulted in great success in the industrial applications. Fuzzy PID control can be viewed as a heuristic, or model-free, control approach. The controller is designed and tuned without any explicit model knowledge about the process. It also can be used when an explicit process model or model-based fuzzy control is available.

A fuzzy PID controller can be obtained with the following rules (Årzén et al, 1999):

- *IF e is NL AND ė is NL   THEN u is NL*
- *IF e is ZE AND ė is NL   THEN u is NS*
- *IF e is PL AND ė is NL   THEN u is ZE*
- *IF e is NL AND ė is ZE   THEN u is NS*
- *IF e is ZE AND ė is ZE   THEN u is ZE*
- *IF e is PL AND ė is ZE   THEN u is PS*
- *IF e is NL AND ė is PL   THEN u is ZE*
- *IF e is ZE AND ė is PL   THEN u is PS*
- *IF e is PL AND ė is PL   THEN u is PL*

where *NL* means negative large, *NS* negative small, *ZE* zero, *PS* positive small, and *PL* positive large. These rules can be presented in a table that has the typical stripe-diagonal form:

		ė		
		NL	ZE	PL
	PL	ZE	PS	PL
e	ZE	NS	ZE	PS
	NL	NL	NS	ZE

Fuzzy Logic Control of a Biscuit-Baking Process

Perrot et al (2000) developed a methodology for the closed-loop control of food product quality using fuzzy sets applied it to a baking process.

The feedback quality control was achieved using a programmable logic controller (PLC) in a closed loop. The air temperature (T_{zi}) in each zone and the airflow rate were controlled around the setpoint set by the operator or by the quality feedback system (classical PI controller). Five numerical inputs were considered: color (L, a, b) measured by an on-line infrared sensor, final product moisture content (using an on-line near-infrared sensor), and product thickness (measured every 10 min off-line). Information in the quality feedback system was processed in terms of linguistic grades. Defuzzification of this information was performed only at the end of the decision module.

The control system consisted of different modules containing different fuzzy algorithms (Figure 8–16). The first module, *evaluation*, reproduced the sensory evaluation of biscuit quality (color, moisture content, thickness) as given by an operator. Fuzzy functions were used to transform crisp values to noncrisp quality values (ranging from overcooked to undercooked). Color was evaluated using a fuzzy *k*-nearest neighbor algorithm (Keller et al, 1985), while the classical notion of membership function was used to estimate moisture and thickness. On-line and real-time sensory evaluation of color was achieved by coupling a sensor (outputs L, a, b) and a fuzzy *k*-nearest neighbor classification algorithm (resulting in five degrees for color). Moisture was associated with the linguistic terms *dry*, *normal*, and *wet*; and thickness with *low*, *normal*, and *high*. Figure 8–17 illustrates the color model (8–17a) and the membership functions for moisture (8–17b) and thickness (8–17c) that are used to control biscuit quality.

The fuzzy evaluations were passed through the decision system, which consisted of two modules: diagnosis and decision. The diagnosis module evaluates the state of the process, starting from sensory evaluations for color, moisture, and thickness. This diagnostic knowledge is expressed as

Oven too hot = (moisture normal OR dry) *AND* color overcooked
Oven not hot enough = (moisture normal OR wet) *AND* color a little undercooked

(24)

Fuzzy operators were used to represent logic combinations of the results obtained by the sensory evaluation module. Classical *AND* and *OR* operators were then implemented, resulting in 11 combinations of expertise.

In the feedback decision module a control resolution was made based on the diagnosis of the state of the process. The fuzzy controller used was a Takagi-Sugeno constant output type (Takagi & Sugeno, 1985). Numerical outputs (temperature setpoints of the three zones of the oven, T_{zic}) were calculated directly with the activation grades (α_j) and constant output parameters (p_j) for each rule j:

$$T_{zic} = \frac{\sum_j \alpha_j p_j}{\sum_j \alpha_j}$$

(25)

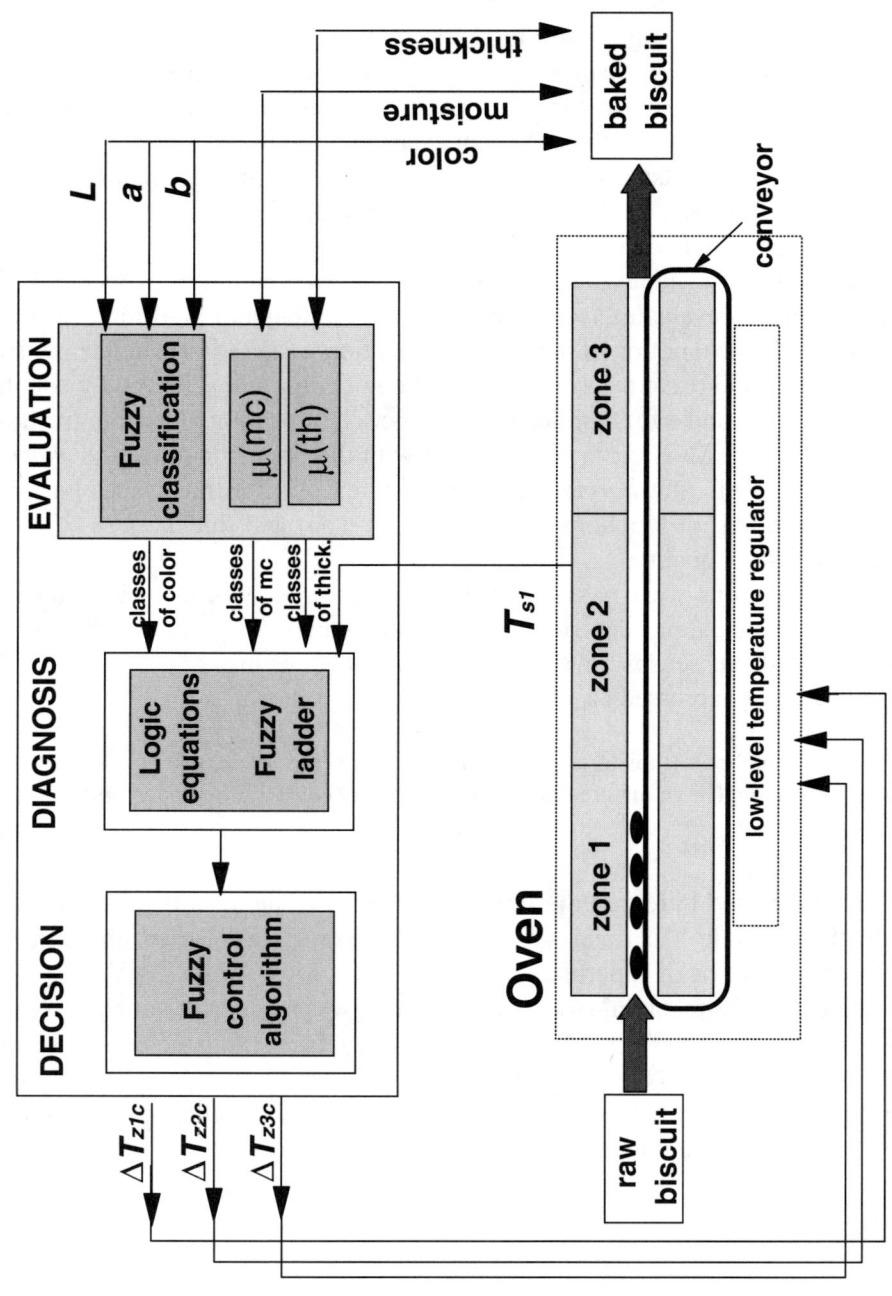

Figure 8–16 The biscuit-baking control system.

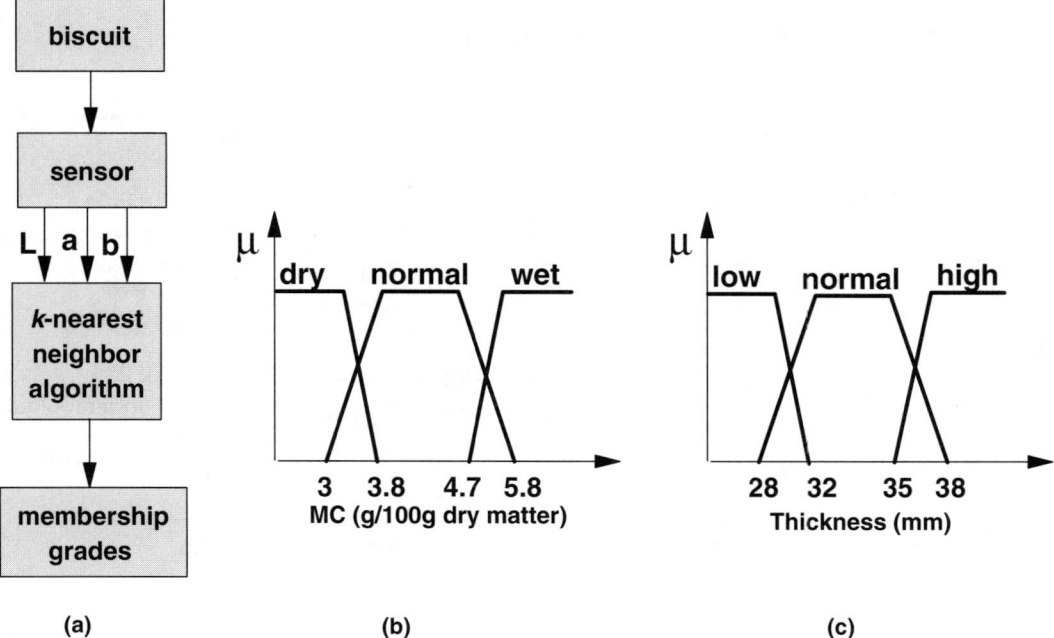

Figure 8–17 (a) Model of human sensory evaluation of color; (b) membership function for moisture content; (c) membership function for thickness.

This knowledge base contains 11 rules of the type *IF the oven is too hot THEN decrease temperature of section 3 by 10 °C.*

Figure 8–18 shows the control system behavior compared to an operator manual control. Here, the controller was able to keep the biscuit quality at the setpoint range after disturbances were applied to the system. Color retuned to the setpoint within 50 and 60 minutes, respectively. It is also shown that the automatic controller provided a good reproduction of the human evaluation of color (in a more gradual way, however). The automatic controller was tolerant of human measurement and therefore able to cooperate robustly with the operator. Basically, it has been shown that the control system processes information at each stage in the range of operator reasoning.

NEURAL NETWORK CONTROL

An artificial neural network (ANN) is an information-processing paradigm inspired by the way the densely interconnected, parallel structure of the human brain processes information. Artificial neural networks are collections of mathematical models that emulate some of the observed properties of nervous systems and draw on analogies of adaptive learning.

Neural networks can be viewed as multivariate nonlinear nonparametric estimation methods. They are used to approximate a function $y = f(x)$, where the functional form is

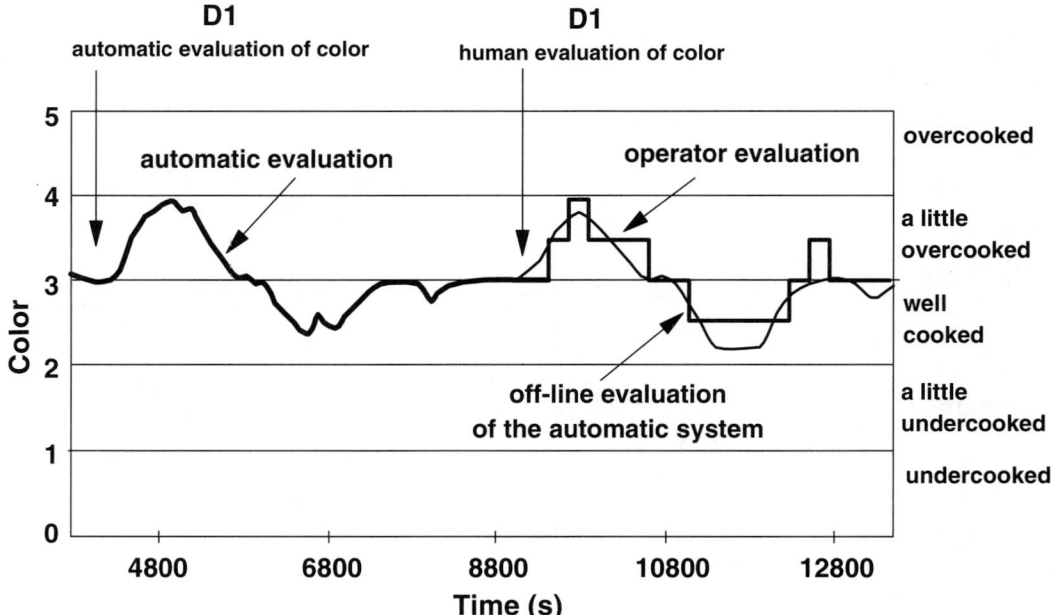

Figure 8–18 Comparison between manual and automatic evaluation of color: Control response to a disturbance, D1.

unknown. Neural networks are universal function approximators that typically work much better in practical applications than more traditional (polynomial) function approximation methods (Ungar et al, 1996).

Neural network methods are inspired by biology, as they are represented as networks of simple neuronlike processors. A typical "neuron" takes in a set of inputs, sums them, takes some function of them, and passes the output through a weighted connection to another neuron. The neuron is just a predictor variable, or a function of a nonlinear combination of predictor variables. The connection weights serve as adjustable parameters that are set by a "training method": that is, they are estimated from part of the experimental data.

The Biological Neuron

The human brain is based on small individual analog processing units called *neurons*. There are about 10^{11} of these neurons in the average brain, and each connects to 10^4 others. All of these neurons operate relatively slowly (around 100 Hz), but there is vast computing power because they all operate in parallel.

A biological neuron consists of a body, called the soma, that receives signals from other neurons via the dendrites. Figure 8–19 shows the essential features of a biological neuron. The signals arriving at the neuron are summed, and if the result is greater than the threshold value the neuron fires.

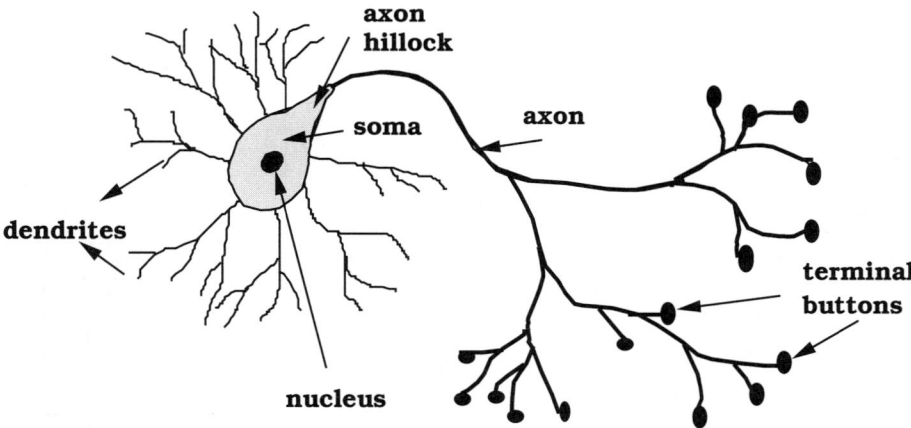

Figure 8–19 Schematic of a biological neuron

The output from the neurons appears on axons. These terminate as synapses that link the output and a dendrite input to another neuron. There is no actual connection between the synapse and the dendrite; the data transfer takes place by the release of chemicals called neurotransmitters.

Artificial Neural Networks (ANNs)

Figure 8–20 shows a schematic of a standard and unifying model for neural networks. The main components of the system are

- A weighted summer
- A linear dynamic single-input, single-output (SISO) system
- A nondynamic function

Weighted Summer

The weighted summer is described by

$$q(t) = Ay(t) + Bu(t) + v$$

(26)

where q is a $N \times 1$ vector, y a $N \times 1$ output vector, u the $M \times 1$ input vector, v a constant $N \times 1$ vector, A an $N \times N$ matrix, and B an $M \times N$ matrix. The ijth element of the matrix A is a_{ij}, and the ipth element of the matrix B is b_{ip}.

Linear Dynamic System

The linear dynamic SISO system can be described, in transfer function form, as

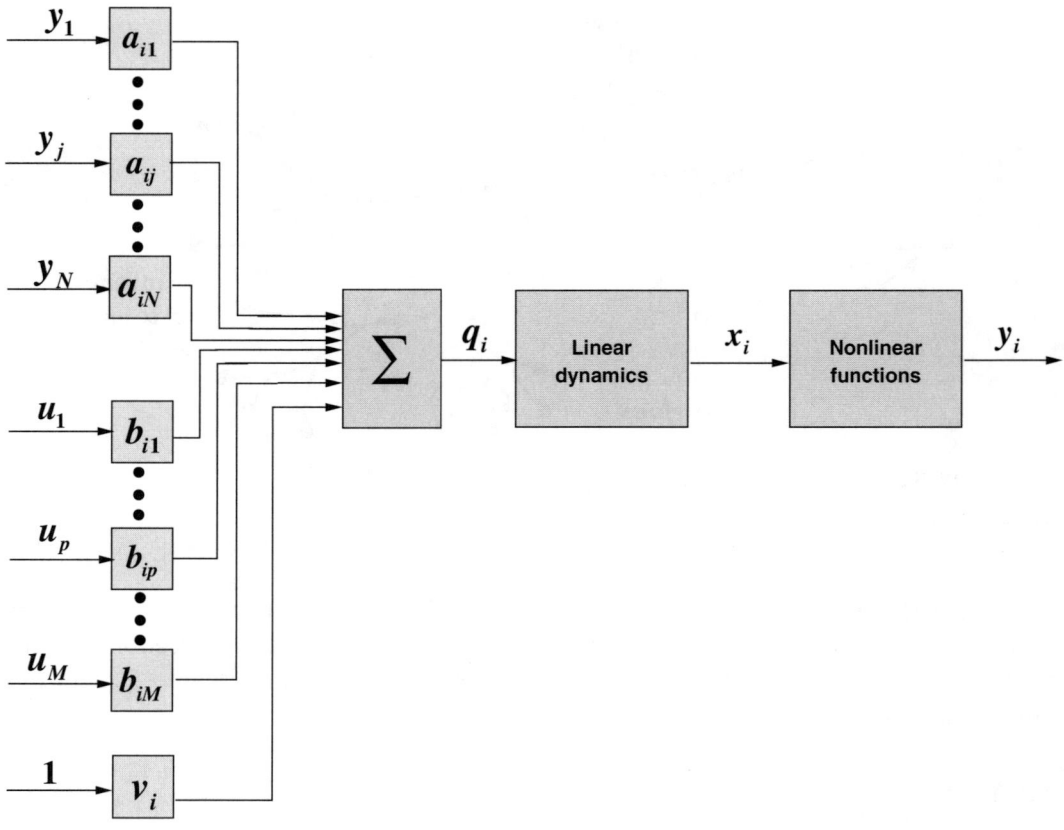

Figure 8–20 The basic model of a neuron

$$X_i(s) = H(s)Q_i(s)$$

(27)

Common examples of $H(s)$ functions include unit impulse, unit step, first-order low pass, and lag. For example, the input-output relation for a first-order low-pass function $(H(s) = 1/(\tau s = 1))$ is

$$\tau \dot{x}_i(t) + x_i(t) = q_i(t)$$

(28)

Nondynamic Nonlinear Function

The nonlinear function $g(\cdot)$ gives the element output y_i in terms of the transfer function output x_i :

$$y_i = g(x_i)$$

(29)

Examples of twofold classifications of these functions include (1) differentiable/nondifferentiable; (2) pulselike, steplike; and (3) positive/zero-mean. Some standard functions are presented in Table 8–2.

Connections

The three components of the neuron discussed above can be combined in several ways. If they are all nondynamic or static ($H(s) = 1$), for example, then an assembly of neurons can be written as the set of algebraic equations

$$x(k) = Ay(t) + Bu(t) + v$$
$$y(t) = g[x(t)]$$

(30)

where x is a vector of N x_i and $g(x)$ is a vector whose components are $g(x_i)$. Otherwise, if each neuron has first-order low-pass dynamics, an assembly of neurons can be written as the set of differential equations

$$t\dot{x}(t) + x(t) = Ay(t) + Bu(t) + v$$
$$y(t) = g[x(t)]$$

(31)

The behavior of this network depends on the interconnection matrix A and the form of $H(s)$.

An ANN is built up in layers, the most common being the three-layer structure shown in Figure 8–21. This consists of an input layer, a hidden layer, and an output layer. Information in these simple networks flows in one direction only, from input to output.

In this network, the connection matrix A is such that the outputs are partitioned into layers so that a neuron in one layer receives inputs only from neurons in the previous layer.

Table 8–2 Nonlinear Functions g(x)

Name	Function	Characteristics
Threshold	+1 if $x > 0$ or else 0	Nondifferentiable, step-like, positive
Threshold	+1 if $x > 0$ or else −1	Nondifferentiable, step-like, zero-mean
Sigmoid	$1/1 + e^{-x}$	Differential, step-like, positive
Hyperbolic tangent	$\tanh(x)$	Differential, step-like, zero-mean
Gausssian	$e^{(-x^2/\sigma^2)}$	Differential, pulse-like

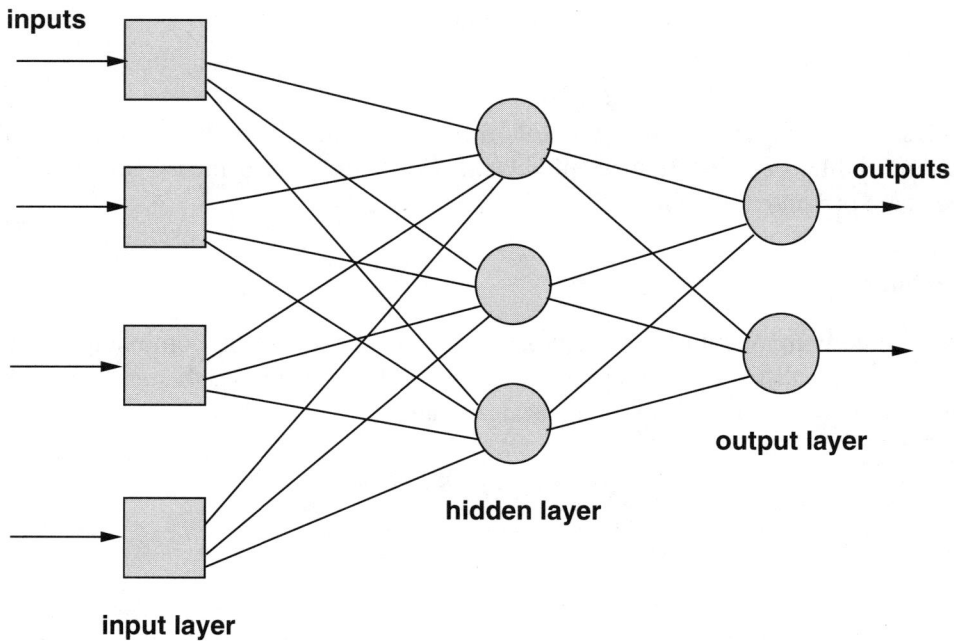

inputs

outputs

output layer

hidden layer

input layer

Figure 8–21 Multilayer neural network

There is no feedback in such network. They are referred to as feedforward networks. For example, in a three-layer network, with each layer containing N neurons, we may partition the network x, y, u, and v vectors from Equation 30 as (Hunt et al, 1992):

$$\begin{bmatrix} x^1(t) \\ x^2(t) \\ x^3(t) \end{bmatrix} = A \begin{bmatrix} y^1(t) \\ y^2(t) \\ y^3(t) \end{bmatrix} + B \begin{bmatrix} u^1(t) \\ u^2(t) \\ u^3(t) \end{bmatrix} + \begin{bmatrix} v^1(t) \\ v^2(t) \\ v^3(t) \end{bmatrix}$$

(32)

where the superscripts denote the corresponding layer in the network. The structure of the A and B matrices for this network are

$$A = \begin{bmatrix} 0_{NN} & 0_{NN} & 0_{NN} \\ A^2 & 0_{NN} & 0_{NN} \\ 0_{NN} & A^3 & 0_{NN} \end{bmatrix}; \quad B = \begin{bmatrix} B^1 & 0_{NM} & 0_{NM} \\ 0_{NM} & 0_{NM} & 0_{NM} \\ 0_{NM} & 0_{NM} & 0_{NM} \end{bmatrix}$$

(33)

where 0_{NN} is the $N \times N$ zero matrix and 0_{NM} the $N \times M$ zero matrix. A^2 and A^3 are the $N \times N$ matrices of weights, while B^1 is a $N \times M$ matrix of weights. For the first layers we have

$$x^1(t) = B^1u^1(t) + v^1$$
$$y^1(t) = g[x^1(t)]$$

(34)

and for the second and third layers,

$$x^a(t) = A^1y^{r-1}(t) + v^r$$
$$y^r(t) = g(x^r(t))$$

(35)

where $r = 2,3$. Different characteristics are obtained using different nonlinearities g ($\bullet$) from Table 8–3. The most common is the sigmoid function (Rumelhart & McClelland, 1986).

Many different network architectures are used, typically with hundreds or thousands of parameters. The resulting equation forms are general enough to solve a large class of nonlinear classification and estimation problems (Hunt et al, 1992). The most widely used of these is the multilayer network with sigmoidal activation functions, called a *back propagation network* (Rumelhart & McClelland, 1986).

Learning in Static and Dynamic Networks

The application of ANN normally proceeds in two phases: developing the network architecture and training (or learning). Training is achieved by adjusting the weights of the individual neurons in the network. Learning can be regarded as a parametric adaptation algorithm. Learning algorithms can be classified into supervised learning and unsupervised learning. The former incorporates an external reference signal, while the latter relies only on local information and internal signals (it does not incorporate external references).

In Static Networks

The standard static ANN learning problem for a single-layer static architecture (considering Equation 30 with $v = 0$) is:

- The input u and output y can be measured.
- The function g is known (taken from Table 8–3).
- A desired output value (or setpoint), w, is known.
- Find a parameter estimate $\hat{\beta}$ such that the error $e^2(t) = [w(t) - y(t)]^2$ is minimized.

In the linear case, when $g(x) = x$, $\hat{\beta}$ would be typically solved in discrete time as

$$\hat{B}(k+1) = \hat{B}(k) + P(k)u(k)[w(k) - y(k)]$$

(36)

where P can be a gradient algorithm, a normalized gradient algorithm, or a least-squares algorithm.

In the case of a multilayer static ANN, where the input to and reference signals for the hidden layers are known, the back-propagation algorithm is generally used. The back-propagation algorithm is a gradient algorithm applied to a nonlinear optimization problem. It solves the missing information problem as follows (Hunt et al, 1992):

- Inputs to the hidden layers are taken as the inputs to the first layers propagated forward through the network.
- The effective reference signals for the hidden layers are obtained by back-propagation of the error through the network. This is accomplished by taking the partial derivative of the squared error against the parameters.

From a control point of view, consider a general network represented by a nonlinear function f and a set of parameters $\boldsymbol{\theta}$ described as

$$y(t) = f(\boldsymbol{u}(t), \boldsymbol{\theta})$$

(37)

where $\boldsymbol{\theta}$ contain the coefficients of (A, B) in Equation 33. The parameters in the model, the weights, $\hat{\boldsymbol{\theta}}$ are chosen to minimize the residual sum of squares error:

$$E = \sum_{t=1}^{T} \lambda^{T-1} \left\| y(t) - f(\boldsymbol{u}(t), \hat{\boldsymbol{\theta}}) \right\|^2$$

(38)

where λ is the forgetting factor and T the number of presentations of different training sets. The estimates $\hat{\boldsymbol{\theta}}$ can be calculated in discrete time as

$$\hat{\boldsymbol{\theta}}(k+1) = \boldsymbol{\theta}(k) + \boldsymbol{P}^*(k)\tilde{\boldsymbol{x}}(k)[y(k) - f(\boldsymbol{u}(k), \boldsymbol{\theta})]$$

(39)

where $\boldsymbol{P}^*(t)$ is the pseudoinverse of a matrix $\boldsymbol{P}$ given by

$$\boldsymbol{P}(k) = \sum_{\tau=1}^{T} \lambda^{T-\tau} \tilde{\boldsymbol{x}}(\tau)\tilde{\boldsymbol{x}}(\tau)'$$

(40)

and

$$\tilde{x}(t) = \frac{\partial F}{\partial \hat{\theta}}$$

(41)

In Dynamic Networks

Dynamic or recurrent networks are different from feedforward ones because they incorporate feedback. Basically, the output of every neuron is fed back with varying gains (weights) to the inputs of all neurons. The architecture of a dynamic ANN is usually one-layered. One of the basic dynamic network descriptions gives linear output equal to state as

$$s(t) = Ay(t)$$
$$Tx(t+1) = -x(t) + g[s(t)] + u(t)$$
$$y(t) = x(t)$$

(42)

Due to the use of distinct learning algorithms, recurrent networks that have the same structure can exhibit different dynamic behavior. So this network is a composition of two dynamic systems: transmission and adjusting systems. The interaction of both systems produces the overall input-output behavior of the network.

There are two general concepts of recurrent structure training: fixed-point learning and trajectory learning. Fixed-point learning is aimed at making the network reaches an equilibrium or performs steady-state matching. Trajectory learning trains the network to follow the desired trajectory in time. For detailed mathematical description of these two concepts, the reader is referred to Hunt et al (1992).

Neural Networks for Control

Neural networks can be used in process controls in a number of ways (Ungar et al, 1996). In direct control, they provide a mapping from the current state and the desired next state of the plant to control action to be taken. In indirect control, the networks are used as a nonlinear model of the plant, as in model predictive control (MPC).

Neural networks can also be used as a piece of a larger model of the plant. They can be trained on the basis of data collected in the laboratory (or off line) to learn the mapping between the on-line measured variables and the off-line variables and then be used as "virtual analyzers." For example, it is often very difficult to measure product quality attributes on line (in a food extrusion process for example), yet it is important to be able to monitor and control them for accurate control actions. Other much easier variables such as pressure and temperature can be measured, and neural networks can be used to estimate the quality attributes.

In addition, neural networks can be used to detect faults and to compensate for sensor failure or drifts. These applications can be used either as filters for controller inputs or as components of supervisory control schemes.

Model Predictive Control with ANN

Model predictive control with ANN is the most widely used method of incorporating neural networks into controllers within the framework of model predictive control (Chapter 6), where the neural network is used as a process model (Ungar et al, 1996).

In this approach, a neural network model provides prediction of the future plant response over the specific horizon. The predictions supplied by the network are passed to a numerical optimization routine that minimizes a specified performance criterion in the calculation of a suitable control signal.

Another possibility is to train a further network to mimic the action of the optimization routine. This controller network is trained to produce the same control output u, for a given plant output, as the optimization routine. An advantage of this approach is that the outer loop consisting of the plant model and optimization routine is no longer needed when training is complete (Hunt et al, 1992).

Neural Network Control Systems for a Continuous Frying Process

Huang et al (1998) used a neural network to develop a prediction model to deal with the complexity of a snack food–frying process. Two inputs, oil temperature and residence time, and two outputs, product outlet color and moisture content, were used in the multiple-input, multiple output (MIMO) neural network process prediction model–based IMC loop for quality control of the continuous frying process.

The continuous process was mathematically described by the following 2×2 discrete-time, nonlinear, autoregressive with exogenous input (NARX) model:

$$y(t) = f[y(k-1), y(k-2); u(k-d-1), u(k-d-2); \boldsymbol{\theta}] + \varepsilon(k)$$

(43)

where d was the time delay that was equal to 20 units for temperature input and 16 units for the residence time input. The resulting smallest structure of the network for the MIMO neural network model was $8 \times 3 \times 2$: that is, 8 inputs by 3 hidden nodes by 2 outputs.

As described before, neural networks are "model-free" estimators that can be used to directly approximate the function $f(\ )$ as follows:

$$\hat{y}(t) = \hat{f}[y(k-1), y(k-2); u(k-d-1), u(k-d-2); \hat{W}]$$

(44)

where $\hat{f}(\ )$ is the approximation of the function $f(\ )$ and $\hat{W}$ is the set of weights and bias terms in the network model. This modeling approach has no network output feedback and takes the structure of a feedforward neural network. It can provide the one-step-ahead predictor for IMC as

$$\hat{y}(k + d + 1) = \hat{f}[y(k + d), y(k + d - 1); u(k), u(k - 1); \hat{W}]$$

(45)

The IMC framework (Chapter 6) provides a good example of how neural networks can be incorporated into controllers. In conventional IMC, a model of the plant, typically linear, is partially inverted to determine the control action. A neural network can be used as the controller (for direct control) or as the plant model (for indirect control) or as both (Ungar et al, 1996). In the current example, the ANN was used for indirect control: that is, as the process model only. For indirect control, the model of the process is learned by training a neural network so that its output $\hat{y}(t)$ in Equation 37 approximates the function $y(t)$ in Equation 45.

Note that Equation 38 contains the input vector, $u(k)$, so that it can be used to compute the control action for the unique solution of $u(k)$ as

$$F[u(k)] = tr(k) - \hat{f}[\hat{y}(k + d), \hat{y}(k + d - 1); u(k), u(k - 1); \hat{W}]$$

(46)

where $tr(k)$ is the tracking signal of the network model predicted output $\hat{y}(k + d + 1)$. Therefore, the control actions can be computed with the inverse of the function $F[u(k)]$. However, the future process outputs $y(k + d - 1)$, $y(k + d)$ need to be estimated to realize the computation.

Because the neural network prediction model equation is nonlinear, an analytical solution cannot be obtained. Thus, the control law needs to be evaluated by solving the nonlinear function $F( )$, the process prediction model, at each time interval iteratively:

$$\hat{u}^k(k) = \hat{u}^{k-1}(k) + \Delta^{k-1}\hat{u}(k)$$

(47)

where $\hat{u}^j(k)$ is the computed $u(k)$ at the kth iteration and $\Delta^{j-1}\hat{u}(k)$ is the updating increment of $\hat{u}(k)$ at the k–1th iteration.

The process model and the control action set up a neural network process one-step-ahead prediction model–based IMC loop for a continuous snack–frying process. The structure of the neural network IMC is shown in Figure 8–22. Here, n is the system noise, M^p the neural process model, M^i the inverse of the neural process model, F the filter for control robustness, and w the setpoints of the process.

EXPERT CONTROL

Expert control was introduced with the idea of obtaining a strict separation of signal processing and logic where the logic was implemented in an expert system (Åström et al, 1986). The system structure processes both signals and symbols. Some of the advantages of expert control are that the structure of the system is nicer, knowledge representation becomes transparent, and debugging is simplified.

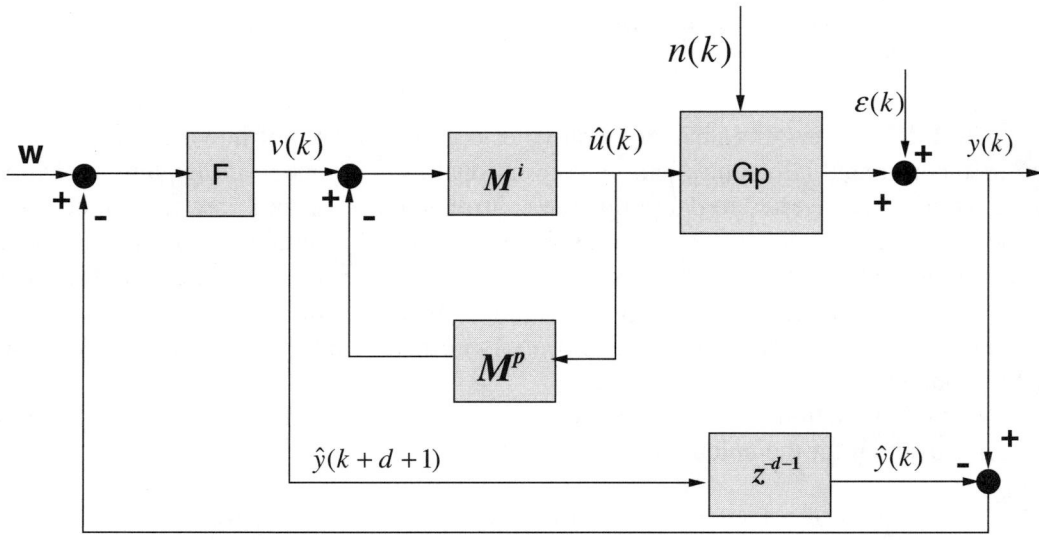

Figure 8–22 Block diagram of the neural network IMC for a continuous frying process

An example of an expert controller is shown in Figure 8–23 (Årzén & Åström, 1996). The system contains the following components: a collection of signal process algorithms for control, control design, parameter estimation, excitation, diagnosis, and logging. A knowledge-based system is also used to coordinate the operation of the algorithms. The expert system (or knowledge-based system) is usually written as part of a *shell*: that is, a general package software designed to facilitate implementation. The shell contains the following components: (1) *a database and a rule base*—data can be entered at system start-up or via the knowledge-based system, in real time, and rules are structured with "if-then" statements; (2) *an inference engine* that provides a means of scanning the available rules to draw conclusions or select appropriate action to be taken; and (3) *a user interface* that displays information, asks questions of the user, and so forth.

The system has an ordinary feedback loop with a process and a controller. There are also a number of auxiliary functions for parameter estimation, control design, supervision, fault detection, and diagnosis. There may be several alternative algorithms for the same task. There are also algorithms for generating perturbation signals to excite the process. The fault and diagnosis tasks are aimed at finding faults that are local to the control loop. The signal-processing algorithms indicated by the boxes can communicate with the expert system by sending and receiving data. The parameters of the algorithms can be changed and algorithms can be replaced. These are coordinated by an expert system, or a knowledge-based system, that decides what algorithm to use and when.

Many other systems can be implemented in this framework: for example, a gain schedule controller can be obtained by having one controller whose parameters are changed on the basis of a measured signal. Another possibility is to implement a controller with automatic tuning by using two control algorithms, a PID controller and a relay feedback

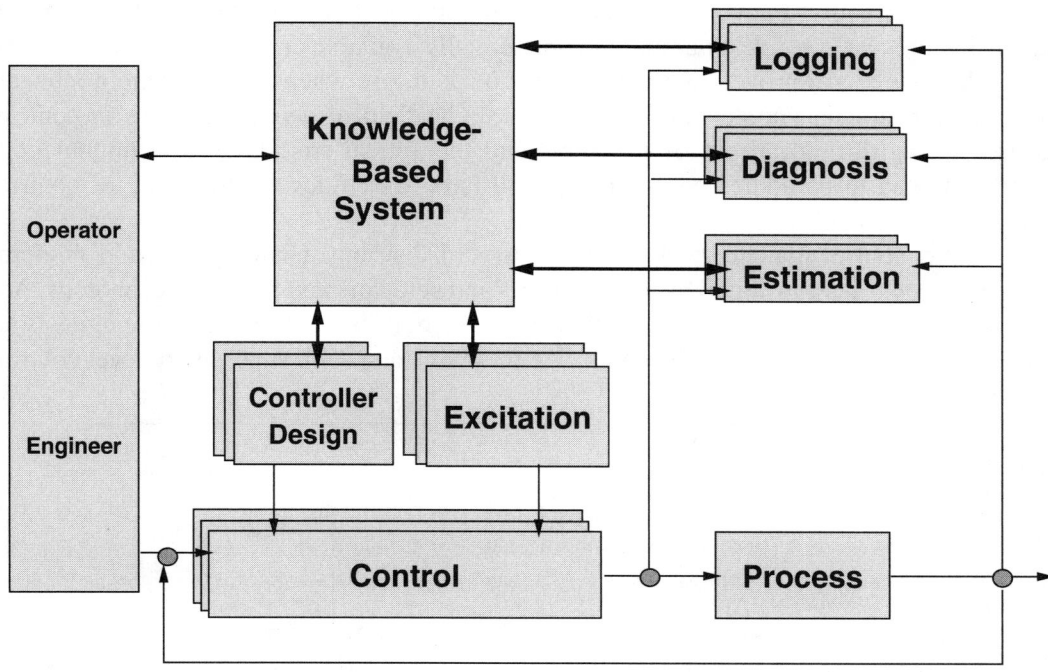

Figure 8–23 Expert control architecture.

with sequencing and logic (Årzén, 1993). Also, an adaptive controller based on recursive parameter estimation and control design is obtained by combining the functions of estimation, excitation, controller design, and controller selection (Åström & Wittenmark, 1995). The supervisory level can also be very well implemented in the system.

CONCLUSION

One type of advanced control discussed in this chapter is statistical process control. This technique uses analysis tools to assist the operator or engineer in decision making. The process controller analyzes the data and presents them in a form that helps the plant operator make decisions. It is expected that this technique will be used more frequently in processing plants in the near future.

In a broad definition, intelligent controllers are viewed as copying human mental faculties such as adaptation and learning, planning under situations of great uncertainty, and coping with large amounts of data to effectively control complex processes. Artificial intelligence, for example, is defined as the science of enabling computer systems to learn, reason, and make judgments.

Fuzzy control and expert control are two paradigms in intelligent control. A major goal of fuzzy control is to model the control actions of the operator. Expert control instead focuses on the generic control knowledge possessed by an experienced control engineer.

Artificial neural networks are another emerging technology that can be fruitfully applied to process control. A number of commercially available control systems use neural networks. They have proved to be attractive models to use when processes are nonlinear or when fundamental models are not available. The field of neural networks in control is proceeding through a mix of empirical and theoretical studies. Many fundamental questions need to be addressed before we will have a well-developed theory of control using neural networks.

Expert control matches well the trend toward distributed intelligence and modern control systems, demonstrated by smart sensors and actuators and field bus technology. An expert controller can be a smart controller that is responsible for control and diagnosis at the local feedback loop level. The controller communicates with supervisory level control and diagnosis functions.

REFERENCES

Åström, K.J., Anton, J.J. & Årzén, K. (1986). Expert control. *Automatica* 22, 277–286.

Åström, K.L. & Wittenmark, B. (1995). *Adaptive Control*. Reading, MA: Addison-Wesley.

Årzén, K. (1993). Expert control: intelligent tuning of PID controllers. In *Applied Control: Current Trends and Modern Methodologies*. New York, NY: Marcel Dekker.

Årzén, K. & Åström, K.J. (1996). Expert control and fuzzy control. *AIChE Proc* 312(92), 47–56.

Årzén, K.E., Jahanson, M. & Babuska, R. (1999). Fuzzy control versus conventional control. In *Fuzzy Algorithms for Control*. Edited by H.B. Verbruggen, H.J. Zimmerman, & R. Babuska. Boston, MA: Klewer Academic Publishers.

Bhat, N.V. & McAvoy, T.J. (1990). Uses of neural nets for dynamic modeling and control of chemical process systems. *Comput Chem Eng* 14, 573–583.

Büller, H. (1994). *Reglage par Logic Floue*. Lausane. Switzerland: Presses Polytechniques et Universitaires Romandes.

Davidson, V.J., Brown, R.B. & Landman, J.J. (1999). Fuzzy control system for peanut roasting. *J Food Eng* 41, 141–146.

Davidson, V.J., Martneau, S. & Brown, R.B. (1996). Quality-based control for drying of food materials. In *Proceedings of the NAFIPS'96*. Biennial Conference of the North American Fuzzy Information Processing Society, Berkely, CA: June 20–22. pp. 750–840.

Deming, W.E. (1982). *Quality, Productivity, and Competitive Position*. Cambridge, MA: MIT Publications.

Eerikainen, T., Linko, S. & Linko, P. (1988). The potential of fuzzy logic in optimization and control: fuzzy reasoning in extrusion cooker control. In *Automatic Control and Optimization of Food Processes*. Edited by M. Renard and J.J. Bimbenet. New York, NY: Elsevier.

Fitzgerald, N. McGrath, M.J., O'Connor, J.F. & Phelan, N. (1998). Integration of on-line quality control into the process control environment for cheese manufacturing. *Food Control* 9, 309–317.

Gaafar, L. & Keats, B. (1992). Statistical process control: a guide for implementation. *Int J Quality Reliability Manage* 9(4), 9–20.

Grant, E.L. & Leavenworth, R.S. (1988). *Statistical Quality Control*. New York, NY: McGraw-Hill.

Grigg, N.P., Daly, J. & Stewart, M. (1998). Case study: the use of statistical process control in fish product packaging. *Food Control* 9(5), 289–297.

Harris, T.J. & Ross, W.H. (1991). Statistical process control procedures for correlated observations. *Can J Chem Eng* 69, 48–57.

Hayes, G.D., Scallan, A.J. & Wong, J.H.F. (1997). Applying statistical process control to monitor and evaluate the hazard analysis critical control point hygiene data. *Food Control* 8(4), 173–176.

Huang, Y., Whittaker, A.D. & Lacey, R.E. (1998). Internal model control for a continuous, snack food frying process using neural networks. *Trans ASAE* 41, 1519–1525.

Hunt, K.J., Sbarbaro, D., Zbikowiski, R. & Gawthrop, P.J. (1992). Neural networks for control systems: survey. *Automatica* 28, 1083–1112.

Kear, F.W. (1998). *Statistical Process Control in Manufacturing Practice*. New York, NY: Marcel Dekker.

Keller, J, Gray, M. & Givens, J. (1985). A fuzzy k-nearest neighbor algorithm. *IEEE Transactions on System, Man and Cybernetics* 15, 580–585.

Kresta, J., MacGregor, J.F. & Marlin, T.E. (1991). Multivariate statistical monitoring of process operating performance. *Can J Chem Eng* 69, 35–47.

Landman, J.J. (1994). Modelling and control of colour development in Virginia peanuts during cross-flow dry roasting. Guelph, Ontario: University of Guelph. PhD dissertation.

Montgomery, D.C. (1985). *Introduction to Statistical Quality Control*. New York, NY: John Wiley.

Negiz, A. & Çinar, A. (1995). A parametric approach to statistical monitoring of processes with autocorrelated observations. Presented at the AIChE Annual Meeting, Miami, FL.

Negiz, A. & Çinar, A. (1995). Statistical process control of autocorrelated observations by monitoring model parameters. *Technometrics*.

Negiz, A., Ramanauskas, P., Cinar, A., Schlesser, J.E. & Armstrong, D.J. (1998a). Modeling, monitoring and control strategies for high temperature, short time pasteurization systems—1: empirical model development. 9(1), 29–47.

Negiz, A., Ramanauskas, P., Cinar, A., Schlesser, J.E. & Armstrong, D.J. (1998b). Modeling, monitoring and control strategies for high temperature, short time pasteurization systems—3: statistical monitoring and process sensor reliability. 9(1), 29–47.

Perrot, N.L., Trystam, G., Guely, F., Chevrie, F., Schoesetters, N. & Dugre, E. (2000). Feedback quality control in the baking industry using fuzzy sets. *J Food Processing Engineering*. In Press.

Perrot, N., Me, L., Trystam, G., Trichard, J.M, & Decloux, M. (1998) Optimal control of the microfiltration of a sugar product using a controller combining fuzzy and generic approaches. *Fuzzy Sets Syst* 94, 309–322.

Perrot, N., Trystam, G., Leguennec, D. & Guely, F. (1996). Sensor fusion for real time quality evaluation of biscuits during baking: comparison between Bayesian and fuzzy approaches. *J Food Eng* 29, 301–315.

Rumelhart, D.E. & McClelland, J.L. (1986). *Parallel Distributed Processing: Explorations in the Microstructures of Cognition*. Cambridge, MA: MIT Press.

Schonauer, S. & Moreira, R.G. (1997). Dynamics analysis of on-line product quality attributes for automation of food extruders. *Food Sci Technol Int* 3, 413–421.

Schonauer, S.L. & Moreira, R.G. (1995). Development of a fixed-GPC controller for a food extruder based on PQA—Part II: control development, implementation and analysis. *Trans Inst Chem Eng* 73(C), 200–210.

Schonauer, S.L. (1995). Product quality adaptive control system for a food extruder. College Station, TX: Department of Agricultural Engineering, Texas A&M University, PhD dissertation.

Seborg, D.E., Edgar, T.F. & Mellichamp, D.A. (1989). *Process Dynamics and Control*. New York, NY: John Wiley.

Shewhart, W.A. (1931). Economic Control of Quality. New York, NY: Van Nostrand.

Takagi, T. & Sugeno, M. (1985). Fuzzy identification of systems and its application to modelling and control. *IEEE Transactions on System, Man and Cybernetics* 15, 116–132.

Ungar, L.H., Hartman, E.J., Keeler, J.D. & Martin, G.D. (1996). Process modeling and control using neural networks. *AIChE Proc* 312(92), 57–67.

Verbruggen, H.B., Zimmermann, H.J. & Babuska, R. (1999). Fuzzy Algorithms for Control. Boston, MA: Kluwer Academic Publishers.

Zadeh, L. (1965). Fuzzy sets. *Inf Control*. 8, 338–353.

Zadeh, L. (1979). A theory of approximation reasoning. *Machine Intell* 9, 149–174.

Zimmermann, H.J. (1991). *Fuzzy Sets Theory—and Its Applications*. Boston: Kluwer Academic Publishers.

CHAPTER 9

Instrumentation

A control system (feedback or feedforward) performs essentially three operations: (1) measurement, (2) manipulation, and (3) signal transmission. Measurement is an integral part of a control system. To control a variable, its values need to be measured and converted to suitable signals.

Consider, for example, the simplified system shown in Figure 9–1, where a feedback control loop is used to aerate a grain bin. For simplification, it is assumed that the ambient temperature is below the setpoint temperature and the relative humidity above 90%. The control objective is to maintain the grain temperature at a desired level (7°C) by turning the fan on or off. The system uses a thermocouple to convert the temperature of the grain into a millivolts signal representing the temperature. The measurement is then sent to a

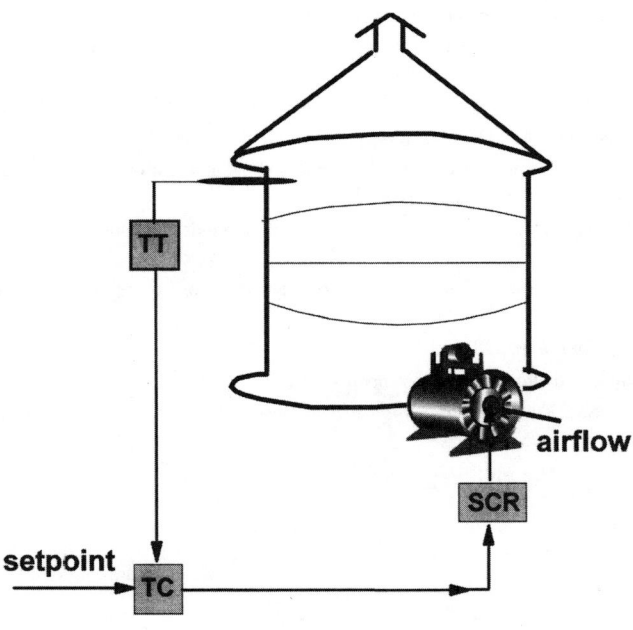

Figure 9–1 Feedback process control for aeration of stored agricultural products

temperature transducer (TT), where it is amplified and then sent to the controller (TC). The output signal (volts) from the controller is sent to a silicon-controller rectifier (SCR), which converts this signal to a form compatible with the electrical fan.

The controller/process interconnection for an analog controller and analog instrumentation of the type shown in Figure 9–1 is referred as an *interface*. The interface element converts information from one form to another—for example, from temperature to voltage.

For many years, food-processing systems such as food extruders used only crude or open-loop control of flow rates (solid and liquid raw material), an ampere meter to estimate the power draw by the drive motor, and a thermocouple and visual observation at the extruder discharge to maintain product specifications (Harper, 1989). New applications nowadays require precise control over the input variables to achieve a consistent product within narrow specification limits and to maximize extrusion throughput. Today, extrusion operations are monitored and controlled by computer systems, using accurate information of input and output variables and advanced control algorithms (Schonauer & Moreira, 1995a, 1995b).

Intelligent selection and use of measurement instrumentation depend on a broad knowledge of what is available and how the performance of the equipment may be best described in terms of the job to be done. New sensors are continually being developed, but certain basic devices have proved their usefulness in broad areas and will be widely used for many years. For detailed information on instrumentation and measurement systems, the reader is referred to Doebelin and Johnson (1999).

MEASUREMENT TRANSDUCER

A measurement transducer consists of a sensing element combined with a transmitter (a driving element). Transmitters are used to convert the sensor output signal to a form compatible with the controller input and to convey the signal through transmission lines connecting the two.

Transducers are devices that convert physical or chemical information in one form into an alternative physical form. Transducers for process control measurement convert the magnitude of a process variable, such as pressure, temperature, or flow rate into a signal that can be sent directly to the controller. A sensing device is required to convert the process variable into a value that can be processed mechanically or electronically within the transducer. Some process variables are easily measured (pressure, temperature, liquid flow rate), but other variables such as product composition, solid flow rate, and product quality are more difficult to measure on line.

MEASUREMENT DEVICES OR SENSORS

A summary of the most used sensors in the food industry is presented below.

Temperature

The most common temperature sensors used in process control are thermocouples, resistance thermal detectors (RTDs), thermistors, integrated circuit (IC) sensors, and pyrometers.

Thermocouples generate emf (electromotive force) by a hot junction. Their major disadvantages are the small voltage produced, which requires a sensitive monitoring instrument; the need for a constant temperature device to serve as a reference junction; and corrosion of connectors and wiring, which affects their resistance and accuracy.

Copper-constantan (T-type) or chromel-alumel (K-type) thermocouples are preferred to overcome corrosion problems instead of the iron-constantan (J-type) ones. Thermocouples of various sizes, construction, and design are available commercially. The most commonly used and available thermocouples have been standardized and are shown in Table 9–1.

Resistance thermal detectors are based on the principle that electrical resistance of various materials changes with temperature. They are capable of measuring very high temperatures accurately. RTDs use a Wheatstone bridge potentiometer and a power supply to obtain a reading through a change in electrical resistance. Thus, they are more expensive than thermocouples. They are especially useful in applications that require extreme levels of accuracy or high and wide temperature capabilities (–240 to 649°C).

Thermistors are semiconductor temperature sensors whose resistance changes inversely with temperature. Most thermistor applications are in the range from –80 to 150°C. Some advantages of these sensors include small size, fast response, narrow span, and high sensitivity. However, they show poor high-temperature stability, are not suitable for large spans, and have high impedance.

Integrated circuit sensors are precision solid-state devices with an accuracy of 0.25% of span, low voltage requirements, low current draw, and low self-heating errors. In an IC sensor, the output is linearly proportional to any temperature scale (ie, °C, K, R, or °F). For example, the LM35 IC sensor produces an output voltage that is equal to 10 mV times the temperature in degrees Celsius over a range from –55 to 155°C.

Table 9–1 American National Standard Thermocouple Type Designations

ANSI Type	Thermocouple Materials	Temperature Range [C]
B	Platinum–6% rhodium/platinum–30% rhodium	0 to 1800
E	Chromel/constantan	–190 to 1000
J	Iron/constantan	–190 to 800
K	Chromel/alumel	–190 to 1370
R	Platinum/platinum–13% rhodium	0 to 1700
S	Platinum/platinum–10% rhodium	0 to 1756
T	Copper/constantan	–190 to 400

ANSI Standard MC 96.1, Temperature Measurement Thermocouples (Instrument Society of America)

Pyrometers are devices based on thermal radiation and are used for very high temperature ranges (up to 3000°C). An optical system collects the visible and infrared energy coming from the object and focuses it on a detector. The detector then converts the energy into an electrical signal. The advantages of a pyrometer include no physical contact with the object whose temperature is being measured, fast response, the ability to measure the temperature of small objects, and the ability to measure high temperatures. Some disadvantages include high cost, a nonlinear response approaching the fourth power of the temperature, variations in the emittance of the object causing error, and the relatively wide temperature span required.

Pressure

A wide range of pressure (0.1 Pa to 100 MPa) is measured and controlled in the food industry. A great variety of primary elements that convert pressure into displacement and force have been developed to measure pressure. A signal conditioner converts the force or displacement into voltage, current, or air pressure signals suitable for use by a controller. Pressure sensors can be classified into strain gauge pressure and deflection type.

Strain gauge pressure sensors are based on the fact that stretching a metal wire changes its resistance. Strain gauges are divided into bounded and unbounded types. In the case of an unbounded strain gauge, the displacement is transferred to the strain wire by a mechanical linkage. Bounded strain gauges are cemented directly onto the body of the transducer.

Deflection-type pressure sensors consist of a primary element (that converts a pressure signal to displacement), a secondary element (that converts displacement into a change of an electrical element), and a signal conditioner.

One example of a common element used in deflection-type pressure transducers is the *Bourdon tube*. It consists of a flattened tube that is shaped into an incomplete circle, spiral, or helix. The tube straightens out as the internal pressure increases, providing a displacement proportional to the pressure. Bourdon tubes are available in bronze, steel, or stainless steel, covering a range from 0 to 10,000 psi or greater.

Another element used in deflection-type pressure sensors is a thin-walled cylinder with corrugated sides called a *bellows*, used to measure pressure up to 100 psi.

The third primary element type is called the *diaphragm*. It can be flat or corrugated, and it allows sufficient movement to balance the pressure against a calibrated spring or force transducer.

Secondary elements are resistance, capacitance, or inductance. A displacement is used to adjust one of the three electric circuit elements so that the signal conditioner can use it to produce a usable signal.

A number of *force transducers* have been developed and used in the food industry. An example of such a transducer has the sensing tip covered with a diaphragm. A force equal to the pressure on the product times the tip area is transmitted through the diaphragm to a strain gauge, LVTD, or piezoelectric transducer to sense its magnitude.

Flow Meters

The flow rate of liquids, gases, and solids is an important variable in the food-processing industry. Volumetric flow rate is generally measured by positive-displacement or rate meters. Some examples of flow meters are differential pressure flow, turbine, ultrasonic, and magnetic flow meters.

Differential pressure flow meters operate on the principle that a restriction placed on a flow line produces a pressure drop proportional to the flow rate squared. The restriction most often used for flow measurement is the orifice plate. The orifice is the primary element and the differential pressure transmitter the secondary element. The orifice converts the flow rate into a differential pressure signal, and the transmitter converts the differential pressure signal into an electric current signal. Advantages of such device include the fact that it is simple and easy to fabricate and has no moving parts. However, an orifice does not work well with slurries.

A *turbine flow meter* consists of a small permanent magnet embedded in one of the turbine blades. A magnetic sensing coil generates a pulse each time the magnet passes by. The number of pulses is related to the volume of liquid passing through the meter. The pulse output of the turbine flow meter is ideally suited for digital counting and control techniques.

Level

The measurement of level can be done by determining the difference in pressure between two fluids (gas and liquid) or by converting the buoyancy effects on a fixed float to level for a liquid of known density. Level meters can be classified as (1) float actuated—limited to liquid-gas interface, (2) hydrostatic head, (3) electrical—based on the conductivity and dielectric constant characteristics between two phases, (4) thermal—based on the thermal characteristics between fluids, and (5) based on the sonic-propagation characteristics between liquid-gas interfaces.

Physical Properties

On-line measurement of properties of food materials such as density, viscosity, composition, color, and pH is crucial for a well-designed process control system.

Density

For liquids, gases, or solutions of solid in a solvent, density is a function of composition for a given pressure and temperature. Some examples of measuring devices are (1) liquid column—based on the measuring gauge pressure at the bottom of a fixed-height liquid column, (2) displacement or hygrometer—based on the weight of liquid displacement, (3) direct mass measurement—based on the amplitude of natural vibration frequency; the change in natural resonant frequency of the system due to the effect of fluid mass is correlated with fluid density, and (4) radiation—based on the absorption of gamma radiation as it passes through the process material; absorption is proportional to changes in material density.

Bulk Density Measurements

On-line automated bulk density monitoring systems for solid materials are available commercially. One of these sensors has three components: (1) a compact sampler, (2) an amplifier module that contains a tare board and signal conditioner, and (3) the controller, which mounts remotely for operator convenience.

Viscosity

On-line viscometers are based on the measurement of either the drag flow or the torque produced by the movement of an element through the fluid. One example of a device uses a bob (cylindrical spindle) attached to the main body and includes temperature measurement. The bob oscillates in the flowing sample material, and the number of oscillations is correlated to viscosity.

Composition (Moisture and Fat Content)

On-line moisture and/or fat sensors are essential in the food industry. They are necessary not just for continuous product quality monitoring but also for process control by providing continuous and reliable signals to a feedback controller system. Some of these sensors use near-infrared reflectance (NIR) technique and do not touch the food during measurement, being particularly suitable to continuous processing lines. The sensor measures the product surface moisture and needs to be carefully calibrated (eg, selecting the correct wavelengths) for correct reading. The principle of NIR involves energy absorbance by chemical bonds. Specific bonds vibrate at a particular frequency and absorb photons of coincident energy. This energy leads to elevation of the energy level of the molecule. Some of this energy is transmitted, and the remainder is reflected. The amount of energy reflected at different wavelengths can be correlated with the number and types of functional groups present. Fundamental absorption bands for organic molecules occur primarily in the infrared region.

Color

On-line colorimeters have been used to measure color of snack food products in processing lines. The signal from the sensor is used to automatically control a continuous process. An example of such a device uses the L^*a^*b color scale. Generally, this sensor is positioned after the cooling section of the processing line, and the b value, which is positive when the product is yellow and negative when the product is blue, is used over a and L for modeling and control because it is affected the most when process variables are varied.

PH

The pH value of liquid food materials is determined by measurement of hydrogen ion concentration, using a glass electrode.

On-Line Sensors

Sensors of the measurement of product physical attributes such as temperature, pressure, flow rate, viscosity, and liquid level have been commonly used in the food industry. Physical or quality attributes of food products are important elements of process control; however, food composition is a critical factor in food processing.

The most important advances in sensor technology for the food industry will occur in the measurement of attributes such as moisture, fat, and protein content. On-line measurement of these parameters is now available commercially (Giese, 1993). On-line instruments and sensors can provide real-time product information, thus linking the process and its control.

Chemical and Biological Sensors

Chemical and biological sensors are of the most interest to the food industry. They monitor chemical attributes and are classified as ion-selective electrodes, ion-selective field effect transistors, and metal oxide gas sensors (Datta, 1992). Biological or biosensors incorporate some type of biological element into their composition.

Ion-Selective Electrodes (ISEs). These sensors measure the concentration of ion species. An example is the ISE pH meter. These sensors have a membrane that selectively allows only certain ions to pass through. The movement of ions from higher to lower concentration results in a uneven charge distribution across the membrane, with a building up of a potential across the membrane in the direction that opposes further movement of ions and achieves an equilibrium. At equilibrium, the resulting potential would exactly balance the net ionic flow in each direction, and no further net flow of charge would occur. In a system in which no current can flow, the total number of ions required to develop this opposing potential is very small and can be measured. However, on-line applications of ISE sensors are limited because the food sample needs to be pretreated before measuring to exclude ion interference.

Ion-Selective Field-Effect Transistors (ISFETs). These transistors are semiconductor chips that have an ion-selective membrane built into them. Similar in principle to the ISE sensor, the charge in the ions at the surface causes a variation of the potential (and thus conductance) in the ISFET semiconductor chip. A direct relationship exists between conductance and pH of the solution. A constant current through the ISFET will cause a change in voltage when the conductance changes and may be measured.

ISFETs are more flexible than the ISE sensors because they are small, rugged, and easy to calibrate and operate (Datta, 1992). An on-line ISFET sensor consists of a pH ISFET mounted in tandem with a reference ISFET (REFET) that has an ion-retarding layer with the ability to retain an established pH for a long period. Figure 9–2 illustrates a schematic of a pH ISFET.

Metal Oxide Gas. These are semiconductor gas sensors based on the surface properties of oxides of tin (SnO_2) or zinc (ZnO). The operating principle of these sensors is based on the fact that the surface conductivity of semiconductors can be markedly changed by the adsorption and subsequent reaction of gases with already adsorbed atmospheric oxygen. Because of reduction or oxidation processes, electrons become available or unavailable for conduction, and the resistance of the surface layers decreases or increases (Datta, 1992). The conductivity versus gas concentration can then be measured. Gas sensors have been used on line to measure ethanol in a fermentation process (Mandenius & Mattiasson, 1983).

The advantages of this type of sensor include small size, low-voltage power supply, and high sensitivity. The disadvantages of metal oxide gas sensors are continuous power drain (about 0.5 W) for sensor heating and sensitivity to ambient conditions (temperature and humidity and the presence of long-term drift).

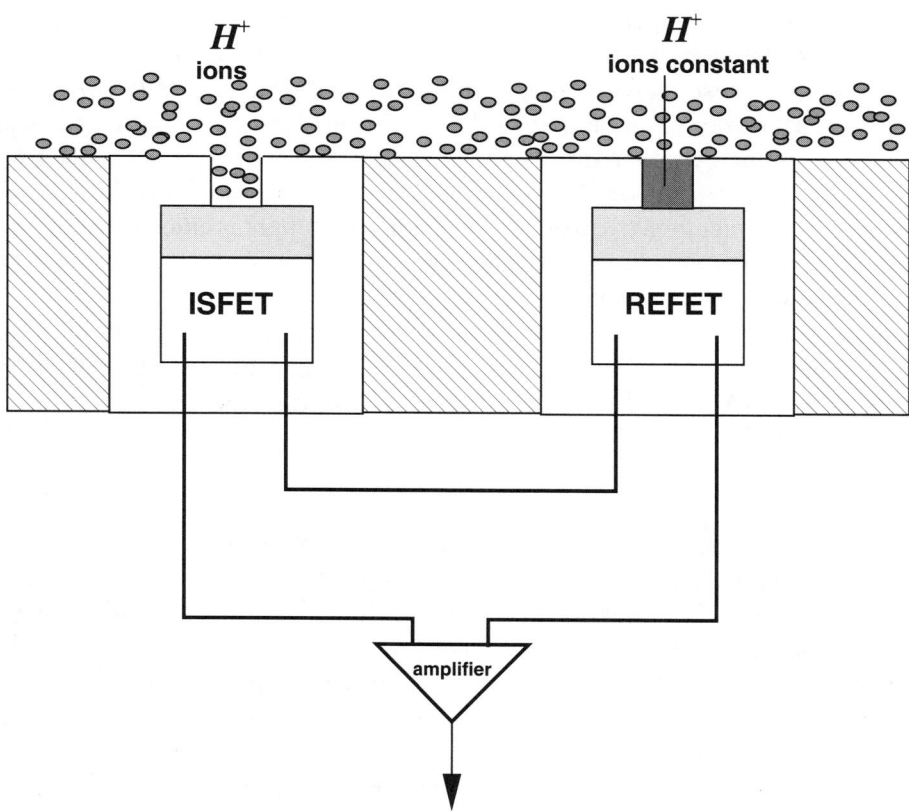

Figure 9–2 Schematic of a pH ISFET

Biosensors. These are biologically sensitive materials immobilized in intimate contact with a suitable transducing system that converts the biochemical signal into a quantifiable and processable electrical signal. Examples of biological sensitive materials include enzymes, antibodies, membrane components, organelles, bacterial or other cells, and slices of mammalian and plant tissues.

Advantages of biosensors include great selectivity and ability to detect extremely small amounts of the desired substance, such as acetic acid, glutamic acid, cholesterol, sugars, and organic acids. Response times for biosensors are on the order of 0.5 to 2 minutes.

The use of biosensors for on-line measurements is limited. Because they are composed of biological materials, biosensors cannot be sterilized and are inactivated by strong acids, bases, and other harsh conditions encountered in food-processing plants. They also have a limited testing life (Giese, 1993).

One example of a biosensor is a microbial sensor for detection of fish freshness developed by Wantanabe et al (1987). The sensor was prepared by immobilizing spoilage-causing putrefacients on a membrane filter that was fixed at the tip of an oxygen electrode. The extent of the assimilation of organic substances (resulting from deterioration of fish) by these microorganisms could be determined from the respiratory activity of the microorganisms by directly measuring the oxygen consumption in the electrode.

Another application of biological-based technology is ATP bioluminescence. It is the only microbiological test available today that offers real-time results (Griffiths, 1996). The technique makes use of the fact that all living cells contain adenosine triphosphate (ATP), a universal energy donor for metabolic reactions. An enzyme-substrate complex, luciferase-luciferin, present in firefly tails converts the chemical energy associated with ATP into light by a stoichiometric reaction. The amount of light emitted (quantified using a luminometer) is proportional to the concentration of ATP present, which is related to the number of metabolically active cells present in the assay.

There are several ATP hygiene monitoring kits commercially available today to detect microorganisms in the food industry. Other bioluminescence techniques available include phage-based ATP assays, adenylate kinase, chemiluminescence, and molecular bioluminescence (Griffiths, 1996).

Fiber-Optic Sensors

Fiber-optic sensors based on the measurement of spectrum modulation have been used in the food industry for pressure, temperature, and moisture sensing (Henrickson, 1992). Fiber optics is the transmission of light through long, thin, flexible fibers of plastic, glass, or other materials via total internal reflection.

One application of fiber-optic technology is in the measurement of the reflective index (Giese, 1993). This sensor has been applied to measure hydrogenation or degree of saturation in edible oils. In this sensor, the fluid enters a cavity; as its reflective index changes, so does the effective path length and thus the extent of spectral modulation. The reflective index correlates to the extent of hydrogenation.

The main advantages of fiber-optic sensors include the lack of optical components and their small size, which makes them less prone to damage and useful in some applications.

Near-Infrared (NIR) Sensors

Current on-line applications of NIR spectroscopy analysis in the food-processing area include the quantitative determination of moisture, fat, oil, sugars, and protein in products such as coffee, orange juice, milk powder, breakfast cereals, soft drinks, snack foods, processed cheese, and flour (Wilson & Kemsley, 1992).

NIR reflectance instruments, which scan in the 700- to 2500-nm spectral region, are based on the principle that fundamental molecular vibrations excited in the mid-infrared region give information on molecular structure. The instruments have an infrared light source that is collimated and filtered into specific wavelengths. The infrared beam is directed onto the surface of the food material, and the amount of energy absorbed at several frequencies is measured and transformed to concentration data. The frequencies measured are selected for the analysis of interest and are based on chemical functions groups such as N-H or O-H bonds (Kess-Rorgers, 1986).

NIR transmission instruments, which scan in a narrower spectral region from 900 to 1025 nm, can be directly installed in a sanitary pipeline cell with windows manufactured from polysulfone (Honings, 1993).

Refractometers

On-line refractometers, used to measure constituents of foods, are based on the refraction of light. The reflective index, defined as the amount of reflection of an electromagnetic radiation ray stricken in a flat surface, can be used to determine component concentration in a food product. On-line refractometers may be used to monitor °Brix in carbonated soft drinks, dissolved solids in orange juice, and percent solids in milk (Giese, 1993).

Microwave and Radio Frequency

Microwave sensors measure the complex dielectric response from which moisture and density can be derived (King, 1992). The instrument radiates a continuous microwave through the food material. The wave undergoes attenuation and time delay due to the dry density and the moisture. The sensor has been developed to allow measurement from one side of the sample.

Ultrasound

High-frequency ultrasonic methods (1–100 MHz) that operate at low power are used to examine food and related systems. Ultrasound is defined as a longitudinal pressure wave with frequency above the range of human hearing (16 kHz).

The propagation of ultrasound through a system depends upon its response to rapid pressure fluctuations, and, in principle, all mechanical and thermal properties have an effect. The transmission of ultrasound through multiphase materials such as foods is influenced not only by the properties of the various phases in isolation but also by the physical structure and the mismatch in ultrasonic properties between phases (Javanaud & Robins, 1993).

Ultrasonic techniques have several advantages for food materials, including the fact that they are noninvasive, accurate, and inexpensive. Sound waves travel easily through many materials that are opaque to light, but they are strongly attenuated by gases; thus, it is difficult to apply this technique to food materials with a larger air content.

The experimental method that has successfully been used with food materials is a pulse overlap to measure the velocity of sound. This method has the advantage of requiring only a fixed path length for the sound, so it is more easily adapted to on-line situations.

Figure 9–3 shows a schematic diagram of the pulse overlap technique, used to measure group velocity in dispersions as a function of height. A water bath, thermostatically controlled to ± 0.1°C, contains quartz transducers facing each other. Between the transducers lies a perspex cell containing the sample liquid. Measurement can be made at different heights to monitor sedimentation processes in dispersions.

The method consists of exciting the emitting transducers by a single high-voltage pulse of duration less than or equal to 100 nanoseconds. The shape of the wave packet is determined by the electromechanical properties of the emitter. The emitted and received signals are observed on an oscilloscope, and the repeat rate for the transmitted pulse is varied until the received signal overlaps successive transmitted pulses. Thus, the reciprocal of the pulse repetition rate equals the total travel time between the transducers. The travel time depends on the group velocity of sound in the liquid.

Machine Vision

Machines that can "see" have been developed to inspect and manipulate processed food products. Machine vision can be defined as the use of a noncontact sensing device to automatically receive and interpret an image to obtain information and/or control processes (Chan & Batchelor, 1993).

Some products cannot be inspected using existing techniques at the rate at which they are produced. A machine-vision system can perform inspection at much higher speeds than a human operator. In addition, machine vision systems offer several advantages to food inspection, being noncontact and nondestructive and having the ability to incorporate feedforward and feedback loops into control processes.

The main components of a machine-vision system are shown in Figure 9–4. The image formation system consists of optics, some form of illumination, and a sensor (the camera). The following is a short description of the main components of a machine-vision system (Chan & Batchelor, 1993):

- *Lighting* is an important part of the system, since the appearance of an object is critically dependent on its attitude toward the light source.
- The *optical techniques* used in machine-vision systems range from fiber-optic devices to complex optical processing methods, including holography, correlation, and diffraction.

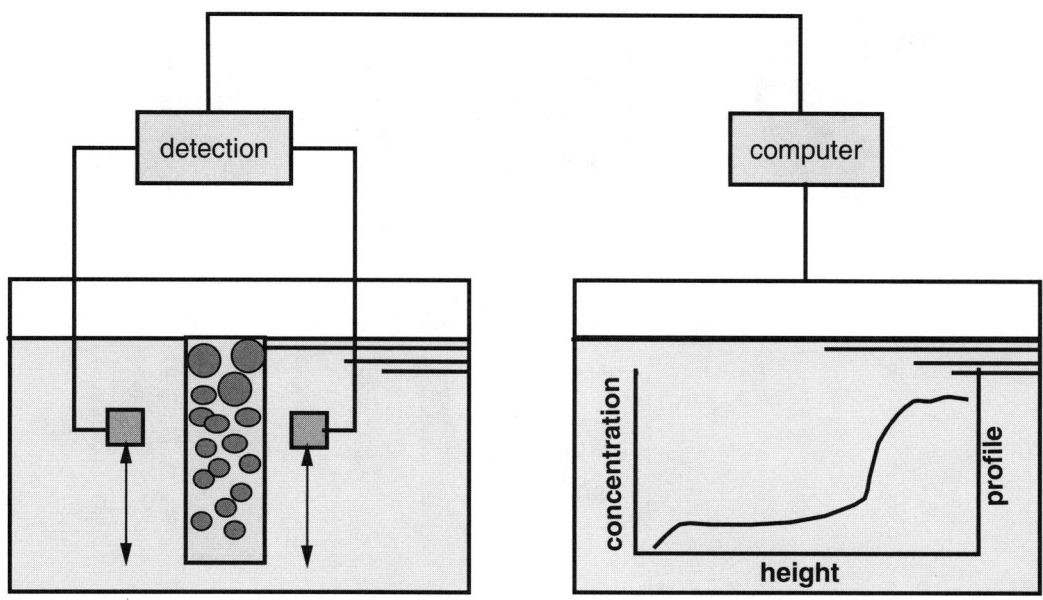

Figure 9–3 Pulse technique to measure group velocity in dispersions as a function of height

- The role of the *image sensor* is to transform certain information related to the physical makeup of a scene into light intensities that can be viewed by the human eye and/or digitized by the vision system. A video camera is the most common form of image sensor.
- *Digital image* acquisition equipment is concerned with the generation of a 2-D array of integer values representing the brightness function of the actual scene at discrete spatial intervals.
- *Image processing* consists of transforming one image into another so that the properties of the image to be analyzed are enhanced.
- *Image analysis* is a technique that transforms the enhanced image into some form of description of the image. The most common techniques used in image analysis are template matching, pattern recognition, and descriptive syntatic processes (Whelan & Batchelor, 1991).
- *Image interpretation* is concerned with making some decision based on the information extracted from the image analysis section. Image understanding requires greater machine intelligence than the previous stages and may incorporate artificial intelligence, a neural network, and/or an expert system.

Final Control Elements and Regulators

The *final control element* serves as an interface between the controller and the process. It is the device that enables a process variable to be manipulated. Figure 9–1 shows the use

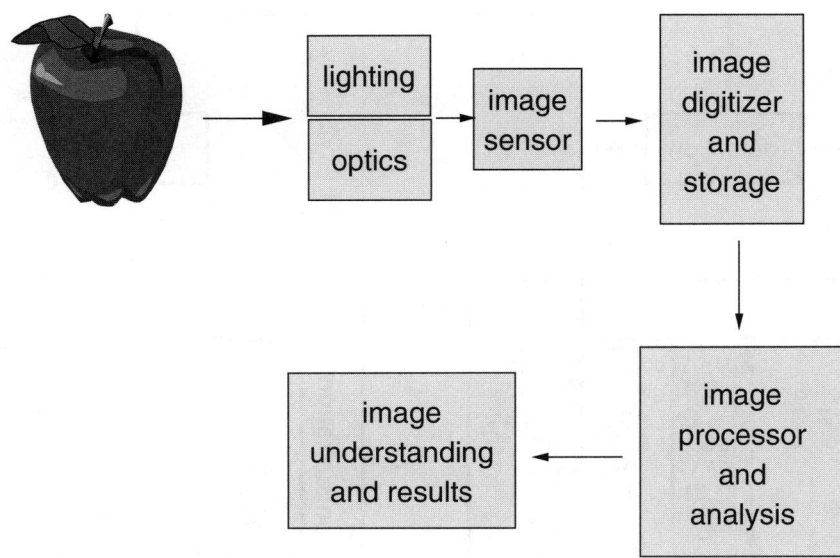

Figure 9–4 Basic components of a machine-vision system

of an electrical fan as the final control element. In this case, a transducer is placed between the controller and the fan. This is needed because the controller output, a voltage signal, cannot be applied directly to the fan, since the controller is not designed to supply the electrical output requirement of the fan. An SCR (silicon-controller rectifier) is designed to provide a linear relationship between the voltage input to its solid-state circuits and its output.

For most food processes, the final control elements adjust raw materials or final product flow rates (liquid, solid, gas), and indirectly energy flows in or out of the process. Examples of ways of varying flow rates include the speed of a screw conveyor, of a valve, of a pump, of a blower, and so on.

The *pneumatic control valve* is the most common final control element used to manipulate the flow rate of a fluid. Figure 9–5 illustrates a typical pneumatic control valve. The input to the valve is air pressure that acts on the diaphragm, compressing the spring, thus, pulling the stem out and finally opening the valve. The actuator of a pneumatic valve can be described as a *direct actuator* (if the actuator spring is located below the diaphragm) or a *reverse actuator* (when the spring is above the diaphragm). The valve position is the displacement plug from the fully seated position. The valve action is in the direction in which the valve plug moves as the air pressure increases. If the valve closes as the pressure increases, an *air-to-close* valve results. If the increased air pressure opens the valve, it becomes an *air-to-open* valve.

Regulators are compact devices that maintain the process variables at a constant value, regardless of disturbances in the load flow. These are self-contained devices that combine the functions of a sensor, controller, and final control element. Regulators are used to control pressure, flow, temperature, level, and other process variables.

TRANSMISSION LINES

Transmission lines are classified according to the signals being transmitted between controllers and instruments: pneumatic pressure signals, electronic signals, and digital signals.

Pneumatic pressure signals are transmitted by tubes, such as PVC-coated copper. The maximum length allowed for dynamic accuracy is limited to 100 or 200 m. Therefore, pneumatic controllers are generally located relatively close to their sensors and actuators.

Electronic controllers, on the other hand, using two-wire current-loop (4 to 20 mA) can be located relatively far from their instruments. Generally, multipair shielded cables are used to improve accuracy. Particular care must be taken, though, when transmitting voltage-level signals, since these signals are better restricted to laboratories.

Signals from *digital instruments and controllers* are transmitted in digital format (on-off pulses) using, for example, a number of parallel wires. New instrumentation systems use a single *data highway* to transmit digital signals to a series of instruments and controllers. These data highways are usually coaxial cables linked serially or in a daisy chain (serially with a complete loop). A microcomputer built into the instrument and controller communicates periodically over the highway by sending or requesting information between different devices installed in the system. Development of digital transmission techniques was responsible for the way modern plants and control rooms are

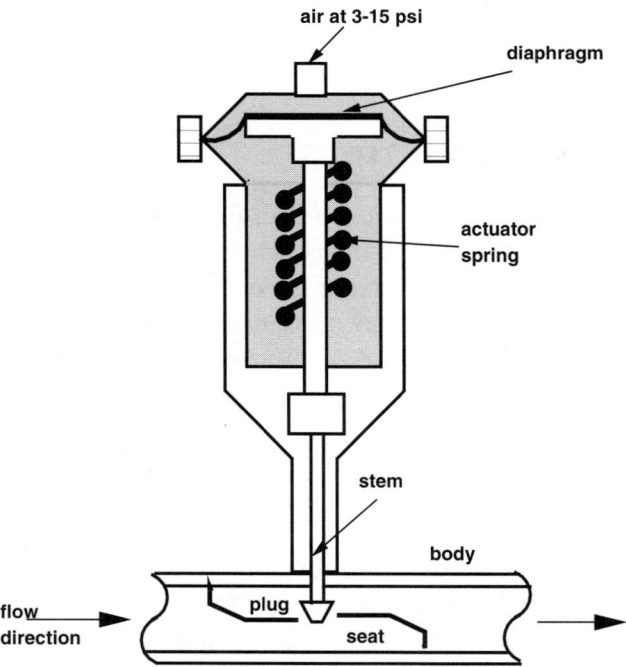

Figure 9–5 Schematic of a pneumatic control valve

wired today. Digital transmission makes installation and maintenance simple by replacing the numbers of wires, cables, and tubes by coaxial or fiber-optic cables.

Examples of data highways are *wide area networks* (*WANs*) and *local area networks* (*LANs*). Networks are communication channels that connects a large number of stations to one or more central stations. WANs are used to connect process systems separated by considerable distances (branch offices in different cities). LAN is a data highway used to connect local processing units with central operator displays and high-level computers. LANs are used to connect processing systems located within a radius of about 1 mile.

Another important advance in communication networks is the *open system interface* (*OSI*). The OSI model is based on several functional levels. The lower levels are responsible for transmitting messages between nodes on the same cable and from one cable to another. The upper levels handle data formatting and access security. A number of LANs are based on the OSI model, such as the *manufacturing automation protocol* (*MAP*) and the *Ethernet*.

The MAP is a multilevel communication network whose objective is to link together all the computers, controllers, equipment, devices, and offices in an entire factory. The Ethernet is able to handle error detection and addressing of the source of destination.

The technology of instrumentation has changed rapidly in the last 20 years. The trend is for more digital and microcomputer-based instrumentation and thus, more data highway transmission lines. Another increasing trend is the integration of sensing elements into silicon chip microcircuitry, or development of smart sensors. Process variables such as temperature, pressure, and composition can be measured with these sensors, which directly incorporate all circuitry needed to self-compensate for the environmental disturbances and to produce a linear amplified output signal suitable for transmission to standard electronic controllers.

COMPUTER-CONTROLLED SYSTEMS

In a computer-controlled system, the output from the process, $y(t)$, is a continuous-time signal. The output (in analog form) is converted into digital form by an analog-to-digital (A/D) converter. The conversion is done at sampling times k. The computer interprets the converted signal, $y(k)$, as a sequence of numbers, uses an algorithm to process the measurements, and generates a new sequence of numbers, $u(k)$. This sequence is then converted to an analog signal by a digital-to-analog converter (D/A). The system runs in an open loop between the A/D and D/A conversion interval. A real-time clock in the computer is used to synchronize the events. A computer-controlled system contains both continuous-time signals and discrete-time (sampled) signals and is often called a *sampled-data system*.

The first full application of computers for process control was made in the early 1960s by the Imperial Chemical Industry, in England. The system was called *direct digital control* (*DDC*). The advantages of the DDC system were (1) lower cost for larger installation than the analog system; (2) easier operator communication (a digital display and a few buttons replacing the large wall of analog instruments); (3) flexibility (i.e., analog systems are changed by rewiring, while computer-controlled systems are changed by reprogramming); and (4) capability of multiloop interactions. The main drawback of the DDC systems at that time was that it was difficult to do unconventional control strategies.

Digital computers underwent substantial development during the 1960s, becoming smaller, faster, more reliable, and cheaper. At that time, it was possible to design efficient process-control systems by using the newly developed *minicomputers*. The combination of minicomputer technology with DDC technology resulted in a rapid increase in the applications of computer control by that time. However, the minicomputer was still a fairly large system, and the price of a unit was still too high for a large number of control problems.

By 1972, microcomputers were developed. By the early 1980s, the price of micro-computer systems had dropped substantially, making the application of computer control technology more affordable. Some of the impacts of the microcomputer on process control technology were (1) replacement of analog hardware even as single-loop controllers, (2) development of smaller DDCs, (3) improvement of operator communication by the introduction of color-video graphics displays, (4) development of hierarchical control systems, and (5) design of special-purpose regulators.

Another important impact of digital control technology was the development of a new system architecture, called *distributed control systems* (*DCSs*). Figure 9–6 shows the elements of a DCS. The system is composed of many digital computers that perform different control functions, such as feedback control, logic, and sequencing during start-up, shutdown, and human-machine operation. The different computers communicate over a computer network. There is also a database that may be either central or distributed where all system variables are stored.

Digital Control Hardware

Knowledge of digital control fundamentals is very helpful in understanding digital control applications. Microprocessors are used in every part of a control system today. They facilitate communication between various sensors and control loops in a processing plant and increase the capabilities and accuracies of field sensors (smart transmitters, control valves).

A *digital computer* consists of five major elements: input, output, memory, arithmetic-logic, and control units. The *input unit* receives analog and digital signals that are condi-tioned to match the internal computer requirements. Computers work only with binary signals, so the analog signals are converted into binary numbers by an A/D converter. For the digital signals, already binary signals, conditioning consists of just producing the correct voltage levels for use in the computer. These input units have a memory, a buffer, to hold the conditioned data until the processor accepts them.

In the *output unit*, binary signals received from the memory unit are conditioned to use in external devices such as a printer, a control valve, a digital indicator, and other final control elements. Analog signals, if presented, are converted from digital to analog by a D/A converter.

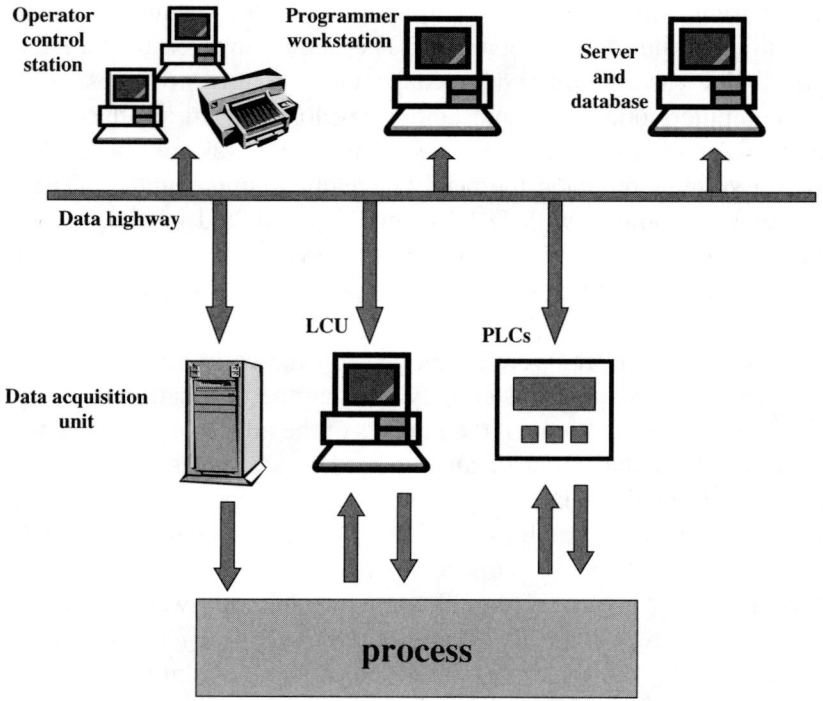

Figure 9–6 A distributed digital control system (DCS)

The *memory unit* consists of a larger number of storage cells, each capable of storing one binary number. Each cell has a unique number called a memory address (MA), identified by its row and column. The memory unit is characterized by its length (number of binary digits that can be stored in each memory cell) and by the total number of memory cells that can be addressed (addressable memory capacity).

The objective of the *arithmetic-logic unit* is to perform the arithmetic and logic operations. These operations can be done with binary numbers or by using binary code to represent decimal digits to perform decimal arithmetic, referred to as *binary-coded-decimal digits* (*BCDs*).

In the *control unit*, the program instructions are taken one at a time from the memory unit, each instruction is interpreted, and then the operations are executed. The program instructions are in the form of binary numbers.

A *microprocessor* is a single electronic component that handles all the processing and control functions of a computer. It operates in a repeating cycle, taking and executing instructions in its program. Microprocessors use buses (series of conductors) to communicate with other components (internally or externally). According to the number of lines in the external data bus, microprocessors can be classified as 8, 16, 32, or 64 bit. Another important characteristic of a microprocessor is the number of memory bytes that it can select, determined by the number of external address bits. A microprocessor with 16-bit address bus, for example, has a physical address space of 2^{16} (64 kilobytes or 64 kB).

Any modern digital *microcomputer* with sufficient input/output (I/O) capability can be used in the application of process control. A microcomputer is composed basically of four units: a microprocessor, a section of read-only memory (ROM), a section of random-access memory (RAM), and an I/O interface. RAM is the working memory of the computer. It can be used for storage of programs and data. RAM can be written in or read out at any time. The term *random access* refers to the ability of the microprocessor to go directly from any location in the memory unit for a READ or WRITE operation. The contents of ROM, on the other hand, cannot be changed, so ROM is used to store permanent system programs and data.

The I/O interface links the microcomputer to the "outside world": that is, to all its peripheral equipment or even to other elements in the network if a data highway is used. The interface contains the hardware logic necessary to detect and respond to external events. The most common means of input or output is through parallel I/O ports.

One important aspect of computer control is its real-time computing application. It allows initialization of data acquisition operations and other tasks such as control output calculations, supervision, and scheduling. *Real time* means that an operation is to be completed within a specified time constraint. An example of a hardware timing device technique is the real-time clock that interrupts the computer periodically (every 0.1 second) and identifies itself as the interrupting device. The system programs within the computer are required to update memory registers (having the time of day) and to execute operations (programmed by the user) at a particular time. Any real-time computer system must be able to respond to interrupts from external devices.

Another important part of a digital control computer is a *terminal* that allows communication between the system and the operator. This terminal is generally a color display monitor that allows visualization of key information. Graphical information is usually used to depict the behavior of different variables around the setpoint or to display distribution in the form of bar graphs. A printer is important to obtain a hard copy of process data containing important updates of key process variables. Moving-head disks store system and user programs. Magnetic tapes or disks are used for long-term storage of system and user programs.

Programming Language

Digital control systems involve software development. Programming is an important aspect of digital process control. The effort required and the approaches used depend on the available software and the nature of the control problem.

Many high-level languages may be used, including BASIC, FORTRAN, C, and PASCAL, that have been extended to permit real-time operations. There is also software in the form of control-oriented programming languages, usually supplied by a vendor of process computer systems. There are many advantages to using high-level languages: it requires less time to develop a system, and the code is easy to read and modify. However, the code required for operator communication is often larger than the pure control code.

The control software can be divided into (1) operating system and executive programs and (2) higher-level programming languages.

A digital computer operates sequentially in time, and ordinary programming languages can represent only sequential activities. Therefore, to run control loops as concurrent activities in parallel, it is necessary to map the concurrent activities into a sequential program. Special-purpose software, a *real-time operating system*, is used to execute this task. The main objectives of the operating system are (1) to provide efficient use of the hardware resources, (2) to coordinate the execution of multiple-use programs, (3) to ensure that external events are dealt with in a real-time manner, and (4) to provide security to the users. The operating system consists of a set of routines that can be used in programming and in executing the use's programs.

The majority of real-time programs developed by the user are written in BASIC, FORTRAN, C, and Pascal. BASIC is an *interpreted* language. The interpreter program scans the user program, one line a time, and executes arithmetic and logical operations using library subroutines. BASIC programs are easy to write, and interpreted BASIC programs are easy to debug and to modify. However, they run more slowly than *executed* programs like FORTRAN.

FORTRAN is a *compiled* language. The FORTRAN compiler modifies the user program to a machine language code that is then loaded and executed directly. So a FORTRAN program will run faster than any interpreted program.

Programmable Logic Controllers

A *programmable logic controller* (*PLC*) is a digital electronic device designed to control machines and processes by performing sequential operations. PLCs use microprocessors programmed to execute Boolean logic (AND, OR, NOT) and to implement sequencing. The PLC started as a reusable, inexpensive, flexible, reliable replacement of hard-wired relay panels. Modern PLCs perform a series of functions, including logic, timing, counting, sequencing, PID control, and fuzzy logic. In addition, PLCs can perform arithmetic operations, analyze data, and communicate with other PLCs and computers.

Most of the control literature deals exclusively with applications involving processes that operate continuously. Processes that are not continuous or that contain equipment that operates discontinuously involve the application of logic decisions in implementing control. PLCs have been very useful in applications such as noncontinuous operations (eg, batch processes) and during start-up and shutdown of continuous processes, when many elements must be correctly sequenced (i.e., upper-levels must be established before downstream devices can be turned on). PLCs are also used for implementing interlocks.

PLCs are programmed using ladder logic diagrams. Ladder diagram programming language uses contact and coil symbols to construct diagrams that are very similar to the ladder diagrams used in relay logic.

The general characteristics of PLCs are described below:

- *Inputs/outputs:* PLCs can handle up to several thousand discrete inputs and outputs, and up to several hundred analog inputs and outputs.
- *Logic:* PLCs can handle combinational binary logic operations and sequential logic. A PLC's capability is measured by its memory scan rate or by the average time required to scan each step in a logic ladder diagram (typically from 50 to 10 milliseconds).
- *Continuous control:* PLCs can incorporate most of the commonly used control functions, such as PID, PID with on/off outputs, cascade control, and lead-lag elements.
- *Operator communication:* Small PLCs are not provided with operator interface, but larger PLC systems are generally provided with sophisticated operator displays and keyboards. Generally, small PLCs are used in a DCS as one element, with the I/O provided by a separate unit in the network. Large PLCs can provide all the control functions in a system.

Figure 9–7 shows an example of a touch screen used for interfacing with PLCs. These interfaces are capable of storing and displaying up to 245 screens that may include graphics and text as well as real-time data from controller. In addition, they can be configured to emulate multistage switches when pressed. Figure 9–7a shows how components such as pushbuttons, switches, displays, counters, and lamps can be replaced using touch screen interfaces. As with a Windows-based program, the operator can open a pop-up window by simply pressing a button on the screen (Figure 9–7b).

Hardware Components

A typical PLC can be divided into three parts (Figure 9–8): the central processing unit (CPU), the input/output (I/O) devices, and the memory devices. In addition, a PLC has a connection for the programming and monitoring unit, printer, and program.

The CPU is the brain of the system. Internally it contains various logic gate circuits. It is a microprocessor-based system that replaces control relays, counters, timers, and sequencers. The CPU is designed in such a way that the user can enter the desired circuit in relay ladder logic. It reads input data from various sensing devices, executes the stored user program from memory, and sends output commands to control devices. A direct current (DC) power source is required to produce the low-level voltage used by the processor and I/O modules.

Memory in a PLC system is divided into the program memory, which is usually stored in EPROM/ROM, and the operating memory (RAM). RAM memory is needed for operating programs and for temporarily storing input and output data. Typical memory size of PLC systems varies from 1 kb to 20 kb.

The input/output (I/O) system consists of modules. It forms the interface between the internal PLC system and the external processes to be monitored and controlled. Typical I/O operating voltages are 5 V–240 V DC (or AC) and currents from 0.1 A up to several amperes. The I/O modules are designed to eliminate the need for any circuitry between the PLC and the process system to be controlled. The number of I/O connections in a PLC unit

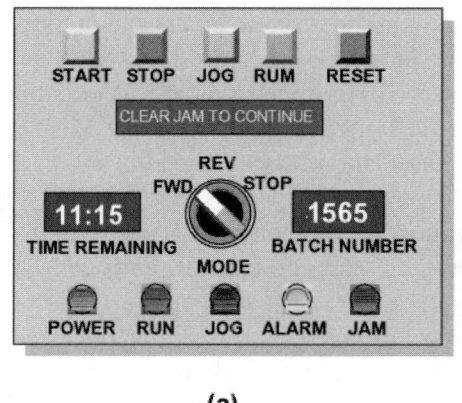

(a)

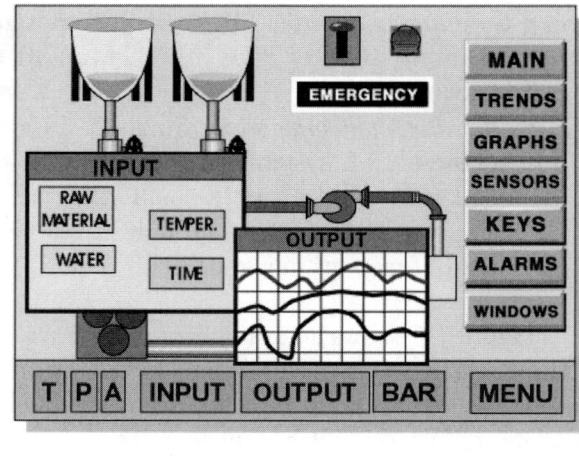

(b)

Figure 9–7 Touch screen used with PLCs where (a) shows how many components can be replaced with operator interface terminals and (b) shows a window operating system

can vary from 40 to more than 128, with either local or remote connections and extensive upgrade capabilities.

The programming units are used to enter the desired program into the memory of the processor. This program is entered using *relay ladder logic*. The programming unit can be a dedicated device or a personal computer. It allows graphical display of the program (ladder diagram). Once the unit is connected to the PLC, it can download the program, which allows real-time monitoring of its operation.

One of the main advantages of a PLC is that it is a programmable device. Unlike the relay logic, it makes it possible to design and modify the program easily without any changes in the wiring. The programming approach as well as programming language are standard for all PLC platforms on the market. This makes programming easy and efficient when operating with PLCs.

Some capabilities of PLC systems that are not present in early relay logic systems include the analog I/O, PID control, and interfaces to a central PLC or computer.

Principles of Operation

A PLC's internal operation is very similar to that of computers. The inputs are continuously monitored and copied from the I/O module into the RAM memory (which

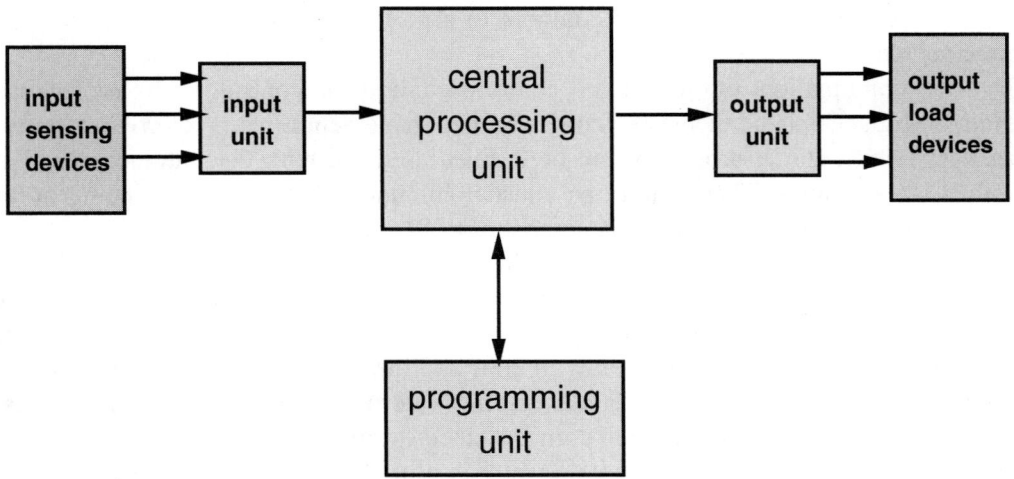

Figure 9–8 Components of a programming logic control (PLC)

is divided into input and output sections). The CPU uses the control program in another section of the memory and takes the input variables from the input RAM. Depending on the program and inputs, the output RAM is filled with control variables, which are then copied into the I/O module so that the process can be controlled.

PLC versus Computer

The architecture of a PLC is basically the same as that of a general-purpose computer. The main differences are the following:

- Unlike computers, PLCs are units of rugged design to operate in an industrial environment and are shielded for improved electrical noise immunity.
- PLCs are designed for easy use by plant technicians and are programmed in relay ladder logic or other easily learned languages. Computers are capable of executing several programs or tasks simultaneously and in any order. Most PLCs execute a single program in an orderly and sequential fashion from first to last instruction.
- PLCs are designed for installation and maintenance by plant electricians who are not required to have advanced computer skills.

Sequential Control

The step-by-step execution of time-ordered events is referred to as sequential control. Conceptually, sequential control can be simple and clear, but real applications often get complex and confusing. The simplest sequential installation involves electromechanical relays. Such relays use series and/or parallel input contact closures to energize a coil that then opens or closes one or more output contacts. With appropriate serial and parallel interconnections of several input contacts to a coil, Boolean logic relationships are implemented.

The combination of sequential control, continuous control such as feedback/feedforward, specialized displays, and database management tools is referred to as *batch process control*. Batch processes can be defined as those with discontinuous feed and product stream flows. They require an unusual amount of logic and sequencing in their control and are thus suited to the characteristics of PLCs.

Consider the simple batch process shown in Figure 9–9 where tortilla chips are manufactured from fresh masa made from corn cooked in the simmering tanks and then soaked in lime solution in the soaking tanks. The following are assumed: (1) the steam-jacketed cooking kettles are charged with corn and lime solution; (2) the corn is cooked by heat supplied by the steam jacket; (3) cooling water is introduced to cooking kettles to cool the cooked corn; (4) the cooked corn is soaked overnight and is mixed every 2 hours; (5) the soaked corn is washed with cold water in a tumbler; (6) the washed corn is dried and ground; (7) the masa is produced with the desired moisture content; (8) the masa is sheeted and cut into triangular shapes; (9) the raw chips are baked for 15 to 20 min; (10) the baked chips are fried for 45 to 120 sec; and (11) the tortilla chips are seasoned and packaged. The basic characteristics of the batch process are as follows:

- Batch processes proceed in sequential steps or phases of operation through the units. The phases are a function of the process and products being produced, not of the control technique used. Considerable communication must exist to successfully process a batch through the process units. In the tortilla chip processing example, when the charge unit completes a charge to each unit operation, the unit controller must be notified so that the processing phase can start.
- Batch processes use large quantities of two-state devices to drive and sense the process. In the example above, two-state valves are used for material transfer operations. Other devices include two-state pumps and motors (conveyors).
- Many timed operations exist in batch processes. In the tortilla chips processing example, the cooking kettles are charged and heated to the trigger point and the temperature is maintained at this level for a specific period of time.
- The same equipment might be used to produce different products in a batch process. For example, the batch process presented above can be used to produce corn chips by changing some of the units or by controlling the steps.
- A batch process uses a number of continuous control loops that operate within the phases of the process. The temperature in the soaking units, for example, is controlled during the soaking phase to a setpoint by manipulating the inlet water temperature using P and PI controllers.

Some of the advantages (Murrill, 1988) of operating such a system under batch control instead of manual control are

- increased production due to higher yields and reduced cycle times
- increased consistency of final product quality

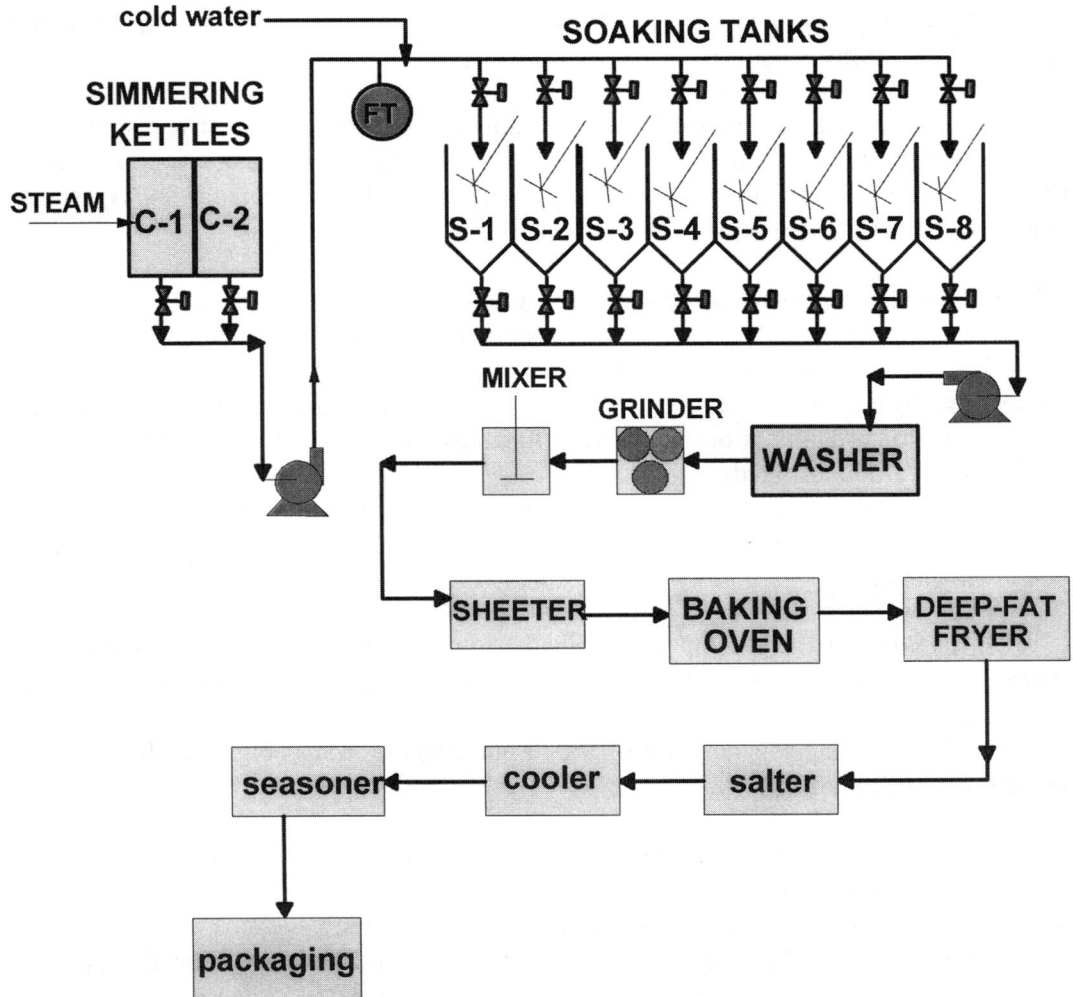

Figure 9–9 Schematic diagram of a batch system for manufacturing tortilla chips

- cost reduction by better use of materials, energy, and equipment
- more flexibility
- better safety

In addition, batch control systems have the following capabilities (Ghosh, 1980):

- They have alarm-handling systems capable of invoking alarm logic and/or falling back to the manual backup.
- They can return automatically to normal operation after an alarm condition has passed.
- They can produce process reports for discrete events.
- Process parameters can be easily changed on line.

- Expansion and alteration of software are easy with plant hardware expansion and modifications.

The design of batch control systems requires the development of a structure for the logic sequential operations. Let us consider the tortilla chips batch system described above. For that example, the sequential logic can be divided into *control states* defined as (Murrill, 1988) (Figure 9–10):

- *Normal*—process operation according to prescribed procedure
- *Hold*—partial shutdown; conditions maintained near operating level
- *Emergency Shutdown*—complete sudden shutdown
- *Standby Shutdown*—similar to *Hold*, but implies a semicontinuous or continuous system in total reflux or total recycle, in which operating conditions are maintained but no product is made
- *Restart*—transition logic from a *Hold* or *Shutdown* condition back to *Normal*
- *Unavailable*—isolated from other systems and scheduling programs for extensive maintainance

On the basis of the system performance, the controller is transferred among those control states. For example, a high-temperature alarm could send you from *Normal* to *Emergency Shutdown*.

The *Normal* state can further be divided into *process states* (Figure 9–10), defined as the following (Murrill, 1988):

- *Prestart*—makes sure that enough raw materials are available to proceed
- *Charge*—charges materials A and B into the reactor
- *Run*—maintains the desired temperature for proper time
- *Discharge*—discharges reactor contents to storage and vents and cools the reactor
- *Cleanup*—prepares the equipment for a new operation

The operator always initiates return to *Normal* from other control states by the panel. When *Restart* logic takes over, it usually returns to a preceding process state to reestablish conditions that existed at the time of interruption. For example, if the system was in *Hold*, and if the soaking tank contents cool below the reaction temperature, the system must return to the heating state before the reaction can finish. The reentry point always depends upon the types of actions or degree of shutdown that takes place in other control states.

Each process state, in turn, has the sequential logic for the basic action of the state.

The entire logic structure shown above represents the batch control program for one system. In a given food-processing plant, there may be many such systems: for instance, one system for mixing of raw materials and another for purifying. Therefore, the final control system needs the hardware and software to execute these various programs and to handle communications among them.

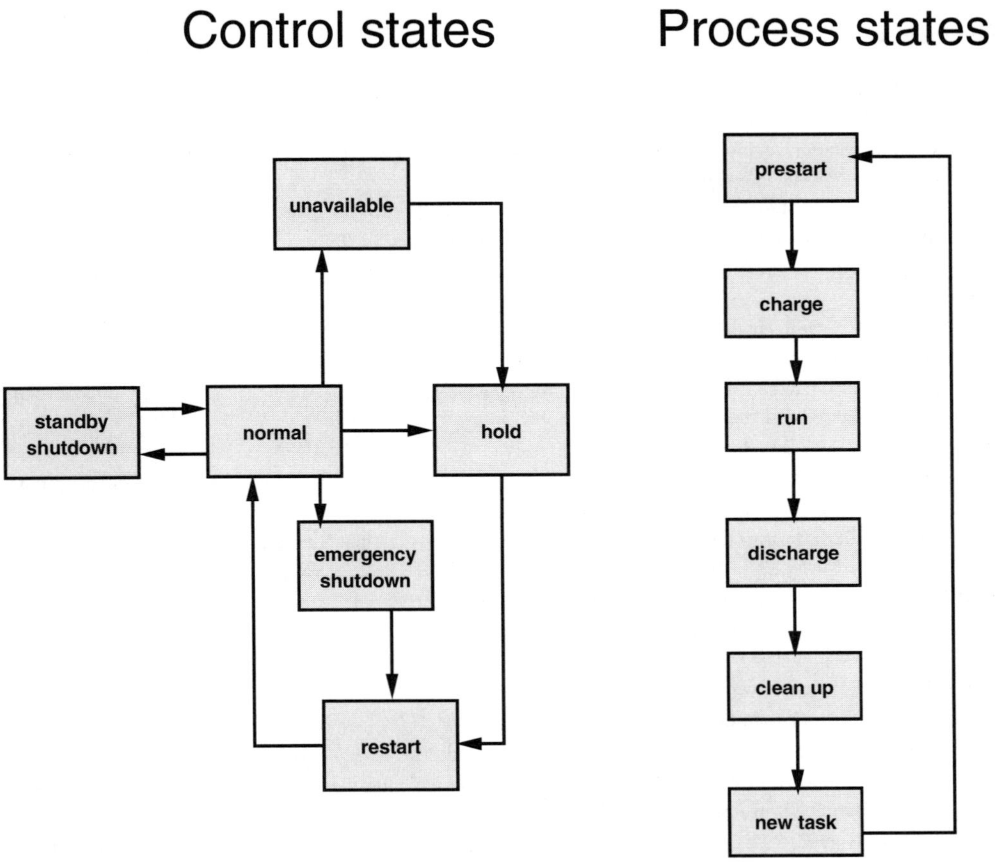

Control states

Process states

Figure 9–10 Control states and process states for a batch control system.

CONCLUSION

This chapter presents an overview of the area of instrumentation and control as applied to food-processing systems.

Sensors and measuring devices have improved significantly over the last decade. Today, smart sensors can measure, self-compensate for disturbances, and even amplify signals. Advances in machine-vision technology will contribute to better characterization of product properties on line.

Another important area of continuing improvement is digital computer control. The integration of PCSs and DCSs to produce LC hybrids will make possible efficient handling of the complex combination of batch and continuous controls required in the food industry. New programming tools are also being developed to improve control applications and communications.

REFERENCES

Chan, J.P. & Batchelor, B.G. (1993). Machine vision for the food industry. In *Food Process Monitoring Systems*. Edited by Pinder & Godfrey. New York. NY: Blackie Academic & Professional.

Datta, A.K. (1992). Sensors and food processing operations. In *Encyclopedia of Food Science and Technology*. Vol 3. Edited by Y.H. Hui. New York, NY: John Wiley.

Ghosh, A. (1980). Checklist for batch process computer control. *Chem Eng* 2, 101–105.

Giese, J. (1993). On-line sensors for food processing. *Food Technol* 47(5), 88–95.

Griffiths, M.W. (1996). The role of ATP bioluminescence in the food industry: new light on old problems. *Food Technol* 50(6), 62–72.

Harper, J.M. (1989). Instrumentation for extrusion processes. In *Extrusion Cooking*. Edited by Mercier, Linko and Harper. St Paul, MN: American Association of Cereal Chemists.

Henrickson, F.W. (1992). *Real-Time Moisture Sensing for the Food Industry*. San Antonio, TX: Research and Development Associates, Military Food and Packaging Systems, Inc. Activities Rept. 44(1).

Honings, D.E. (1993). *Product Application Notes*. Hagerstown, MD: Katrina, Inc.

Javanaud, C. & Robins, M.M. (1993). Ultrasonic methods. In *Food Process Monitoring Systems*. Edited by Pinder & Godfrey. New York, NY: Blackie Academic & Professional.

Kess-Rorgers, E. (1986). Instrumentation in the food industry. I: chemical, biochemical, and immunological determinants. *J Phys E: Sci Instrum* 19(1), 13–21.

King, R.J. (1992). Microwave sensors for process control. *Sensors* 9(9), 68–74.

Mandenius, C.F. & P. Martensson. (1983). Improved membrane gas sensor systems for on-line analysis of ethanol and other volatile organic compounds in fermentation media. *Eur J Appl Microbiol Biotechnol* 18, 197–200.

Martin, P.G. (1984). Computer control of batch processes. *Meas Control* 9, 213–220.

Murrill, P.W. (1988). *Application Concepts of Process Control*. Raleigh, NC: Instrument Society of America.

Pretuzella, F.D. (1989). *Programmable Logic Controllers*. New York, NY: McGraw-Hill.

Schonauer, S.L. & Moreira, R.G. (1995a). Development of a fixed-GPC controller for a food extruder based on PQA. part I: system identification. *Trans Inst Chem Eng* 73(C), 189–199.

Schonauer, S.L. & Moreira, R.G. (1995b). Development of a fixed-GPC controller for a food extruder based on PQA. part II: control development, implementation and analysis. *Trans Inst Chem Eng* 73(C), 200–210.

Wantanabe, E., Nagumo, A., Hoshi, M., Konagaya, S. & Tanaka, M. (1987). Microbial sensors for detection of fish freshness. *J Food Sci* 52, 592–595.

Whelan, P. & Batchelor, B.G. (1991). Automatic packing of arbitrary shapes. *Proceedings of the SPIE. Machine Vision Systems: Integration and Applications*. Boston, MA: International Society of Optical Engineering.

Wilson, R.H. & Kemsley, E.K. (1992). On-line process monitoring using infrared techniques. In *Food Processing Automation II. Proceedings of the American Society of Agricultural Engineers*. Lexington, KY: American Society of Agricultural Engineers.

List of Sources

CHAPTER 1

Figure 1–1 Reprinted with permission from E. Morris, The State of Food Manufacturing, *Food Engineering*, September, pp. 66–82, © 1998, Cahners Business Information.

Figure 1–2 Reprinted from U.S. Department of Commerce, 1998.

Figure 1–3 Reprinted with permission from E. Morris, The State of Food Manufacturing, *Food Engineering*, September, pp. 66–82, © 1998, Cahners Business Information.

CHAPTER 3

Figure 3–20 Adapted from D. Platt, A. Palazoglu and T.R. Rumsey, *Drying Technology*, Vol. 10, No. 2, pp. 333–363, © 1992, by courtesy of Marcel Dekker, Inc., N.Y.

Figure 3–40 Adapted with permission from W. Mann, Digital Control of a Rotary Dryer in the Sugar Industry, *6th IFAC/IFIP Conference on Digital Computer Applications*, © 1982.

Figure 3–41 Adapted with permission from W. Mann, Digital Control of a Rotary Dryer in the Sugar Industry*, 6th IFAC/IFIP Conference on Digital Computer Applications*, © 1982.

Figure 3–42 Adapted with permission from W. Mann, Digital Control of a Rotary Dryer in the Sugar Industry, *6th IFAC/IFIP Conference on Digital Computer Applications*, © 1982.

CHAPTER 5

Figure 5–8 Adapted with permission from W. Mann, Digital Control of a Rotary Dryer in the Sugar Industry, *6th IFAC/IFIP Conference on Digital Computer Applications*, © 1982.

Table 5–1 Reprinted from S. Schonauer and R.G. Moreira, Dynamic Analysis of On-line Product Quality Attributes for Automation of Food Extruders, *Food Science Technology International*, Vol. 3, pp. 413–421, © 1997, Aspen Publishers, Inc.

Table 5–2 Reprinted from S. Schonauer and R.G. Moreira, Dynamic Analysis of On-line Product Quality Attributes for Automation of Food Extruders, *Food Science Technology International*, Vol. 3, pp. 413–421, © 1997, Aspen Publishers, Inc.

Table 5–3 Reprinted from S. Schonauer and R.G. Moreira, Dynamic Analysis of On-line Product Quality Attributes for Automation of Food Extruders, *Food Science Technology International*, Vol. 3, pp. 413–421, © 1997, Aspen Publishers, Inc.

CHAPTER 6

Figure 6–10 Reprinted with permission from S. Shonauer and R.G. Moreira, Development of a Fixed-GPC Controller for a Food Extruder Based on PQA- Part II: Control Development, Implementation and Analysis, *Transactions of the Institution of Chemical Engineers*, Vol. 73, No. c, pp. 200–210, © 1995, Institution of Chemical Engineers.

Figure 6–11 Reprinted with permission from S. Shonauer and R.G. Moreira, Development of a Fixed-GPC Controller for a Food Extruder Based on PQA- Part II: Control Development, Implementation and Analysis, *Transactions of the Institution of Chemical Engineers*, Vol. 73, No. c, pp. 200–210, © 1995, Institution of Chemical Engineers.

Figure 6–12 Reprinted with permission from S. Shonauer and R.G. Moreira, Development of a Fixed-GPC Controller for a Food Extruder Based on PQA- Part II: Control Development, Implementation and Analysis, *Transactions of the Institution of Chemical Engineers*, Vol. 73, No. c, pp. 200–210, © 1995, Institution of Chemical Engineers.

Figure 6–13 Reprinted with permission from S. Shonauer and R.G. Moreira, Development of a Fixed-GPC Controller for a Food Extruder Based on PQA- Part II: Control Development, Implementation and Analysis, *Transactions of the Institution of Chemical Engineers*, Vol. 73, No. c, pp. 200–210, © 1995, Institution of Chemical Engineers.

Figure 6–17 Reprinted with permission from Q. Liu, *Stochastic Modeling and Automatic Control of Grain Dryers: Optimizing Grain Quality*, Michigan State University, Ph.D Dissertation, © 1998.

Figure 6–18 Reprinted with permission from Q. Liu, *Stochastic Modeling and Automatic Control of Grain Dryers: Optimizing Grain Quality*, Michigan State University, Ph.D Dissertation, © 1998.

Figure 6–19 Reprinted with permission from Q. Liu, *Stochastic Modeling and Automatic Control of Grain Dryers: Optimizing Grain Quality*, Michigan State University, Ph.D Dissertation, © 1998.

Figure 6–20 Reprinted with permission from Q. Liu, *Stochastic Modeling and Automatic Control of Grain Dryers: Optimizing Grain Quality*, Michigan State University, Ph.D Dissertation, © 1998.

Table 6–2 Reprinted from S. Schonauer and R.G. Moreira, Dynamic Analysis of On-line Product Quality Attributes for Automation of Food Extruders, *Food Science Technology International*, Vol. 3, pp. 413–421, © 1997, Aspen Publishers, Inc.

CHAPTER 7

Figure 7–7 Reprinted with permission from Nybrant and Regner, ASAE Paper No. 85–3011, © 1985, American Society of Agricultural Engineers.

Figure 7–10 Reprinted from *Automatica*, Vol. 21, R. Isermann and K.H. Lachmann, Parameter-Adaptive Control with Configuration Aids and Supervision Functions, pp. 625–638, Copyright 1985, with permission from Elsevier Science.

Figure 7–11 Reprinted from *Automatica*, Vol. 21, R. Isermann and K.H. Lachmann, Parameter-Adaptive Control with Configuration Aids and Supervision Functions, pp. 625–638, Copyright 1985, with permission from Elsevier Science.

Figure 7–12 Reprinted with permission from Z. Chang and J. Tan, *Adaptive Control of Food Processes with Time Delay*, Paper No. 92–6544, © 1992, American Society of Agricultural Engineers.

Figure 7–13 Reprinted with permission from Z. Chang and J. Tan, *Adaptive Control of Food Processes with Time Delay*, Paper No. 92–6544, © 1992, American Society of Agricultural Engineers.

CHAPTER 8

Figure 8–4 Reprinted from *Statistical Monitoring and Process Sensor Reliability*, Vol. 9, No. 1, A. Negiz, et al., Modeling, Monitoring and Control Strategies for High Temperature Short Time Pasteurization Systems, pp. 29–47, Copyright 1998, with permission from Elsevier Science.

Figure 8–5 Reprinted with permission from S. Shonauer and R.G. Moreira, Development of a Fixed-GPC Controller for a Food Extruder Based on PQA- Part II: Control Development, Implementation and Analysis, *Transactions of the Institution of Chemical Engineers*, Vol. 73, No. c, pp. 200–210, © 1995, Institution of Chemical Engineers.

Figure 8–10 From K. Arzen and K.J. Astrom, Expert Control and Fuzzy Control, in *International Conference on Intelligent Systems in Process Engineering*, Davis, Stephanopolous, and Venkatasubramanian, eds., Vol. 312, No. 92, pp. 47–56. Reproduced with permission of the American Institute of Chemical Engineers. Copyright © 1996 American Institute of Chemical Engineers (AIChE) and Computer Aids for Chemical Engineering Education (CACHE). All rights reserved.

Figure 8–12 Adapted with permission from K.E. Arzen, et al., Fuzzy Control Versus Conventional Control, in *Fuzzy Algorithms for Control*, Vergruggen, Zimmerman, and Rabuska, eds., © 1999, Kluwer Academic Publishers.

Figure 8–16 Reprinted with permission from N.L. Perrot, et al., Feedback Quality Control in the Baking Industry Using Fuzzy Sets, *Journal of Food Process Engineering*, 2000, in press, Food and Nutrition Press, Inc.

Figure 8–17 Reprinted with permission from N.L. Perrot, et al., Feedback Quality Control in the Baking Industry Using Fuzzy Sets, *Journal of Food Process Engineering*, 2000, in press, Food and Nutrition Press, Inc.

Figure 8–18 Reprinted with permission from N.L. Perrot, et al., Feedback Quality Control in the Baking Industry Using Fuzzy Sets, *Journal of Food Process Engineering*, 2000, in press, Food and Nutrition Press, Inc.

Figure 8–23 From K. Arzen and K.J. Astrom, Expert Control and Fuzzy Control, in *International Conference on Intelligent Systems in Process Engineering*, Davis, Stephanopolous, and Venkatasubramanian, eds., Vol. 312, No. 92, pp. 47–56. Reproduced with permission of the American Institute of Chemical Engineers. Copyright © 1996 American Institute of Chemical Engineers (AIChE) and Computer Aids for Chemical Engineering Education (CACHE). All rights reserved.

CHAPTER 9

Figure 9–14 From P.W. Murrill, *Application Concepts of Process Control*, Copyright © 1988, ISA. Reprinted by permission. All rights reserved.

Table 9–1 From ISA-MC96.1–1982, Temperature Measurement Thermocouples, Copyright by ISA. All rights reserved. To obtain a copy, contact ISA, PO Box 12277, Research Triangle Park, NC 27709, (919) 549–8411 or www.isa.org.

Index